GROWTH AND DIFFERENTIATION IN PLANTS

P. F. WAREING and **I. D. J. PHILLIPS**

School of Biological Sciences
University College of Wales

Department of Biological Sciences
University of Exeter

THIRD EDITION

PERGAMON PRESS

OXFORD · NEW YORK · BEIJING · FRANKFURT
SÃO PAULO · SYDNEY · TOKYO · TORONTO

U.K.	Pergamon Press, Headington Hill Hall, Oxford OX3 0BW, England
U.S.A.	Pergamon Press, Maxwell House, Fairview Park, Elmsford, New York 10523, U.S.A.
PEOPLE'S REPUBLIC OF CHINA	Pergamon Press, Qianmen Hotel, Beijing, People's Republic of China
FEDERAL REPUBLIC OF GERMANY	Pergamon Press, Hammerweg 6, D-6242 Kronberg, Federal Republic of Germany
BRAZIL	Pergamon Editora, Rua Eça de Queiros, 346, CEP 04011, São Paulo, Brazil
AUSTRALIA	Pergamon Press Australia, P.O. Box 544, Potts Point, N.S.W. 2011, Australia
JAPAN	Pergamon Press, 8th Floor, Matsuoka Central Building, 1-7-1 Nishishinjuku, Shinjuku-ku, Tokyo 160, Japan
CANADA	Pergamon Press Canada, Suite 104, 150 Consumers Road, Willowdale, Ontario M2J 1P9, Canada

First edition 1970
Reprinted 1971, 1973, 1975, 1976, 1977
Second edition 1978
Third edition 1981
Reprinted 1982, 1985, 1986

British Library Cataloguing in Publication Data

Wareing, Philip Frank
Growth and differentiation in plants.—3rd ed—
(Pergamon international library).
1. Growth (Plants)
I. Title II. Phillips, Irving David James
III. Control of growth and differentiation
in plants
581.3'1   QK731   80–41659

ISBN 0-08-026351-8 (Hardcover)
ISBN 0-08-026350-X (Flexicover)

Printed in Great Britain by A. Wheaton & Co. Ltd., Exeter

Preface to the 3rd Edition

The preparation of this 3rd edition of *Growth and Differentiation in Plants* has involved extensive revision and rearrangement of the material of the preceding edition. The subject matter has now been grouped into four main sections. Section I provides an introduction to the structural aspects of development, at various levels of organization, to provide a basis for the biochemical and physiological approaches presented in the subsequent sections. Section II deals with the major classes of plant growth substances and their role in the internal control of development. The material has been extensively rearranged and the chapters dealing with the biochemistry and mode of action of growth substances have been expanded significantly. The various aspects of environmental control of development, dealt with in Section III, have been brought up to date and the chapter on Growth Movements has been extensively rewritten. Finally, Section IV deals with more general aspects of the control of development, especially at the molecular level. Since development can be regarded essentially as a process involving selective gene expression, the present state of knowledge regarding the structure of the plant genome and the control of gene expression in eukaryotes is summarized briefly, although at present it is not possible to relate this information directly to the vast body of information on plant development which has been accumulated by other approaches. Indeed, although the different approaches represented by the four sections of the book have each contributed substantially to our understanding of development, it is not yet possible to present an overall synthesis of these various facets of knowledge. It is true that plant hormones can be seen to play a vital role, on the one hand, in the growth and differentiation of cells and tissues, and on the other hand, in plant responses to environmental factors, but until we understand how hormones act at the molecular and sub-cellular levels our understanding of their role in developmental processes will necessarily be incomplete. Moreover, although the paramount importance of hormones in the control and co-ordination of *growth* is not in question, the extent of their role in *differentiation* is still unclear, since each major class of hormone has a wide spectrum of effects in different parts of the plant, and the specificity of the response depends on the "programming" ("competence") of the target tissue. We know very little as to the processes which determine the specificity of the response, as compared with the mass of information on hormone chemistry and biochemistry and hormone-induced effects. Thus, our emphasis on the role of hormones reflects, in fact, current lack of knowledge of other factors controlling development.

We are indebted to Dr. J. Ingle and Professor H. Smith, and also our colleagues, Dr. R. Horgan, Dr M. A. Hall and Dr P. F. Saunders, for reading various sections of the revised manuscript and for their helpful and constructive criticisms.

E/14.50
9L

PERGAMON INTERNATIONAL LIBRARY
of Science, Technology, Engineering and Social Studies

The 1000-volume original paperback library in aid of education,
industrial training and the enjoyment of leisure

Publisher: Robert Maxwell, M.C.

GROWTH AND DIFFERENTIATION IN PLANTS

Titles of Related Interest

COOMBES *et al.*
Techniques in Bioproductivity and Photosynthesis, 2nd Edition

FAHN
Plant Anatomy, 3rd Edition

MAYER & POLJAKOFF-MAYBER
The Germination of Seeds, 3rd Edition

VOSE & BLIXT
Crop Breeding: A Contemporary Basis

Contents

SECTION I

Structural and Morphological Aspects of Development

Introduction

The orderly succession of changes leading from the simple structure of the embryo to the highly complex organization of the mature plant presents some of the most fascinating and challenging outstanding problems of biology. In this book we shall be concerned primarily with describing and examining what is known about the processes underlying and controlling plant development.

Development is applied in its broadest sense to the whole series of changes which an organism goes through during its life cycle, but it may equally be applied to individual organs, to tissues or even to cells. Development is most clearly manifest in changes in the form of an organism, as when it changes from a vegetative to a flowering condition. Similarly, we may speak of the development of a leaf, from a simple primordium to a complex, mature organ.

The problems of development can be studied in a number of different ways, but basically there are two major types of approach, viz. (1) the morphological and (2) the physiological and biochemical. Developmental morphology and anatomy were formerly largely concerned with describing the visible changes occurring during development, but current interest is mainly directed to trying to understand the factors and processes determining plant form, using experimental techniques, such as surgery, tissue culture, autoradiography and so on. However, development cannot be fully understood without a study of the manifold biochemical and physiological processes underlying and determining the morphological changes, and it is these latter aspects of development which form the main subject of this work.

The experimental morphologist often uses the term *morphogenesis,* which, in the literal sense, is concerned with the origin of form in living organisms. However, by the term "form" should be understood not only the gross external morphology of the plant, but its whole organization, which may be recognized as existing at several different levels; thus, we may recognize (1) the structural organization of the individual cell, as shown by electron microscopy, (2) the organization of cells to form tissues, and (3) the organization of the plant body at the macroscopic level. Moreover, in the study of morphogenesis we are concerned not only with observable changes in form and structure but also with the underlying processes controlling the development of organs and tissues, and insofar as these processes must ultimately be explainable in terms of physics and chemistry, this aspect of morphogenesis is identical with developmental physiology and biochemistry. However, at the present time our knowledge of the molecular basis of morphogenesis is very fragmentary and we know very little about the physiological and biochemical processes regulating, for example, the initiation and development of leaves.

When we come to consider the physiology of development, we find a further dichotomy of approach. On the one hand, a considerable body of knowledge has been acquired about the role of hormones as "internal" factors controlling growth and differentiation; on the other hand, the profound importance of environmental factors, such as day length and temperature, in the regulation of some of the major phases in the plant life cycle has been clearly demonstrated, although there is considerable evidence that a number of environmental influences are mediated through effects on the levels and distribution of hormones within the plant.

In the present section we shall consider briefly the structural and morphological aspects of development, first at the cellular and then at the organ and whole plant level.

1

The Plant Cell in Development

1.1 INTRODUCTION

Plant development involves both *growth* and *differentiation*. The term growth is applied to *quantitative* changes occurring during development and it may be defined as an irreversible change in the size of a cell, organ or whole organism. The external form of an organ is primarily the result of *differential growth* along certain axes. However, during development there appear not only quantitative differences in the numbers and arrangement of cells within different organs, but also *qualitative* differences between cells, tissues and organs, to which the term *differentiation* is applied. Differentiation at the cell and tissue level is well known and is the primary object of study in plant anatomy. However, we may also speak of differentiation of the plant body into shoot and root. Similarly, the change from the vegetative to the reproductive phase may

also be regarded as another example of differentiation. We shall, therefore, apply the term differentiation in a very broad sense to any situation in which meristematic cells give rise to two or more types of cell, tissue or organ which are qualitatively different from each other.

Thus, we may say that *growth and differentiation are the two major developmental processes.* Usually growth and differentiation take place concurrently during development, but under certain conditions we may obtain growth without differentiation, as in the growth of a mass of callus cells (Chapter 6).

In the present chapter we shall discuss some general aspects of cell growth and differentiation, but we shall consider later, in more detail, the mechanism of cell elongation growth (Chapter 4) and the molecular aspects of differentiation (Chapter 13).

1.2 THE LOCALIZATION OF GROWTH

One of the essential characteristics of organisms is that they are able to take up relatively simple substances from their environment and use them in the synthesis of the varied and complex substances of which cells are composed. It is this increase in the amount of living material which is basically what we mean by growth. At the cellular level the increase in living material normally leads to an increase in cell size and ultimately to cell division. These two aspects of growth are seen in their simplest form in unicellular organisms such as bacteria, unicellular algae and protozoa, where growth leads to enlargement of each cell which then divides and the process is repeated.

When we come to consider the growth of multicellular organisms, such as the higher plants, the situation is much more complex. It is true that here, also, growth ultimately depends on the enlargement and division of individual cells, but not all cells of the plant body contribute to the growth of the organisms as a whole, for growth is restricted to certain embryonic regions, the *meristems.* This restriction of the growing regions is probably related to the fact that mature plant cells are normally surrounded by relatively thick and rigid cell walls, and many cells of mechanical and vascular tissues are, of course, non-living. These facts would probably render co-ordinated growth, involving both cell division and cell enlargement, difficult in an organ, such as a stem, once a certain stage of differentiation had been reached. We shall see later that most living plant cells retain the capacity to divide under certain conditions, but even if they do divide the daughter cells do not necessarily increase in size, unless they are relatively thin-walled cells which are able to revert to the embryonic or "meristematic" condition. In having rather strictly localized embryonic regions higher plants differ from animals, where growth typically occurs throughout the organism as a whole.

This difference between higher plants and animals is no doubt related to the basic differences in the modes of nutrition of the two groups. Because they have to take up water and mineral salts from the soil, the autotrophic land plants must necessarily be rooted and sessile, whereas most animals have to forage for their food, whether they are herbivorous or carnivorous, and they need, therefore, to be mobile. This requirement for mobility in animals which forage for their food, in turn, demands that they should have flexible bodies, whereas the plant body can be much more rigid and indeed it needs to be so in erect-growing plants, especially in large forest trees. This rigidity and firmness of the plant body depends upon the presence of relatively thick and

firm cell walls, whether in the living cells of the leaf, for example, or in the non-living cells of mechanical tissue of the stem. (The rigidity of those tissues consisting mainly of living cells depends, of course, on the turgidity of the cells and not simply on the mechanical properties of the walls, but even in such tissues a cell wall is an essential requirement for the attainment of the turgid condition.) On the other hand, in aquatic plants, whether they are lower plants or angiosperms, nutrients may be absorbed from the surrounding water directly into the shoot, so that they may be free-floating, and the mechanical tissues are usually less well developed than in land plants.

A number of different types of meristem may be recognized in the plant body. The axial organs, the stems and roots, have *apical meristems,* i.e. growth in length is restricted to the tip regions and the new tissue is added to the plant body on the proximal side, so that the pattern of growth may be described as *accretionary.* The apical meristems of the stem and root usually remain permanently embryonic and capable of growth over long periods—for hundreds of years in some trees. Consequently we may describe these as *indeterminate* meristems.

On the other hand, other parts of the plant, particularly the leaves, flowers and fruits, show rather different patterns of growth and they are embryonic for only a limited period before the whole organ attains maturity. Thus, the growing regions of such organs are sometimes referred to as *determinate* meristems. In such organs the pattern of growth resembles that of animals in that, firstly, there is an embryonic phase of limited duration and secondly, in such organs growth is more generalized than in stems and roots.

The presence of indeterminate meristems, together with the capacity for forming branches, each with its apical meristem, gives the plant body a much less precise and definite form than is the case for the animal body. Indeed, the general form of the plant body resembles a colony of coelenterates, such as corals, rather than that of an individual higher animal. On the other hand, the organs showing determinate growth, such as leaves and flowers, generally show much more precise morphology and may have fairly precise numbers of parts, such as petals.

In addition to classifying meristems as indeterminate and determinate we may classify them in various other ways. For example, we may distinguish the apical meristems of stems and roots, from the *lateral meristems,* comprising the cambium and phellogen (cork cambium). In some plants there are *intercalary meristems,* inserted between regions of differentiated tissues. One of the best known examples of this type of meristem is seen in grasses, where the internodes and leaf sheaths continue growth in the basal region, after the upper parts have become differentiated. The structure of some of these meristems will be described in more detail in Chapter 2.

1.3 CELL DIVISION AND CELL VACUOLATION

The growth of a multicellular plant involves both increase in cell number, by cell division, and increase in cell size. These two aspects of growth have no sharp spatial boundaries; however, in the apical regions of shoots and roots cell division occurs most intensively towards the extreme tip of both organs, whereas the region of most rapid increase in cell size is in a zone a few millimetres back from the tip (Fig. 2.17). In organs of determinate growth, such as leaves and fruits, these two aspects of growth tend to be separated in time, so that there is an early phase in which cell division is predominant, followed later by a phase when cell division ceases and there is active increase in cell size. The greater part of this increase in size is due to vacuolation, i.e. by water uptake, and as a result the cytoplasm may come to be limited to a thin boundary layer against the cell wall.

Cell division involves the replication of all cell organelles, and of these the nucleus is the most conspicuous and has been most intensively studied. Successive divisions of the nucleus, involving the formation of chromosomes and the process of mitosis alternate with periods when the nucleus appears to be in a resting state referred to as "interphase".

Mitosis involves a doubling in the number of chromosomes, which are divided equally between the two daughter cells. Consequently the amount of DNA must be duplicated at some stage before the doubled number of chromosomes becomes visible during prophase. The precise stage of DNA synthesis can be determined either by feeding plant tissues, such as root tips, with radioactive thymidine (one of the bases of DNA) at various stages and determining the time at which it becomes incorporated into DNA; alternatively, the amount of DNA per nucleus can be followed spectrophotometrically, to determine the time at which the total amount of DNA per nucleus is doubled. These techniques indicate the stage of DNA synthesis, referred to as the "S" phase. The remaining periods of interphase preceding and following the S phase, during which there is no DNA synthesis, are referred to as G1 ("Gap 1") and G2 ("Gap 2"), respectively (Fig. 1.1). The durations of G1 and G2 vary considerably in different types of tissue.

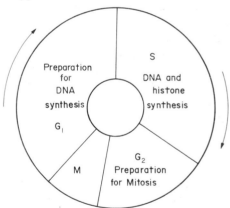

Fig. 1.1. Scheme of the mitotic cell cycle. After mitosis (M) the daughter cells enter the G1 phase, towards the end of which there is preparation for DNA synthesis. DNA and histone synthesis take place in the following S phase, and the chromosomes are replicated so that the cell DNA content is doubled. The cell then enters the G2 phase, which represents a period of preparation for mitosis involving condensation of the chromosomes.

During interphase most of the DNA is relatively extended, but during the prophase of mitosis it becomes compacted and coiled as a result of changes in associated proteins which are described later (p. 304). Conversely, after mitosis, the chromosomal material undergoes a process of un-coiling which leads to the extended condition in the interphase nucleus.

Cell division also involves the replication of various organelles, including the plastids and mitochondria. The simplest type of plastid is the proplastid, from which other types of plastid, including the chloroplast, develop. The plastids are semi-autonomous organelles which are capable of duplication by division or by budding. Most higher plant cells contain several to many plastids, but there are large differences in plastid number between cells of different types. In general, the numbers of plastids in a given cell type remains approximately constant, suggesting that plastid replication keeps roughly in step with cell division. However, the partition of the plastid population of the mother cell between the two daughter cells appears to be a random process.

In the tip regions of roots and shoots in which cell division predominates, the cells are relatively small, and have prominent, spherical nuclei lying towards the centre of the cytoplasm, which is non-vacuolated and tends to be densely staining; the cell walls are thin (Figs. 2.3, 2.5). As a result of division, each of the two daughter cells is only half the size of the parent cell. These cells then proceed to enlarge, but such cell growth involves the synthesis of cytoplasm and cell wall material and not vacuolation.

Since the number of cells in the zone of cell division tends to remain fairly constant (at least over limited periods), it is clear that not all the daughter cells formed in this zone retain the capacity for unlimited further division. The situation is perhaps best illustrated by reference to plants which grow by a single apical cell, such as certain algae and the bryophytes and some pteridophytes (Fig. 13.13) where it can clearly be seen that division of the apical cell results in one cell on the outside which becomes the new apical cell, and a second daughter cell, on the proximal side, which gives rise to the differentiated tissue of the thallus or shoot. This latter daughter cell usually undergoes several further divisions but ultimately the derivative cells lose their capacity for division. Thus, whereas the apical cell remains permanently

meristematic, the derivative cells are capable of only a limited number of further divisions. The situation must be analogous in the more complex apices of gymnosperms and angiosperms, where there is normally a number of initial cells, i.e cells which remain meristematic and undergo repeated division, but it is more difficult to recognize which of the daughter cells is destined to remain meristematic and which will give rise to mature tissue. The problem of why cells in the initial zone remain permanently embryonic or meristematic, whereas the derivative cells on the proximal side are capable of only a limited number of further divisions is an intriguing one, but it remains unsolved at the present time.

At a certain distance from the apex, in both shoots and roots, the process of vacuolation commences and, as a result of this process, the root cells of *Allium* may increase in length from $17\,\mu$m to $30\,\mu$m and in volume by 30-fold. In other tissues the cells may increase up to 150-fold in total volume during vacuolation. It appears that this great uptake of water during cell extension is essentially governed by osmosis, and if we apply the usual concept relating to water uptake by cells, then, in general, the ability of the cell to take up water is given by its water potential (ψ) which is equal to the osmotic potential (π) of the vacuolar solution plus the wall or turgor pressure (p). That is, $\psi = p + \pi$. Now clearly water uptake may involve changes in either the osmotic potential or in the wall pressure, or both. Studies on the changes in osmotic relations of the vacuolar solution during growth have yielded no evidence of a change in osmotic potential. Indeed, since the vacuolar sap becomes greatly diluted during growth, considerable amounts of additional osmotically active substances, such as sugars, salts, organic acids, etc., must pass into the vacuole during growth, in order simply to maintain the osmotic potential at a steady value. In some organs the osmotic potential of the vacuole may actually rise, due to dilution, during this phase of growth. Thus, in the petioles of the water lily, *Victoria regia,* which may increase in length from 9 cm to 68 cm in 24 hours, the osmotic potential may rise to less than half its original value during the extension phase. On the other hand, there is considerable evidence that in vacuolating cells the wall pressure is reduced by increased plasticity of the cell wall at this time (p. 88). As a result of its increased plastic extensibility the wall undergoes irreversible elongation during vacuolation.

Although the greater part of the increase in cell volume during vacuolation is due to water uptake, the synthesis of new cytoplasm and cell wall material proceeds actively during this period, so that the cell increases considerably in dry weight. Thus, the processes of cell growth initiated before vacuolation commences are continued during this latter phase. Moreover, the zones of cell division and cell vacuolation are not sharply demarcated, and in both shoots and roots of many species cell division occurs in cells which have started to undergo considerable vacuolation. Division may also occur in vacuolated cells in wound tissues. In root tips the separation of the zones of division and vacuolation are somewhat sharper and division in vacuolated cells is less frequent.

Since growth involves various endergonic, i.e. energy-requiring, processes, including protein synthesis, it is not surprising to find that rapidly elongating tissues of the root have a high respiration rate, when compared with mature tissues on the basis of equal *volumes* of tissue, although when expressed *per cell* the respiration rate of mature cells may be greater than that of meristematic cells, since the latter are smaller and contain less cytoplasm. Moreover, growth requires aerobic conditions and an adequate supply of carbohydrate, both as an energy source and as structural material.

The role of growth hormones in cell division and cell extension will be discussed later.

1.4 THE GROWTH OF CELL WALLS

Electron microscopic studies have shown that the most prominent structural element of the wall in higher plants is a framework of cellulose *microfibrils*. Cellulose occurs as long linear chains of ß, 1-4 linked glucose units, which are aggregated to form the microfibrils. The chains lie parallel to each other in a very regular arrangement so that the microfibril has a para-crystalline structure. The chains are hydrogen bonded through the hydroxyl groups at position 6 in one chain and the glycosidic oxygens (O_5) in an adjacent chain; there is also intra-molecular bonding between hydroxyl groups at position 3 and bridging oxygens of adjacent glucose units of the same chain (Fig. 1.2). These structural features confer great strength on the microfibrils.

The microfibrils are embedded in a matrix of *non-cellulosic polysaccharides* which include fractions composed of the residues of the pentoses (5 carbon sugars) *arabinose* and *xylose,* and the hexoses (6 carbon sugars) *glucose, galactose* and *mannose.* The xyloglucans consist of ß, 1-4 linked

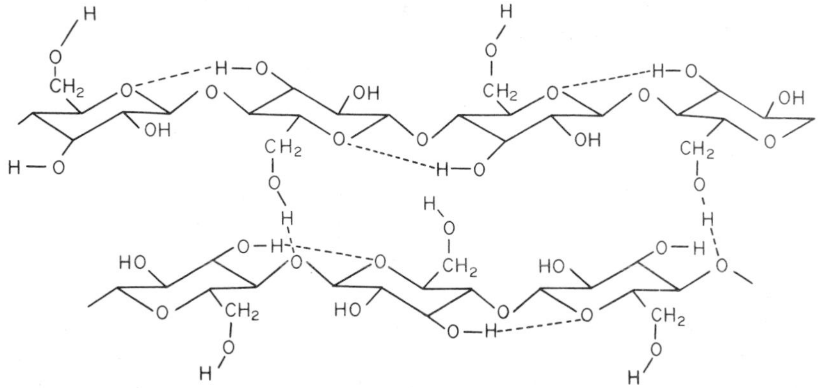

Fig. 1.2. ß-1,4 linked chains of cellulose, showing the hydrogen bonding between adjacent chains, and the intra-molecular bonding between adjacent glucose units in the same chain. (From W. D. Bauer *in The Molecular Biology of Plant Cells,* (Ed.) H. Smith, Blackwell Scientific Publications, Oxford, 1977.)

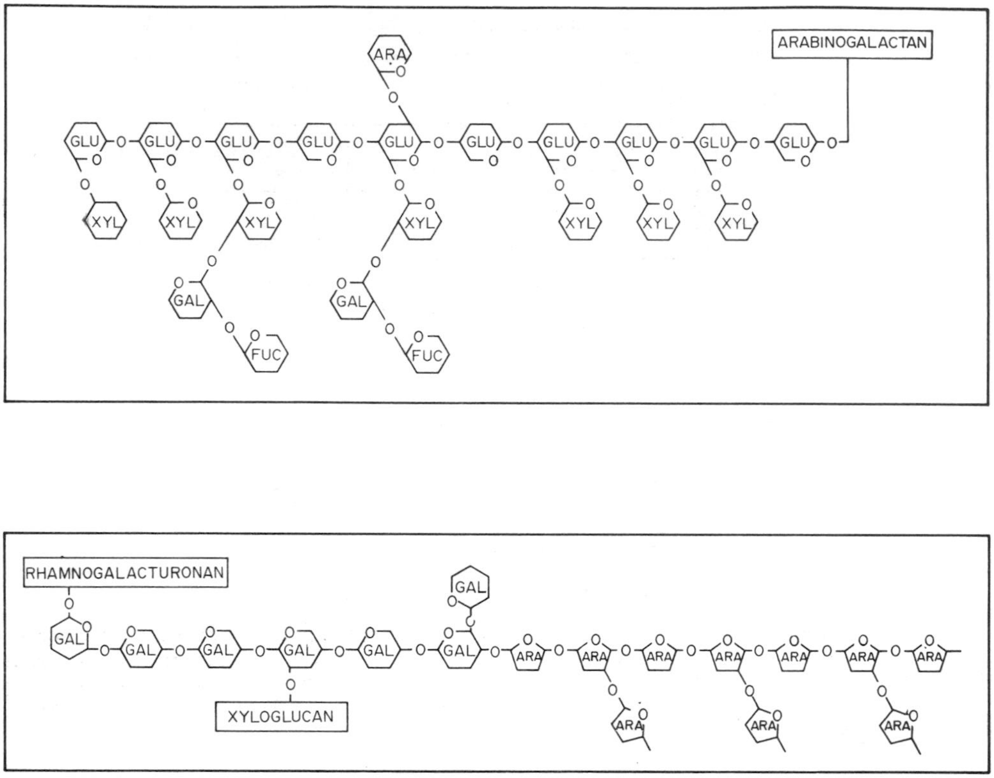

Fig. 1.3. Proposed structure for sycamore callus xyloglucan and arabinogalactan. X—xylose; F—fucose; GAL—galactose; A—arabinose; G—glucose (after Albersheim.)

glucose chains, with frequent xylose side chains, while the arabinogalactans may have linear or branched chains (Fig. 1.3). As well as these polysaccharides, which were formerly referred to as "hemicelluloses", there are other fractions containing a high proportion of galacturonic acid, previously referred to as "pectins". Primary cell walls also contain a structural protein, which contains a high content of the unusual amino acid hydroxyproline.

Although the way in which the various components of the wall associate is still a matter of some controversy it is clear that both covalent and non-covalent interactions between components are involved. The xyloglucans are hydrogen bonded to cellulose chains and in turn may be connected via arabinogalactan to rhamnogalacturanan. The rhamnogalacturanan may also be linked to the cell wall protein via another type of arabinogalactan. A schematic representation of such a cell wall is shown in Fig. 1.4. The foregoing description applies to dicotyledons; the cell walls of monocotyledons, although similar, do differ somewhat in the nature of their matrix components.

The formation of a new cell wall between the two daughter cells following mitotic division of the nucleus begins with the appearance of large numbers of vesicles in the plane of the equator of the spindle. These vesicles are formed by Golgi bodies and apparently contain polysaccharides from which the first stage of the new cell wall, known as the cell plate, is formed by coalescence of the vesicles (Fig. 1.5). The cell plate forms first in the centre of the cell and its edge extends outwards, apparently by the addition of vesicles and their contents from the Golgi bodies which lie on the periphery of the plate, until it joins up with the lateral walls. The membranes of the Golgi vesicles fuse to form the new plasmalemma of each

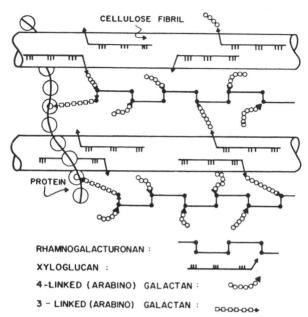

Fig. 1.4. Schematic representation of the polymeric components of sycamore primary cell walls and their interconnections. Xyloglucans are co-crystallized with the cellulose glucan chains on the surface of the microfibrils. Some of the reducing ends of the xyloglucans are glycosidically attached to ß-1,4 linked arabinogalactan side chains of the rhamnogalacturonan chains, which themselves are attached to the structural proteins. (From W. D. Bauer, *in The Molecular Biology of Plant Cells,* (Ed.) H. Smith, Blackwell Scientific Publications, Oxford, 1977.)

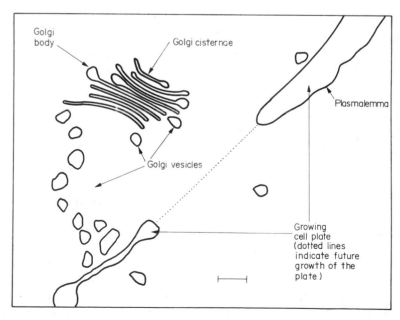

Fig. 1.5. Association of Golgi bodies with cell plate. Scale = 200 nm. (From Bryant, 1976.)

daughter cell. The protoplasm of the two daughter cells remains connected by membranous strands through the wall, forming the plasmadesmata.

It seems clear that the polysaccharides forming the first formed layer of the new cell wall (the middle lamella), which contain a high proportion of galacturonic acid, are synthesized in the Golgi bodies. After the laying down of the middle lamella, cellulose microfibrils are formed on each side of it to give the primary cell wall.

As in the formation of the cell plate, much of the non-cellulosic material laid down during the further formation of the primary cell wall is synthesized in the Golgi apparatus and this material is transported through the cytoplasm and across the plasmalemma into the cell wall. On the other hand, the cellulose microfibrils are formed at the inner surface of the wall, adjacent to the plasmalemma. It appears that the enzymes involved in cellulose biosynthesis are located on the membranes of the Golgi apparatus so that they become functional when the vesicles fuse with the plasmalemma.

In cells which undergo elongation during vacuolation the microfibrils are initially laid down in a plane at right angles to the axis of elongation (i.e. transversely), but during the extension of the walls the microfibrils become re-oriented ("pulled-out"), so that they come to lie predominantly along the longitudinal axis. During growth, however, new transverse microfibrils are added to the inside of the wall, so that in a section through the wall we find a gradual transition from transversely to longitudinally oriented microfibrils, in passing from the inside to the outside (Fig. 1.6). In cells which do not elongate but remain isodiametric during growth the microfibrils are apparently laid down at random.

It is not known what controls the orientation in which the microfibrils are laid down in any cell type, but it is found that they usually lie parallel to certain *microtubules* which, as the name suggests, are elongated cylindrical structures of diameter 23—27 nm, found in the boundary layers of the cytoplasm. Moreover, treatment with colchicine, which disrupts the microtubules, also disorganizes the arrangement of the microfibrils, but does not prevent their deposition. It has been suggested that the microtubules control, in some unknown way, the orientation in which the microfibrils are laid down by the cellulose synthesizing apparatus.

Microtubules may also play a role in determining where the cell plate is formed and fuses with the wall of the parental cell. In resting cells the microtubules lie in the outer cytoplasm, just inside the plasmalemma. In cells which are about to undergo division, but in which the nucleus has not yet entered prophase, the "wall microtubules" just described disappear and a band consisting of a large number of tubules appears in the outer cytoplasm near the longitudinal walls and at right angles to the axis of the cell (Fig. 1.7). This "preprophase band" appears to run right round the outer zone of cytoplasm, in the mid-region of the cell. When the cell plate develops it fuses with the lateral walls of the parent cell at a position which

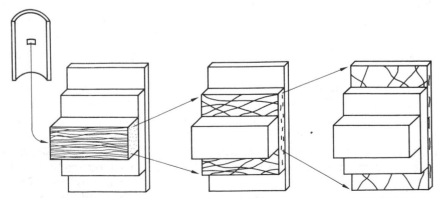

Fig. 1.6. The multinet concept of cell wall growth showing (left to right) re-orientation of microfibrils as successive stages of wall extension. (From P. A. Roelefsen, *Adv. Bot. Res.* **2**, 69–150.)

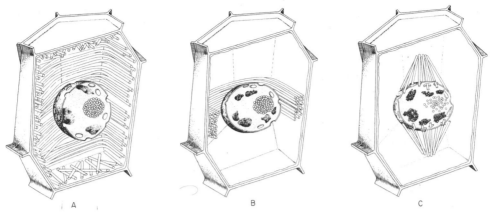

Fig. 1.7. Diagram of changes in microtubules at cell division. (Prints supplied by Dr. Myron Ledbetter, reproduced from *Symposium Int. Soc. Cell. Biol.* **6**, 1967.)

divides the pre-prophase band almost exactly equally between the two daughter cells. It is therefore thought that the pre-prophase band in fact determines the position and orientation of the cell plate.

Growth of the cell wall involves the yielding of the wall to the stress generated by turgor pressure. The increased plasticity of the cell wall during vacuolation, referred to earlier, must indicate that the various types of chemical bond which link the different wall components must be broken during wall growth, possibly as the result of the activities of hydrolytic enzymes (p. 89).

1.5 CELL DIFFERENTIATION

At the cellular level, the term differentiation is sometimes used in two different senses: viz. (1) it may be applied to the development of different specialized types of mature cell within an organ or tissue; or (2) it may be used to refer to the changes which occur during the development of a meristematic cell into a mature cell, usually involving vacuolation and enlargement. In this book we shall use the term in the first sense and we shall use the term *maturation* for the processes which lead to the formation of a mature cell from a meristematic one.

Usually maturation involves cell vacuolation and enlargement, and some aspects of this process

have already been described (pp. 5-6). Cells undergoing maturation may show relatively little other change in structure, as in the formation of parenchymatous tissue, or there may be great changes, as in the formation of xylem and phloem tissue. It is the diverging pathways followed by different cells during maturation which result in differentiation.

In addition to the visible changes involved in differentiation, there are also biochemical differences, some of which will be described later (p. 13).

Between the biochemical differences occurring at the molecular level and the structural changes observable with the optical microscope, one might expect to find changes at the ultrastructural level, to be seen under the electron microscope. Studies on various types of tissue have shown that, in general, living differentiated higher plant cells have the normal cell organelles, including mitochondria, Golgi bodies (dictyosomes), plastids and endoplasmic reticulum, but there are certain exceptions, e.g. sieve tubes, in which most of the organelles disintegrate during differentiation. Mitochondria also vary quite markedly in number and structure in different types of cell and the Golgi bodies show active and quiescent states related to the state of wall growth, secretion and so on. The endoplasmic reticulum varies in abundance and localization in several types of specialized cell, particularly those concerned with secretion. The organelle which shows the most marked differences in various types of tissue is the plastid, the structure of which varies enormously according to whether it occurs in leaf tissue, storage tissue, fruits (as in tomato), or flower parts, such as petals.

1.6 THE PLASTID AND CELL DIFFERENTIATION

Mature plastids are formed from proplastids which are small structures present in meristematic and other cells. Proplastids apparently multiply by division and are distributed between the two daughter cells in cell division. Mature plastids can be of various types, viz. (1) *chloroplasts,* in which chlorophyll is the predominant pigment, although carotenoids, such as xanthophyll, are also present; (2) *chromoplasts,* which contain little or no chlorophyll, so that the yellow, orange or red colours of the carotenoids are visible, and which give the characteristic colours of certain flowers, fruits and autumn leaves, and (3) *leucoplasts,* which are non-pigmented and include plastids which are biochemically specialized to produce storage products such as starch (*amyloplasts*) or fats (*elaioplasts*).

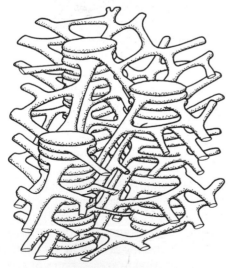

Fig. 1.8. Three-dimensional model of grana with interconnecting membrane system in higher plant chloroplast. (From T. E. Weir *et al., Jour., Ultrastructural Res.,* 8, 122, 1963.)

Although chloroplasts may take many shapes and sizes in the algae, they are fairly uniform in higher plants. They are bounded by a double membrane and consist of a colourless, fluid matrix (*stroma*), within which is a complex system of flattened sacs of membrane, known as *thylakoids,* which are associated in piles (*grana*). The thylakoids, which are circular in outline, are continuous with those in other grana through connecting membranes (Fig. 1.8). The membrane of the thylakoids consists of lipids and proteins, to which the chlorophyll is attached.

Light is normally necessary for the development of a chloroplast from a proplastid by the budding off from the inner membrane of flattened vesicles, which form the collapsed double-membrane lamellae of the thylakoids, and ultimately develop chlorophyll. However, if a seedling is maintained in continuous darkness, instead of the vesicles forming normal thylakoids the material accumulates partly as flattened sheets and partly in a structure known as the *prolamellar body* (Fig. 1.9). The material in the prolamellar body consists of tubules which are arranged in a 3-dimensional para-crystalline lattice. Plastids which have developed in the dark in this way are known as etioplasts. If a dark-grown seedling is exposed to light a remarkable change rapidly takes place, in which the prolamellar body and the membrane sheets break down and the material is reassembled to give the normal membrane structure of the chloroplast of a light-grown leaf and chlorophyll is formed (Fig. 1.9).

1.7 THE CELL WALL AND DIFFERENTIATION

Many of the differences between various types of tissue relate to the cell wall, particularly the *secondary* wall. As we have seen, the primary wall is formed during cell expansion and hence it must have properties which allow stretching, whereas the secondary wall is laid down after elongation has ceased.

Secondary wall formation involves very active synthesis of cellulose, with a lower rate of formation of matrix material. Whereas in the primary

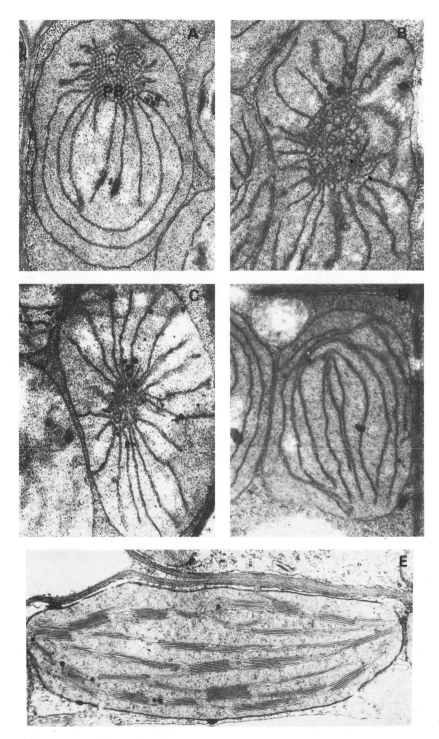

Fig. 1.9. Stages of chloroplast development during the greening of the primary leaves of 14-day old, dark-grown beans under illumination of 3 MW cm^{-2}. A, no illumination; B, 105 minutes illumination; C, 4 hours illumination; D, 5 hours illumination; E, 48 hours continuous illumination. Magnification × 25,000. Note the progressive disappearance of the prolamella body (PB) as the thylakoid system develops to the mature state shown in E. (From J. W. Bradbeer, *in The Molecular Biology of Plant Cells,* (Ed.) H. Smith, Blackwell Scientific Publications, Oxford, 1977.)

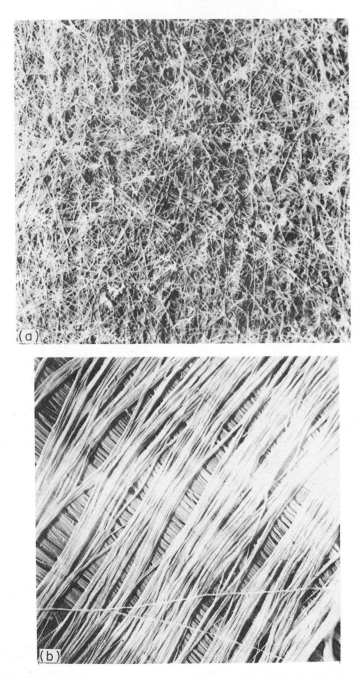

Fig. 1.10. (a) Electron micrograph showing structure of the primary wall of *Valonia* × 8000. (b) As above, but of the secondary wall × 7000. (From F. C. Steward and K. Mühlethaler, *Ann. Bot.* N.S., **17**, 295, 1953.)

wall the microfibrils are frequently laid down in a random manner or are initially orientated transverse to the direction of elongation, in the secondary wall the microfibrils are orientated in precise directions (Fig. 1.10).

In very elongated cells, such as fibres, the microfibrils are oriented parallel to the main axis of the cell, while in others, such as the tracheids of conifers, they are oriented at an angle to the cell axis, and frequently two or three distinct layers of microfibrils can be recognized, which differ in their orientation, sometimes in a "criss-cross" arrangement (Fig. 1.11). In cells of wide diameter, such as vessel elements the microfibrils are formed at right angles to the long axis.

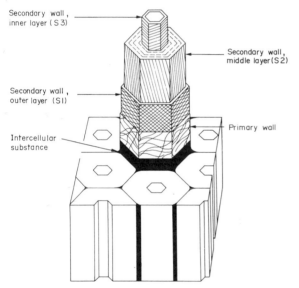

Secondary wall, inner layer (S3)

Secondary wall, middle layer (S2)

Secondary wall, outer layer (S1)

Primary wall

Intercellular substance

Fig. 1.11. Secondary cell wall structure of a wood fibre. (From A. B. Wardrop and D. E. Bland, *Proc. 4th International Congress of Biochem.,* 2, 96, 1959.)

As in the development of the primary wall, the orientation of the microfibrils in the secondary wall appears to be controlled by that of the microtubules lying in the outer layers of the cytoplasm, since there is frequently a close correlation between the orientation of the two structures. The microtubules and the endoplasmic reticulum are involved in the localized deposition of secondary wall which is responsible for the various types of wall thickenings seen in xylem vessel elements. During primary wall growth, the microtubules lie close to the plasmalemma over the whole cell surface, but at the end of primary wall growth the microtubules become localized into groups in cells which are differentiating into vessel elements. In regions of the cell boundary where there are no microtubules the endoplasmic reticulum is concentrated near the plasmalemma (Fig. 1.12). Secondary wall is formed in the regions where the microtubules occur, whereas no secondary wall is laid down in the areas occupied by endoplasmic reticulum, suggesting that the latter prevents the synthesis of secondary wall in these regions. In this way secondary wall thickening is laid down in a regular pattern in xylem cells.

In many types of cell, secondary wall formation is accompanied by lignification. Lignin is a complex polymer, of which the primary building blocks are phenylpropanoid alcohols, such as coniferyl, sinapyl and p-hydroxycinnamyl alcohols. Lignin is deposited throughout the thickness of the secondary wall, and in some cases may even extend to the primary wall.

Communication between adjacent cells after formation of the secondary wall is maintained through the fine strands of cytoplasm which penetrate the walls, known as plasmodesmata. In many cells these are distributed randomly over the cell wall, often in vast numbers (e.g. 20,000 m^2). In other types of cell, plasmodesmata may be closely clustered in areas known as primary pit-fields, when they occur in the primary wall. Secondary wall formation does not occur in these pit-fields, so that communication between cells is maintained. In the tracheids and xylem vessels of some plant species the secondary wall extends beyond the edge of the pit-field, to give a bordered-pit.

1.8 THE ORIGIN OF CELL DIFFERENCES

One of the central problems of development concerns the nature of the processes whereby cells of common lineage are caused to diverge into alternative pathways of differentiation. One such process is by *unequal* or *asymmetric division* of a parent cell, leading to two daughter cells which

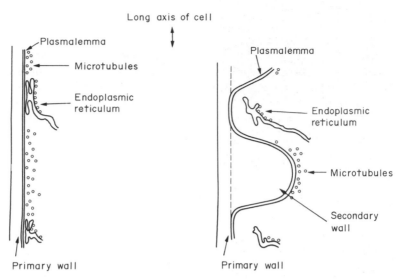

Fig. 1.12. Diagram illustrating the relationship between the endoplasmic reticulum and the pattern of secondary wall deposition in xylem vessel elements. (From Bryant, 1976.)

subsequently follow divergent patterns of differentiation. This process is illustrated by the events leading to the formation of root hairs in certain grasses. During the differentiation of the roots, it can be seen that in certain epidermal cells, which are arranged so that their long axis is parallel to the root axis, the cytoplasm is more dense at the distal end, whereas the proximal end is vacuolated. Mitosis occurs and a transverse wall is formed in a position which gives rise to a smaller daughter cell with dense cytoplasm and a larger cell with less dense cytoplasm (Fig. 1.13). The smaller cell then develops a root hair, whereas the larger daughter cell develops into a normal epidermal cell.

A similar situation is seen in the development of the stomatal guard cells of certain monocotyledons. Here also, certain epidermal cells of the developing leaf show unequal division, and cut off a small cell with densely staining cytoplasm at one end, and a larger cell at the other (Fig. 1.14). The smaller cell becomes a stomatal mother cell and it undergoes a further division, at right angles to the first division, and the two daughter cells so formed are in this case identical and give rise to the guard cells. Thus, where a parent cell shows polarized differences in the cytoplasm along its axis, division is unequal and leads to differentiation between the resulting daughter cells, but where the plane of division appears to divide the cytoplasm equally, the resulting daughter cells are identical.

The fact that unequal division is preceded by polarized differences in the cytoplasm is well illustrated in the development of pollen grains in which the spindle is oriented so that one of the daughter nuclei passes to the end of the cell at which there is a denser cytoplasm, and becomes the generative nucleus, whereas the other daughter nucleus moves to the region of the cell with less dense cytoplasm and becomes the vegetative nucleus (Fig. 1.4B). Occasionally the plane of the spindle accidentally becomes oriented *across* the axis of the pollen mother cell, and in this case two equal cells are formed and the further development of the pollen grain is disturbed.

It is not known how the polarized differences in the cytoplasm in the mother cell arise in the first instance, but the process of division itself is likely to lead to polarization, since the end of a daughter

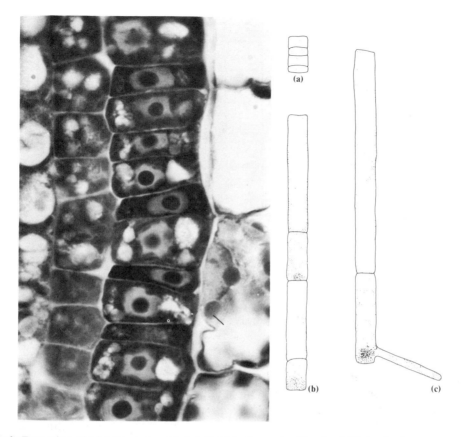

Fig. 1.13. *Left:* Formation of trichoblasts (root hair initials) in the root epidermis of *Hydrocharis morsus-ranae*. Unequal division has given rise in each case to a small cell which is the trichoblast and a large cell which is the epidermal cell (e). (From E. G. Cutter and L. J. Feldman, *Amer. J. Bot.* **57**, 190-201, 1970.)

   Right: Development of root hair initials in the grass *Phleum pratense,* at successive stages of development (a—c). The smaller cells formed by unequal division (a) gives rise to the root hair cell (c). (From E. W. Sinnott and R. Bloch, *Proc. Nat. Acad. Sci., U.S.A.* **26**, 223-7, 1939.)

cell at which the equator was formed is likely to contain a different distribution of organelles from the end at which the pole of the spindle occurred.

In most of the examples we have just considered, the cells derived from unequal division do not themselves undergo further division, but differentiate directly. In some instances, however, the derivative cells from an unequal division undergo several more divisions. For example, in plants of the castor bean (*Ricinus communis*) there are secretory cells which contain tannin and unsaturated fatty acids. These cells rise from an initial unequal division, and one daughter cell undergoes a series of divisions to give rise to a row of cells, each of which becomes a secretory cell.

Thus, the differential state can apparently be transmitted by cell lineage in some instances.

1.9 CELL POLARITY

It is self-evident that plant species show characteristic form and one aspect of this form is that there is typically a well-developed longitudinal *axis,* bearing lateral organs such as leaves and flowers. Differences occur along the axis, so that the two ends are not the same—for example, the plant axis is usually differentiated into a shoot end and a root end. In this respect the axis is said to show *polarity,* which has been defined as "any

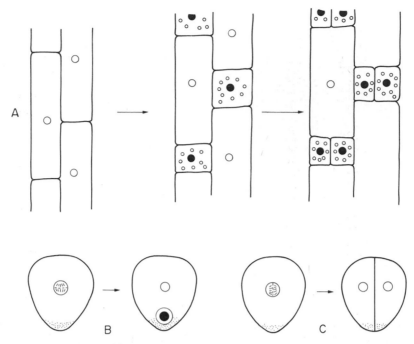

Fig. 1.14. A. Unequal division in formation of stomatal guard cells in leaf of a monocotyledonous plant. Epidermal cells undergo unequal division with the formation of a smaller stomatal mother cell (nucleus shaded) and a larger cell (nucleus not shaded). Chloroplasts develop in the stomatal mother cell, which divides again at right angles to the plane of the first division to produce two daughter cells which develop into the guard cells.

   B, C. Development of pollen grain.

   B. Normal development. Nucleus divides so that one of the daughter nuclei moves to the region of more dense cytoplasm at one end of the cell and becomes one generative nucleus. The other daughter cell nucleus becomes the vegetative nucleus.

   C. Abnormal development. Nucleus divides at right angles to the normal plane. Daughter nuclei thus remains in the same cytoplasmic environment, and two equal cells are formed which distrupts the further normal development of the pollen grain. (Redrawn from E. Bünning, *Handbuch Protoplasmaforschung,* Vol. VII, Vienna, 1958.)

situation where two ends or surfaces in a living system are different". Polarity of the axis of plants is most readily seen from morphological differences, but it is also manifested in several physiological properties, as in the basipetal "polar" transport of auxin in stems, and in the regeneration of buds from the upper end and roots from the lower end of root segments (Fig. 1.15).

Although axial polarity is one of the most striking features of the plant body, it should not be forgotten that there are other forms of polarity. For example, the *dorsiventrality* of leaves involves differences between the upper and lower sides and may be regarded as a form of polarity. There may also be radial polarity in spherical bodies, such as cells of *Chlorella* or apple fruits, where there is a degree of radial symmetry, but there are differences between the inner and outer layers with

respect to both chemical constituents and structure. In the following discussion we shall be concerned almost entirely with axial polarity and the term "polarity" will be taken to refer to this type, unless it is otherwise stated.

Polarization of the plant body into shoot and root in higher plants is initiated by the first unequal division occurring in the zygote (p. 22) and the axis of polarity of the zygote appears to be determined by its position in relation to the surrounding maternal tissue, since the future "root" end is always towards micropyle and the "shoot" end away from it. However, in the brown seaweed, *Fucus,* polarity is not already predetermined in the zygote, which thus provides favourable experimental material, since it is a free living cell in which the plane of the first division can be influenced by various treatments.

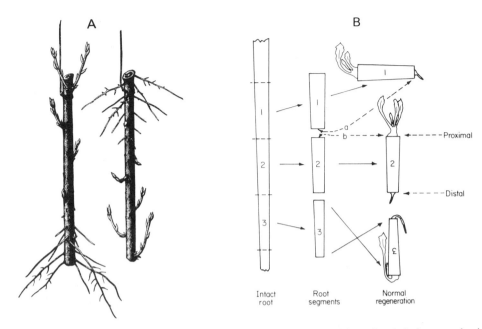

Fig. 1.15. A. Polarity of regeneration in willow stem. *Left*: stem cutting suspended in moist air in its normal orientation. *Right*: stem cutting similarly treated but in inverted position. Roots grew out at the morphologically lower end and shoot buds at the morphologically upper end, regardless of orientation. (Reprinted from *Pfeffer's Physiology of Plants,* 2nd ed., Clarendon Press Oxford, 1903.)
B. Polarity of regeneration in root segments, such as those of *Taraxacum* and *Cichorium*. Shoot buds develop at the proximal end (i.e. that originally furthest from the root tip), and roots at the distal end, regardless of orientation. (Reprinted from H. E. Warmke and G. L. Warmke, *Amer. J. Bot.* **37**, 272, 1950.)

The eggs of *Fucus* are released into the surrounding seawater and after fertilization they settle in a solid stratum, where the development of a localized protuberance is the first sign of a rhizoid, by which the young plant becomes attached to a rock. The position of the rhizoid is determined by the orientation of the spindle in the first mitosis, which in turn determines the plane of the first dividing wall. The orientation of the spindle can be influenced by gradients in various external factors, such as light, heat, pH and osmotic activity. If the eggs are illuminated unilaterally, the shaded side (remote from an incident light) begins to form a protuberance at about 14 hours after fertilization and mitosis takes place so that the axis of the spindle is parallel to the direction of the incident light and a cell wall is formed at right angles to this, cutting off a larger cell which gives rise to a future thallus and a smaller cell which forms the rhizoid (Fig. 1.16). External factors can influence the axis of polarity up to a few hours before the appearance of the rhizoid, after which the position of emergence of the future rhizoid, appears to be irreversibly fixed.

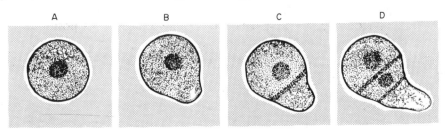

Fig. 1.16. Successive stages in the early development of the embryo of *Fucus*. In C, a wall has been formed giving rise to the rhizoid and thallus cells. In D, further division has occurred.

Before there are any visible signs of emergence of the rhizoid, the nuclear surface becomes highly polarized with finger-like projections (representing membranous extensions of the nuclear envelope) radiating towards the site of rhizoid formation. Mitochondria, ribosomes and fibrillar vesicles are also concentrated in this region of the rhizoidal half of the zygote, which is thus more densely cytoplasmic. At the same time changes occur in the cell wall at the future rhizoid side, with the deposition of the polysaccharide, fucoidin. As a result of these submicroscopic differences between the two halves of the cell, the two daughter nuclei come to lie in different cytoplasmic environments.

These submicroscopic changes are associated with changes in the electrical properties of the eggs. It has been shown that the future rhizoidal side of the cell becomes electronegative with respect to other parts, as a result of which a current is driven through the cell. Moreover, the axis of polarity of an unpolarized egg can be fixed by applying a voltage across it. There is evidence that the electrical polarization leads to a gradient of calcium ions across the egg, as a result of a strong influx of calcium ions in the future rhizoidal region and a calcium efflux in the future thallus region. Indeed, the localized application of a high concentration of calcium ions to one side of the egg of the alga, *Pelvetia,* leads to development of the rhizoid on that side.

Light has a similar effect in determining the plane of the first cell division in the spores of the horsetail, *Equisetum,* and of other lower plants, the first cell wall being at right angles to the incident light.

Once polarity has been induced it becomes extremely difficult or impossible to reverse it. This fact is well illustrated in regeneration experiments. Thus, if we take stem pieces of a plant such as willow and suspend them in a moist atmosphere, they will develop adventitious roots towards the morphologically lower end and the buds will tend to grow out most strongly at the upper end (Fig. 1.15A). Similarly, if we take segments of the roots of chicory, dandelion or dock, and plant them in moist sand, roots will develop mainly from the

morphologically lower (distal) end and buds from the upper (proximal) end (Fig. 1.15B). Thus, although the willow cutting, and still less the root cutting, does not show any morphological differentiation between the upper and lower ends, nevertheless it is clear that these organs possess a marked physiological polarity which affects the pattern of regeneration of roots and buds. This polarity is inherent in the tissues themselves and is not dependent on gravity, illumination or other external conditions, as is shown by the fact that willow cuttings suspended upside down in a moist atmosphere still regenerate roots predominantly at the original, morphologically lower end (Fig. 1.15A), and similar effects are obtained with root segments of chicory which are planted so that the original upper end is now lowermost. Lower animals such as Hydra and planarians show analogous properties in the polarity of their regeneration patterns.

The physiological polarity of pieces of stem and root is a "built-in" property of the tissues. It might be thought that physiological polarity is due to gradients of metabolites or other substances along the stem or root, but this cannot be so, since polarity persists from one season to the next, through the dormant period, when metabolism is proceeding at a low rate, so that concentration gradients of metabolites are unlikely to persist. The basipetal transport of auxin (p. 107) and the occurrence of auxin gradients in the stem must therefore be seen as the *result* of the polarity of the tissues and not the cause of it.

We have seen that a piece of stem shows polarity of regeneration, with a tendency for buds to develop from the upper end and roots from the basal end. If such a stem is divided into two, each half behaves in a similar manner, and this process can be repeated even with very short pieces of stem. If it were possible to carry out this process to the limit, we might conclude that each individual cell exhibits polarity, and indeed there is evidence that this is the case. Thus, the filamentous green alga, *Cladophora,* shows polarity of the plant body, in that the basal end forms a rhizoid. If the cells of a filament of *Cladophora* are plasmolysed, so that the protoplasts are pulled away from the cell walls (which will break protoplasmic connections between cells), and are then deplasmolysed, each cell subsequently regenerates a new filament, developing a rhizoid at the basal end of the cell (Fig. 1.17). This experiment seems to provide clear evidence for polarity of individual cells of *Cladophora*. Comparable evidence for the cells of higher plants is more difficult to obtain, but much indirect evidence supports the hypothesis, such as the occurrence of unequal cell divisions.

Fig. 1.17. Single cell from a filament of *Cladophora* regenerating a thallus from its apical end, and a rhizoid from the basal end. (Redrawn from A. T. Czaja, *Protoplasma,* II, 601, 1930.)

These various observations suggest that (1) polarity of the tissues of a stem or root is remarkably stable and is not easily reversible, (2) polarity persists even throughout dormant periods, when metabolism is proceeding at a low rate, and (3) the polarity of a tissue apparently reflects the polarity of the individual cells. On these grounds, it has been suggested that there must be some permanent, structural basis for cell polarity. The observation that a piece of stem can be divided into several pieces, each of which shows the same polarity of regeneration, is very reminiscent of the fact that a bar magnet can be similarly divided and each piece becomes a small magnet. In a magnet each iron atom possesses magnetic polarity and the atoms become aligned during the

process of magnetization, so that the "North" and "South" pole of each atom becomes oriented along the axis of the bar. By analogy, we might postulate that each cell of polarized organs, such as stems, contains polarized molecules which are oriented along the axis of each cell to form a "cytoskeleton", but attempts to demonstrate such oriented molecules in the cell have so far proved unsuccessful. It is interesting that the only linearly oriented structures that we do know of, the microtubules, lie at *right angles* to the axis of polarization in stems and roots (p. 8). It has been suggested that polar transport of auxin may depend upon polarization of the plasma membranes at the end walls of the cell (p. 110).

1.10 POLARIZED CELL DIVISION

The planes of cell division during the development of an organ play a very important role in determining its final form and shape. Indeed, we may say that without oriented cell divisions there could be no organized form within the plant body, as we see in cultures of callus tissue, where the plane of cell division is at random, and the resulting tissue forms a shapeless and structurally unorganized mass (p. 154). Thus, polarized cell divisions give the plant body its three-dimensional form. For example, in a developing internode the majority of cell divisions are oriented so that the mitotic spindle lies parallel to the axis of the internode. Consequently the internode grows mainly in length and relatively much less in diameter. For gourd fruits of different shape it has been shown that in long, elongated fruits divisions in which the mitotic spindle is oriented parallel to the long axis are much more frequent than divisions in which the spindle is in other planes, whereas in round fruits divisions in one particular plane do not predominate. However, we cannot yet say what determines the planes of cell division along certain axes. It is clear that nuclear genes are involved, since many differences in form, including those affecting fruit shape in gourds, are inherited in a simple Mendelian manner. It would appear that the axis of polarity of the whole organ also has a strong influence on the plane of cell division, by affecting the orientation of the mitotic spindle. How the orientation of the spindle is controlled is not known.

FURTHER READING

General

Clowes, F. A. L. and B. E. Juniper. *Plant Cells,* Blackwell Scientific Publications, Oxford, 1968.
Gunning, B. E. S. and M. W. Steer, *Plant Cell Biology,* Edward Arnold, London, 1975.
Hall, M. A. (Ed.), *Plant Structure, Function and Adaptation,* MacMillan, London, 1976.

More Advanced Reading

Bloch, R. Histological foundations of differentiation and development in plants, *Encyc. Plant Physiol.,* 15(1), 146, 1965.
Gunning, B. E. S. and M. W. Steer, *Ultrastructure and the Biology of Plant Cells,* Edward Arnold, London, 1975.
Quatrano, R. S. Development of cell polarity, *Ann. Rev. Plant Physiol.,* 29, 487, 1978.
Robards, A. W. *Dynamic Aspects of Plant Ultrastructure,* McGraw Hill, London, 1974.
Yeoman, M. M. (Ed.). *Cell Division in Higher Plants,* Academic Press, London, 1976.

2

Patterns of Growth and Differentiation in the Whole Plant

2.1 LEVELS OF DIFFERENTIATION

So far, we have been concerned primarily with growth, rather than differentiation. In the present chapter we shall describe the development of the main organs of the plant and the way in which differentiation arises within the plant.

Now, when we consider the manifold forms of differentiation in the plant it is evident that it occurs at various levels. At the highest level, there is differentiation in the plant body as a whole, as seen in the division into root and shoot. Within the shoot we can observe the differentiation into various organs, such as stems, leaves, buds and flowers, and within each of these organs there is differentiation at the cellular and tissue level. These three levels of differentiation also constitute a series of successive stages in *time*—there is first differentiation into root and shoot in the embryo, and this is followed by the formation of organ primordia, as a result of the activities of the apical meristems. These organ primordia do not at first show differentiation at the cell and tissue level, which occurs during the later stages of their development.

Certain other changes occur during the life cycle of seed plants which must be regarded as aspects of differentiation, of which the most important is the transition to the reproductive phase, which involves a profound change in the structure of the shoot apex (p. 37). We shall see later that in many species the onset of flowering is controlled by environmental factors, but in many other species it appears to be determined more by progressive changes occurring during the development of the plant itself than by environmental factors. Often these progressive physiological changes are reflected in morphological characters, such as leaf shape, in which a gradient up the stem may frequently be seen. These aspects of development will be dealt with later, but it is important to recognize that they represent an aspect of differentiation within the shoot as a whole.

2.2 DIFFERENTIATION INTO ROOT AND SHOOT IN EMBRYO DEVELOPMENT

The first major step in differentiation occurs at a very early stage in the development of the embryo, with the establishment of a shoot end and a root end.

As an example of embryo development in angiosperms we may consider that of the Shepherd's purse, *Capsella bursa-pastoris*. The cytoplasm of the unfertilized egg, of *Capsella* is highly polarized, the half towards the micropyle being filled with a large vacuole, whereas the other "chalazal"

end contains the nucleus and much of the cytoplasm (Fig. 2.1). The first division is unequal and gives rise to a smaller, densely cytoplasmic terminal cell which forms most of the future embryo, and a larger vacuolate basal cell which forms the suspensor (Fig. 2.2). The terminal cell divides by two successive longitudinal divisions, with the plane of the second division at right angles to that of the first, to give four cells; these cells then each divide transversely to form eight cells, which constitute the octant. Each octant cell divides to form an outer protodermal cell, which gives rise to the future epidermis, and an inner cell. The inner cells continue to divide to form the cotyledons and the hypocotyl.

By several successive transverse divisions the basal cell gives rise to a row of cells which forms the suspensor, the end cell of which enlarges and

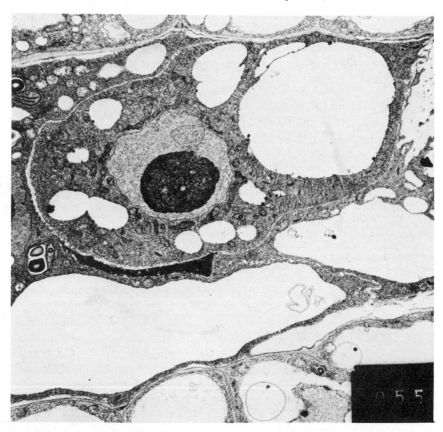

Fig. 2.1. Electron micrograph of the zygote of *Capsella bursa-pastoris,* showing polarization of the cell as manifested by the occurrence of the nucleus lying in the dense cytoplasm at the micropylar end (*left*) and a large vacuole at the chalazal end (*right*). (From Sister Richardis Schultz and W. A. Jensen, *Amer. J. Bot.* **55**, 807-19. 1968.)

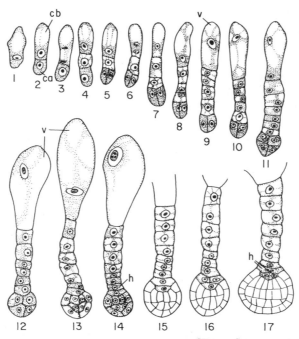

Fig. 2.2. Early stages in the development of the embryo of *Capsella bursa-pastoris*. Note the initial unequal division into a larger basal cell (cb) and smaller terminal cell (ca). The basal cell gives rise to the suspensor. (From A. Fahn, *Plant Anatomy*, Pergamon Press, Oxford, 1967, adapted from Souèges, 1914.)

becomes sac-like. The suspensor cell nearest the embryo undergoes several divisions, to give a group of cells, of which the outer ones form the future root cap and root epidermis, while the inner ones form the remainder of the radicle. The fully developed embryo is formed by further repeated divisions of these various regions.

There is considerable variation from the pattern of development described for *Capsella* among the various groups of angiosperms, but the details do not concern us here. However, whatever the variation in further development, the initial stages have certain features in common, namely that the first division of the zygote gives rise to two unequal cells, and of these, the basal cell is normally the one nearer the micropyle of the ovule, and gives rise to the root end of the embryo, whereas the terminal cell gives rise to the shoot end. Thus, even the very young embryo shows polarity, in that it has a shoot end and a root end. Indeed, the egg itself shows differences in the density of cytoplasm between its two ends, suggesting that the first un-

equal division of the zygote is already predetermined by polarization in the unfertilized egg.

After the embryo has become differentiated into root and shoot regions, apical meristems are established and some organs become differentiated, often while the seed is still developing on the parent plant, so that not only cotyledons but also a rudimentary epicotyl and, in grasses, even several leaf primordia, may be present.

2.3 SHOOT APICAL MERISTEMS

Although this book is primarily concerned with flowering plants, it is useful to consider, briefly, the patterns of growth in the lower plants. In simple filamentous algae, such as *Spirogyra*, every cell appears to be potentially capable of division and growth and is not localized to particular regions. However, in many algae there is marked localization of growth. Thus, the alga, *Chara*, has a single prominent apical cell, which divides repeatedly, giving a larger outer (distal) cell, which continues to function as the apical cell, and a smaller proximal daughter cell, which proceeds to undergo further division, the resulting cells forming the mature tissue of the thallus.

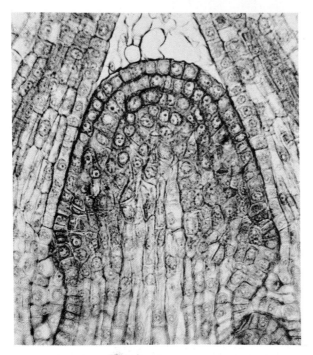

Fig. 2.3. Median section through the shoot apical meristem of *Alternanthera philoxeroides,* showing the two-layered tunica overlying the central corpus. (From E. G. Cutter, *Phytomorph.* **17,** 437, 1967.)

In the bryophytes and many pteridophytes, also, the shoot grows by a single well-marked apical cell, which normally has the form of an inverted tetrahedron, and it divides so that the three "inner" faces cut off daughter cells in succession and these latter cells undergo further division to form the tissues of the shoot.

The early plant anatomists of the nineteenth century were so impressed with the essential unity of structure in vascular plants that they expected to find single apical cells also in gymnosperms and angiosperms and indeed described such cells. Later, however, it became apparent that there is no clearly recognizable *single* apical cell in the shoots of higher plants, but two zones may be

distinguished in shoot apical region of flowering plants: (1) the outer *tunica* or mantle, which surrounds and envelops, (2) the inner *corpus* (Fig. 2.3). These zones can be distinguished fairly sharply by the predominant planes of cell division. In the tunica the divisions are predominantly *anticlinal,* i.e. with the axis of the mitotic spindle parallel to the surface, so that the resulting crosswall separating the two daughter cells is perpendicular to the surface. The corpus, on the other hand, is characterized by the fact that divisions occur in all planes, viz. both anticlinal and *periclinal* (i.e. the spindle is perpendicular and the new wall parallel, to the surface). The thickness of the tunica is somewhat variable and it may consist of one, two or more layers of cells, according to the species. Moreover, even within a single species the number of tunica layers may vary according to the age of the plant, the nutrient status and various other conditions.

It should be noted that the tunica-corpus theory is largely descriptive and simply based on what can be observed—it makes no predictions regarding the future destiny of the tunica and corpus cells. The epidermis does, of course, arise from the outer layer of the tunica, but there is considerable variation between different species in the extent to which the two layers may contribute to the origin of leaves and buds (see p. 26). However, although we should not regard the tunica and corpus as rigidly and permanently demarcated, it is possible to show that in some species the outer tunica layers remain remarkably distinct from the deeper tissue for long periods. One method by which this has been demonstrated is to induce polyploidy in the cells of the shoot apices of *Datura* and of maize by treatment with colchicine. The tetraploid cells so formed can be recognized by their larger nuclei. If tetraploid cells are formed in the outer layer of the tunica, for example, all the resulting daughter cells will show large nuclei and it can be seen that such cells are strictly limited to a single layer of the tunica (Fig. 2.4), indicating that periclinal divisions are very rare and that each layer of the tunica retains its identity remarkably constantly. Each of the layers of the tunica arises from a set of initial cells at the shoot apex, and the

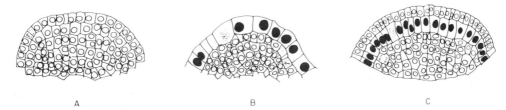

Fig. 2.4. Longitudinal sections of the shoot apex of *Datura*. A. Normal apex, with all cells diploid. B. Apex after treatment with colchicine, showing polyploid nuclei in outer tunica layer. C. After treatment with colchicine, with polyploid nuclei in inner tunica layer. (Adapted from S. Satina, A. F. Blakeslee and A. G. Avery, *Amer. J. Bot* **27**, 895, 1940.)

corpus apparently arises from its own set of initials, although there is some difference of interpretation on this point.

The distinction between tunica and corpus is even less definite in gymnosperm shoot apices, and although the divisions in the outer layers of *Pinus*, for example, are predominantly anticlinal, there are also quite frequent periclinal divisions as well.

In addition to the tunica and corpus and their initials, it is possible to recognize a number of other zones in the apices of some species. One simple form of zonation is shown in Fig. 2.5. At the summit of the apical dome there is a group of

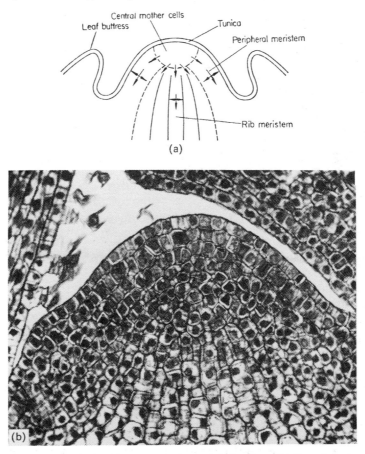

Fig. 2.5. (a) Generalized diagram to illustrate zonation in the shoot apical region of a flowering plant. (b) Median section through the vegetative apex of *Chrysanthemum morifolium*, in which the various zones shown in A may be distinguished. (From R. A. Popham, *Amer. J. Bot.* **37**, 476, 1950.)

initial cells which divide mainly anticlinally, and so give rise to the tunica, but the lower layers may also divide periclinally. Below these initial cells there is a group of larger cells, known as the *central mother cells,* which appear to have a low rate of division. Divisions at the boundary of the central mother zone give rise to (1) a zone of actively dividing cells which form the flanks of the apex and is sometimes called the *flank* or *peripheral meristem*; and (2) a zone of cells which divide mainly along the shoot axis and give rise to the longitudinal rows of cells of the cortex and pith and hence is sometimes called the *rib meristem.* The distinction between these zones, which have been given various names by different workers, is somewhat arbitrary, since their boundaries are ill defined and grade into each other. Moreover, there is considerable variation between species, both with respect to the general shape of the apical region and to the number of zones which can be recognized within it, and yet the apices of all these species produce similar end products, viz. stem, leaves and buds. It would seem, therefore, that not much morphogenetic significance can be attached to the various zones which can be recognized in the shoot apex.

2.4 THE INITIATION OF LEAVES AND BUDS

A leaf originates from the periclinal divisions of a group of cells on the flank of the apex (Fig. 2.10). As a result of these divisions in a localized area, a small protuberance (Figs. 2.9 and 2.10) is formed and gives rise to the future leaf primordium. The number of layers involved in these initial divisions varies considerably in different species. In many grasses, the peridinal divisions commence in the outermost layer of the tunica, and in the layer below. In other monocotyledons and in dicotyledons, periclinal divisions do not take place in the outermost layer, but in the layers below it. Thus, the extent to which the tunica and corpus are involved in the initiation of the primordium varies greatly. In many species the initial divisions involve both the tunica and the corpus, while in others they may occur in only one or other

of these layers. Variations may occur even within a single species.

Lateral buds usually appear somewhat later than leaf primordia, in the sequence of developmental changes seen at the shoot apex. Buds arise in the outer layers of the stem tissues, as a result of cell divisions which may be predominantly anticlinal in the outer layers, or both anticlinal and periclinal in the deeper layers. As with leaves, there is considerable variation between species in the extent to which tunica and corpus are involved in these cell divisions giving rise to lateral buds. As a result of these cell divisions the bud emerges as a protuberance and it soon develops an apical structure similar to that of the main shoot apex for that species.

2.5 THE SITING OF LEAF PRIMORDIA

The siting of leaf primordia at the shoot apex is a rather precisely regulated process, although there are considerable differences from one species to another in the pattern of arrangement of leaf primordia. In considering this problem it has to be remembered that the apex is growing continuously and that as it does so the older leaf primordia are left behind on the flanks of the apex and they steadily increase in size as they do so (Fig. 2.6). As the apex grows, new leaf primordia are being continuously initiated above the existing ones.

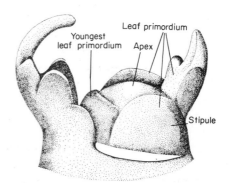

Fig. 2.6. Vegetative shoot apex of *Vitis.* (From A. Fahn, *Plant Anatomy,* Pergamon Press, Oxford. 1967.)

Fig. 2.7. Diagram to illustrate spiral phyllotaxis as seen in cross-section through shoot apical region of *Saxifraga* (genetic spiral shown by broken line). (From F. Clowes, *Apical Meristems*, Blackwell Scientific Publications, Oxford, 1961.)

The siting of leaf primordia at the apex determines the arrangement of leaves on the mature shoot, for which the term *phyllotaxis* is used. The most common type of leaf arrangement is *spiral* phyllotaxis, in which it can be seen that a line drawn through successively older leaf primordia at the shoot apex forms a spiral, known as the *genetic* or *developmental spiral* (Fig. 2.7). The mathematical treatment of spiral phyllotaxis has interested botanists for more than a century, but here we shall consider the problem only so far as it relates to the siting of primordia at the shoot apex.

The fern apex provides very convenient material for studying phyllotaxis, since it is relatively flat and the primordia are well spaced out (Fig. 2.8). In considering this problem it is useful to use a system of nomenclature in which the youngest primordium is called P_1 and the successively older primordia, P_2, P_3, etc. The next primordium to arise is referred to as I_1 (i.e. Initial$_1$) and the successively *younger* primordia as I_2, I_3, etc. If we draw radii from the centre of the apex to two successive leaf primordia, it is found that the angle of divergence between the two lines varies from one species to another, but where there are numerous primordia at the apex, it is found to approach a "limiting" value of 137·5°. It is also found that in many species the radial distance of successive primordia increases in geometric progression, indicating that there is a corresponding increase in

the rate of expansion of the apex with distance from the centre. Thus, we can reproduce the spiral phyllotaxis seen at a shoot apex by marking points consecutively around a centre at a constant divergence of 137·5° and radially at a distance from the centre which increases in geometric progression. if we then join the successive points, we shall obtain a genetic spiral.

What determines that the next primordium will be formed at a position which will cause an angular divergence from the preceding primordium of approximately 137·5°? There has been a great deal of controversy on this question, but at present there are two main theories to account for the observed facts, which may be referred to as the "Repulsion Theory" and the "Available Space Theory", respectively. Both theories postulate that the positions in which the leaf primordia arise are determined by the positions of older primordia. According to the Available Space Theory

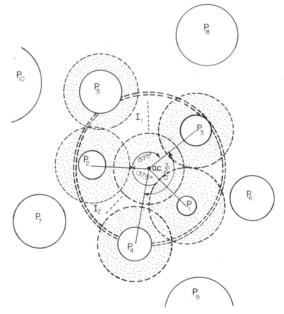

Fig. 2.8. The growth centre and field concept as it may apply to the apex of *Dryopteris*. The apex is seen from above, *ac,* apical cell; P_1, P_2, P_3, P_4, etc., leaf primordia in order of increasing age; I_1, I_2, the next primordia (as yet invisible) to be formed in that order. The large double-dashed circular line indicates the approximate limit of the apical cone. The subapical region lies outside the double-dashed circle. The hypothetical inhibitory-fields around the several growth centres are indicated by stippling. (Adapted from C. W. Wardlaw, *Growth,* **13,** Suppl. 93, 1949.)

(first put forward by Hofmeister in 1865) a certain minium space between existing primordia and the centre of the apex is necessary before a new primordium can arise. As the apex grows, the spaces between existing primordia increase in size and it is postulated that the next primordium (I_1) arises between P_2 and P_3 because this is the first area to reach the necessary minimum size. The next primordium after this (I_2) will arise between P_2 and P_4, and so on.

The Repulsion Theory was put forward by Schoute (1913) who postulated that the centre of a leaf primordium is determined first, and that a specific substance is produced which inhibits the formation of others in the immediate vicinity, so that new primordia again arise in the gaps between older ones, where they will presumably be outside the inhibitory fields of the neighbouring primordia (Fig. 2.8). Inhibition of new centres by the main apex was also postulated, so that new primordia are prevented from forming within a minimum distance from the summit of the apex. There is, indeed, evidence for the existence of mutual inhibition between growth centres.

These two hypotheses are not necessarily exclusive, since the "available space" may not be determined simply by the superficial area between adjacent primordia, but by freedom from their inhibitory influence. Both theories postulate that phyllotaxis is determined primarily by the geometry of the shoot apex. It can be shown, from purely geometrical considerations, that I_1 will arise at a divergence of 137·5° from P_1, if it occurs at a point which divides the angle between P_2 and P_3 in the inverse ratio of their respective ages; that is, the new primordium will not be equidistant from the neighbouring primordia, but will be displaced towards the older (and larger) of these (Fig. 2.8). It is not clear what is the significance of this fact, but it is consistent with the hypothesis that the neighbouring primordia play an important part in determining the position of a new primordium.

Attempts have been made to test the "available space" and "repulsion" theories by surgical experiments on the shoot apex. For example, in the experiments with *Lupinus albus,* R. and M. Snow made radial cuts in the area at which I_2 would be expected to arise, thereby reducing the space available, and it was found that no primordium developed in this space, presumably because it was reduced below the minimum area necessary for primordium development. On the other hand, Wardlaw isolated I_1 of fern apices by two radial cuts and found that it then grew *more rapidly* than normally (Fig. 2.11B), suggesting that it had thereby been released from the inhibitory effects of neighbouring primordia. Although these and other surgical experiments have produced interesting results, they have not given decisive evidence in favour of either of the two theories.

Recently, a mathematical model for phyllotaxis has been put forward by Thornley, the basic assumptions of which are very similar to those postulated in the Repulsion Theory. In this model it is assumed that each leaf primordium is the site for the production of a "morphogen", M. The strength of M as a source of M depends upon its age, size and distance from the apex. The morphogen is postulated to move by diffusion from the leaf primordia and to be degraded at a rate proportional to its concentration. It is assumed that the initiation of new primordia can only take place when the concentration of M drops to a certain critical value, this assumption being identical to that in the earlier Repulsion Theory. The mathematical model based upon these initial postulates predicts that under certain conditions the angle between successive primordia will approach the observed value of 137·5°. Thus, this approach also shows that the observed arrangement of leaf primordia at the shoot apex can be predicted from certain rather simple initial assumptions.

2.6 DEVELOPMENT OF THE LEAF

The overall development of the leaf may be divided into the following steps: (1) formation of the foliar buttress, (2) formation of the leaf axis, and (3) formation of the lamina. The following account is based upon the development of the tobacco leaf.

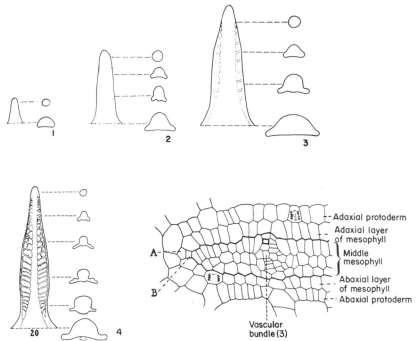

Fig. 2.9. Diagrams of longitudinal and cross-sections of leaf primordia of *Nicotiana tabacum* at different ontogenetic stages. 1. A young, more or less cone-shaped primordium. 2. Primordium in which the narrow margins, from which the lamina will develop, can be seen. 3. Primordium in which the beginning of development of the main lateral veins can be seen. 4. Primordium 5 mm long in which the early development of the provascular system can be seen. 5. Cross-section of the marginal region of tobacco leaf showing submarginal initial and the origin of the mesophyll and a vascular strand. A, B, submarginal initials. (From G. S. Avery, *Amer. J. Bot.* **20**, 565, 1933.)

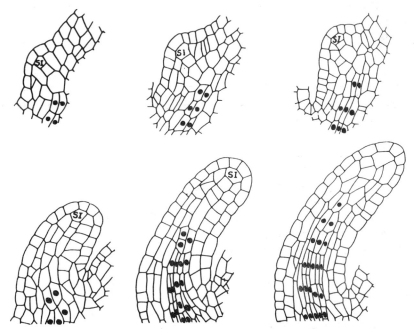

Fig. 2.10. Early stages in development of leaf primordium in flax (Linum), showing development of procambial strand (cells with nuclei indicated). S.I., submarginal initial cell. (From G. Girolami, *Amer. J. Bot.* **41**, 264, 1954.)

As we have seen (p. 26), a small protuberance arises on the surface of the flanks of the meristem (Fig. 2.10) by periclinal divisions in the surface layers. Certain cells towards the centre of this *foliar buttress* now begin to divide actively and a small finger-like protuberance emerges from the buttress (Fig. 2.9). This protuberance proceeds to grow in size by the activities of apical cells until it is about 1 mm long. At this stage the leaf primordium consists of little more than the future axis (midrib and petiole) of the leaf. Soon, however, certain cells on its flanks begin to grow, so that it now acquires a more flattened appearance in cross-section. These dividing cells on the flanks of the mid-rib constitute the *marginal meristems* and they give rise to the future lamina of the leaf. The tip growth of the mid-rib ceases fairly early, when the total leaf length is 2-3 mm and further growth is more generally dispersed.

The marginal meristems consist of superficial *marginal initial* cells, together with the underlying *submarginal initials* (Figs. 2.9, 2.10). In dicotyledons, the marginal initials normally divide only anticlinally, to give rise to the surface layers of the leaf, whereas the submarginal initials form the inner tissues of the future leaf. As a result of these activities of the marginal and submarginal initials a definite number of layers is established in the young lamina and under any given set of environmental conditions this number remains rather constant over most of the leaf, throughout its future development. The relative constancy in the number of layers arises from the fact that the cells continue to divide mainly anticlinally, i.e. at right angles to the surface of the leaf, so that there is a steady increase in area but not in thickness.

The different layers of the leaf are found to cease cell division at different stages. The cells of the upper epidermis usually cease dividing first, when the tobacco leaf is only 6-7 cm long, i.e. when it is only one-fifth to one-sixth of its final size. However, these epidermal cells continue to increase in size until the whole leaf ceases to enlarge. On the other hand, the palisade cells continue to divide and so give rise to a closely packed layer of cells which keeps pace in its growth with the upper epidermal cells. They cease dividing and

enlarging shortly before the epidermal cells do so, so that to keep pace with the latter they are pulled apart slightly in the final stages, thus giving rise to the intercellular spaces between the mature palisade cells. The cells of the future spongy mesophyll cease dividing and enlarging earlier than the cells of the palisade, so that the cells become pulled apart more in the final stages, giving rise to larger intercellular spaces than in the palisade layers and to a more irregular arrangement of the cells. This study of the origin of the different layers of the mesophyll of the leaf provides a good example of how a study of the morphology of development can give a better understanding of the way in which the mature structure of an organ arises.

The first procambium is formed towards the base of the developing leaf primordium at a very early stage (Fig. 2.10). This first procambium forms the future mid-vein and it develops both outwards towards the tip (acropetally) and downward (basipetally) to link up with the procambium of the stem (Fig. 2.14). The first vascular elements to appear are those of the protophloem, followed later by the protoxylem. Shortly after, the lateral meristems start to form the lamina, and the first signs of the lateral veins appear when the primordium is about 1·55 mm long. The connecting veins soon appear, towards the tip (Fig. 2.9).

Complete normal leaf development depends upon exposure to light and is one of the important aspects of "photomorphogenesis" (p. 214). Hormones, especially auxin and cytokinin, also appear to play an important part in leaf growth (p. 125).

2.7 LEAF DETERMINATION

The differentiation of the shoot into leaves, buds and stems commences, as we have seen, with the formation of leaf and bud primordia at the shoot apex.

The early stages of development are very similar in both leaves and buds, but whereas the leaf primordium very early assumes a dorsiventrality (i.e. it becomes flattened, with upper and lower surfaces), the bud remains radially symmetrical. Moreover, whereas the apical meristematic cells of the leaf primordium cease to be active at an early stage and the development of the leaf is determinate, the bud primordium develops a typical shoot apical meristem which shows indeterminate growth.

Although we know from its position and sequence that a given primordium is normally destined to become a bud or a leaf, a very young primordium is not yet irreversibly determined to become one or the other. Indeed, in the early stages of their development, the primordia which, by their position, would normally give rise to leaves, may be converted into buds by certain surgical treatments. These techniques were first developed by R. and M. Snow and have since been used extensively by others, especially by Wardlaw and his associates, using the fern apex. The interconvertibility of leaf and bud primordia has been demonstrated in *Dryopteris dilatata,* by making a deep tangential cut between the shoot apical cell and a very young presumptive leaf primordium, as a result of which it develops into a radially symmetrical bud instead of a dorsiventral leaf (Fig. 2.11A). This conversion of a leaf into a bud can

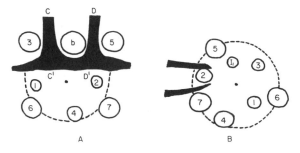

Fig. 2.11. A. Shoot apex of *Dryopteris* in which the I_1 position was isolated from the apical cells (ac) by a tangential incision (AB) and from leaf primordia 3 and 5 by radial incisions (CC[1] and DD[1]). A bud has been formed in what was normally a leaf position. B. Shoot apex in which leaf primordium, P_2, has been isolated from P_5 and P_7 by radial incisions: P_2 grows rapidly and is soon larger than older primordia. (From C. W. Wardlaw, *Proc. Linn. Soc,* **162**, 13, 1950-1.)

only be effected with very young primordia. When a leaf primordium is a little older it remains as a leaf when such surgical treatment is applied. Thus, the primordia produced on the shoot apex are at

first uncommitted, and they only become determined as leaves at a later stage.

A different approach to the problem of leaf determination has been made by Steeves using sterile culture methods. When primordia are removed from the apex of the fern *Osmunda cinnamomea* and placed on a sterile nutrient medium they are able to undergo further growth and development (Fig. 2.12). When the youngest primordia, P_1—P_5, were so tested they developed not as leaves but as shoots which eventually became rooted plants. Progressively older primordia showed an increasing tendency to develop as leaves, and P_{10} always developed as a dorsiventral leaf. From these experiments, Steeves concluded that leaf primordia of *Osmunda* are not irreversibly determined from their inception, but undergo a relatively long period of developing during which they remain undetermined. Determination is gradually imposed on the primordium.

Once determination has occurred, the future development of the complex pattern of the leaf is self-controlled, as shown by the fact that isolated leaf primordia in culture appear to go through all the normal stages of development, even though the resulting leaves are minute.

2.8 LEAF SHAPE

The very wide range of variation in leaf shape in seed plants needs no emphasis. The shape of the mature leaf is determined by three factors: (i) the shape of its primordium; (ii) the number, distribution and orientation of cell divisions; (iii) the amount and distribution of cell enlargement.

The form of the early leaf primordium varies considerably from species to species. As we have seen, in tobacco the young primordium has a simple, finger-like structure, but in maples (*Acer*), the development of the primordium forming the midrib is shortly followed by the appearance of two lateral branches at its base, and these three finger-like structures give rise to the main veins of the leaf. In a compound leaf such as that of ash (*Fraxinus*), a number of lateral lobes are formed from the central primordium, and these give rise to the

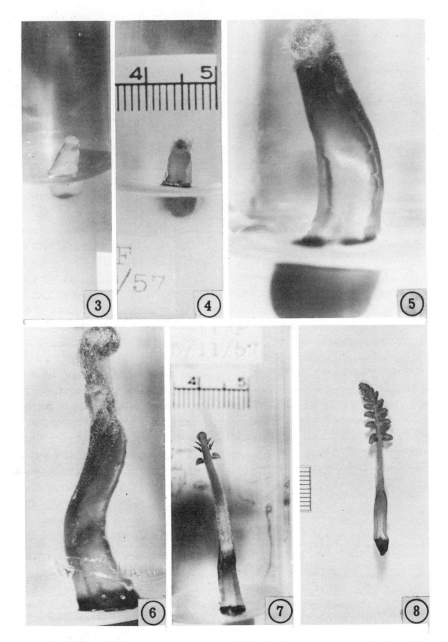

Fig. 2.12. Sterile culture of isolated leaf primordia of the fern, *Osmunda cinnamomea,* on a simple medium. Successive stages in the development of the leaf, from the earliest stage (top left). (From J. D. Caponetti and T. A. Steeves, *Can. J. Bot.* **41**, 545, 1963.)

leaflets of the mature leaf. The subsequent development of each leaflet resembles that of a simple leaf.

The comparative growth rates of the lamina and of the main veins have a profound effect on the ultimate form of the leaf. If lamina growth keeps pace with that of the main veins, then a leaf of simple outline results. On the other hand, if growth is more vigorous near the veins than in the other regions, then a lobed leaf will be formed, the ultimate shape depending also upon the pattern of vein development. In maple, for example, the early localized lamina growth around the main veins is normally followed by a wing-like growth to form a continuous sheet of lamina joining the veins, so that a *palmate* leaf is produced, but in some genetical variants the growth of the lamina is restricted more to the regions adjacent to the veins, so that a more "dissected" type of leaf is formed.

Leaf shape may be profoundly modified by environmental factors; for example, the submerged leaves of some aquatic plants, such as *Sagittaria* and *Ranunculus* spp., have a very different form from that of the aerial leaves. In certain species a considerable number of genes which cause wide variations in leaf shape are known. The successive leaves formed on the stem from the seedling stage onwards very commonly show characteristic changes in shape (p. 253).

2.9 DIFFERENTIATION IN THE STEM

As we have seen, the cells in the apical meristem itself are generally small, densely cytoplasmic, have large nuclei, and are non-vacuolated. As we pass downwards from the apex to the regions in which cell vacuolation and differentiation begins to be apparent in the pith and cortex, we notice that there is a zone of cells between these two latter regions which is characterized by smaller, deeply staining cells, as seen in cross-section (Fig. 2.13).

The latter zone gives rise to the future procambium. The level at which the procambium can be recognized varies considerably from species to species, but it commonly can be recognized in the zone of leaf initiation. At the highest level these densely staining smaller cells may be seen to form a complete ring, but not all the cells of the ring are destined to form the future vascular tissues and lower down the stem it can be seen that the ring has become broken into discrete strands, by vacuolation of some of the intervening cells of the former ring. These strands constitute the first clearly delimited procambium. Although the cells of the strands appear relatively small in transverse section, in longitudinal section they can be seen to be elongated and spindle-shaped.

As we have seen (p. 30), the development of the procambium is closely associated with leaf

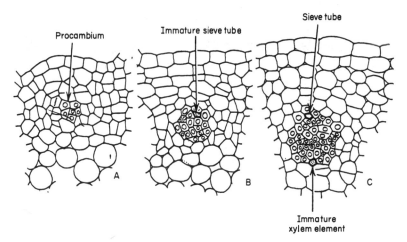

Fig. 2.13. Successive stages in the development of the procambium (cells with nuclei) in transections of a stem of *Linum perenne*. (All × 430. From K. Esau, *Amer. J. Bot.* **29**, 738, 1942.)

development. Very detailed studies on the shoot apex of *Populus* have shown that procambial strands can be observed below the position of a future leaf primordium before the cell divisions leading to its formation have commenced. As the leaf primordium develops, the procambium extends into it; protophloem differentiates acropetally from the older protophloem below, whereas protoxylem differentiates both acropetally and basipetally from a point at the base of the leaf primordium. The basipetal development results in the leaf trace forming connections with other vascular strands in the stem (Fig. 2.14).

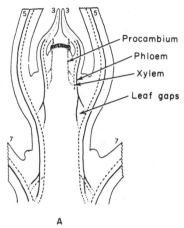

A

Fig. 2.14. Diagram illustrating the initial vascular differentiation in a shoot with a decussate leaf arrangement, as seen in longitudinal section. Continuous lines indicate phloem and broken lines indicate xylem. (From K. Esau, *Plant Anatomy*, John Wiley & Sons, Inc., New York, 1953.)

2.10 THE SHOOT APEX AS A SELF-DETERMINING REGION

The observation that procambial strands can be detected below the position of a leaf primordium before the latter has emerged raises the question as to whether the procambium plays a role in determining the initial siting of leaf primordia. If this were the case, then it would imply that the activities and organization of the shoot apex are influenced and controlled by the already differentiated regions of the shoot. However, several lines of evidence seem to argue against this conclusion.

Firstly, it is clear that an organized shoot apex must arise *de novo* during the development of the embryo, and this may occur in free cell cultures in which the developing embryo is independent of the influence of any surrounding vascular tissues (p. 157). Similarly, shoot buds may arise spontaneously from undifferentiated tissues; for example, if chicory (*Cichorium*) or dandelion (*Taraxacum*) roots are cut in pieces, under appropriate conditions they will regenerate shoot buds from the cut surface at the upper end (Fig. 1.15). Adventitious buds will also develop in callus cultures of various species (p. 157). These observations suggest that the shoot apex represents a stable configuration of cells which will as it were, "crystallize out" from an undifferentiated mass of tissue under appropriate conditions. Thus, the shoot apex appears to be a self-organizing region.

Further evidence that the shoot apex is a self-determining region, and that it behaves as an organized whole, is provided by the observation that when the apex of *Lupinus alba* is divided into four sectors by vertical radial cuts through the centre, each of these sectors regenerates into a normal apex.

Other experiments have included the surgical isolation of the apex from surrounding tissues by four vertical incisions, so that it stands on a plug of parenchymatous tissue (Fig. 2.15). Under these conditions, in which the influence of the vascular tissue of the older parts of the shoot was removed, the apices of both *Lupinus* and the fern, *Dryopteris,* continued to behave in a perfectly normal manner and produced new leaf primordia in the normal phyllotactic sequence, indicating, once again, the self-determining properties of the shoot apex. Indeed, so far from being controlled by the acropetal development of procambial tissue, there is much evidence that developing buds and leaves exert a stimulatory effect on the differentiation of vascular strands in the stem tissue below. For example, if young leaf primordia are removed from a shoot apex at a very young stage, vascular tissue fails to develop in the stem or, in the case of ferns, may be greatly modified or reduced. On the other hand, in callus cultures of chicory or lilac there is no differentiated vascular

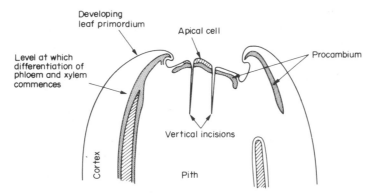

Fig. 2.15. Experiment involving surgical isolation of the shoot apical meristem of *Dryopteris aristata* from existing vascular tissues. Longitudinal section showing two of the vertical incisions which isolated the apical meristem on a plug of parenchymatous pith cells. Consequently, the only connection left between the apical meristem and existing vascular tissues was via parenchyma, yet the apex continued to grow, produced new leaf primordia, and developed normal vascular tissues. (Redrawn from C. W. Wardlaw, 1947.)

tissue, but if a bud is grafted into the callus, then vascular strands develop in the callus below the bud (p. 120).

The stimulatory effect of a bud on the development of vascular tissue appears to be due to the hormones, especially auxins and gibberellins, which it produces (p. 121). There is also evidence that hormones play an important role in the regeneration of vascular tissue (p. 121).

From these various types of evidence it is apparent that in respect of many of its activities the vegetative shoot apex is a self-determining region. On the other hand, when flowers are initiated, so that a vegetative shoot apex is converted into a flowering one, it appears that in many species this transition occurs under the influence of a stimulus arising within the mature leaves in the older parts of the plant (p. 222).

2.11 ROOT APICES

The apical region of roots shows both similarities and differences in comparison with shoot apices. No lateral organs such as leaves are initiated at the root apex, and hence growth is more uniform and there is no division into nodes and internodes. On the other hand, the apex of the root is covered by a root cap, which is not represented in the shoot apex.

It is possible, by careful study of the patterns of division in the root apical region, to trace back the origin of the main zones of the differentiated root, the epidermis, cortex and vascular cylinder, to certain groups of initial cells in the main zone of cell division, the *promeristem* (Fig. 2.16A). It has proved remarkably difficult to identify the initial cells with certainty, however, and there is still considerable difference of opinion as to the number of initial cells in roots. Some authors have claimed that there are relatively few initial cells—perhaps only three, or even one. On the other hand, Clowes has produced evidence that, in the root tip of *Zea mays* and other species, there is a group of rather inert cells which constitute the *quiescent centre* (Fig. 2.16B) and that the actively dividing initial cells occur at the boundary or "surface" of the quiescent centre. The zone of actively dividing cells takes the form of an inverted "cup". Several techniques have been used to study this problem, including autoradiographic studies to show the zones of active DNA synthesis and cell division, using a radioactive DNA precursor, such as ^{3}H-thymidine. Nuclei which show active DNA synthesis will incorporate the ^{3}H-thymidine, whereas inert cells will show no incorporation (Fig. 2.17). In this way, the existence of the quiescent centre can be clearly shown in various species. The function of the quiescent centre is still obscure.

The initial cells and their immediate daughter cells are non-vacuolated and cell division proceeds

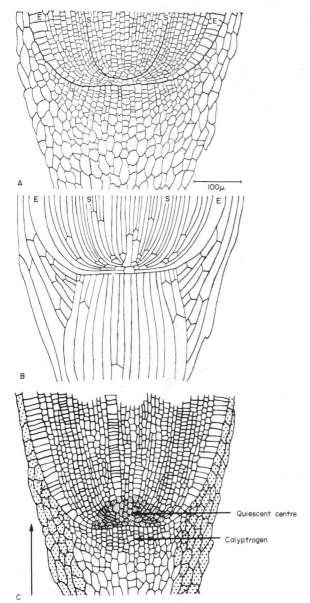

Fig. 2.16. A, B. Median longitudinal section of the root of maize (*Zea mays*). In A the outlines of the individual cells are shown, while in B the cell-lineages which can be traced back to the meristem region are outlines. E, epidermis; S, outer layer of stele. (From Clowes, 1961.) C. Longitudinal section of the root of maize (*Zea mays*), showing the position of the quiescent zone. (From Clowes and Juniper, 1968—see Further Reading.)

sharply separated into regions of cell division and cell extension, but in others (e.g. beech, *Fagus sylvatica*) there is a certain amount of division in cells which are beginning to vacuolate.

The boundaries of the future vascular cylinder, cortex and epidermis of the root become recognizable at a short distance back from the apical initials. These zones can be distinguished by the sizes of the cells and by the planes of division; the cells of the inner layers of the cortex tend to develop by periclinal divisions, whereas the planes of the cell walls of the future vascular tissue are less regular.

The endodermis is recognizable at an early stage, as the innermost layer of the cortex, while the outermost layer of the vascular cylinder, the pericycle, is recognizable within 100μ or less of the apical initial region in some roots.

The procambium develops acropetally and the differentiation of the xylem and phloem follows in the same direction. The first observable changes to occur in the central vascular cylinder is the blocking out of the future xylem groups, by the radial enlargement of certain cells. On the other hand, the first cells to differentiate into mature cells are the sieve elements of the protophloem, which may occur within a distance of 230μ from the promeristem, in slow growing roots. Thus, it is clear that histogenesis may occur at a very short distance from the promeristem itself (Fig. 2.18).

It was shown earlier than, in many respects, the shoot apex is a self-determining region, and this appears to be true also for the root apex. Thus, the pattern of vascular differentiation appears to be controlled by the apex itself. For example, if the apical 2 mm of roots is cut off, and the tip is turned about its longitudinal axis and replaced on the stump, the vascular tissue which later differentiates in the tip is out of line with that of the stump. Again, Torrey cut off the extreme tips of roots of *Pisum sativum* and grew them in a suitable culture medium. It was found that whereas the original roots showed triarch xylem (i.e. it showed three protoxylem groups), a certain proportion of the regenerated roots showed a diarch structure. Thus, the experimental treatment destroyed the original pattern of differentiation

actively in this zone. Further back along the root, however, cell division becomes less frequent and cell vacuolation and extension commence. In the roots of many species (e.g. wheat) growth is fairly

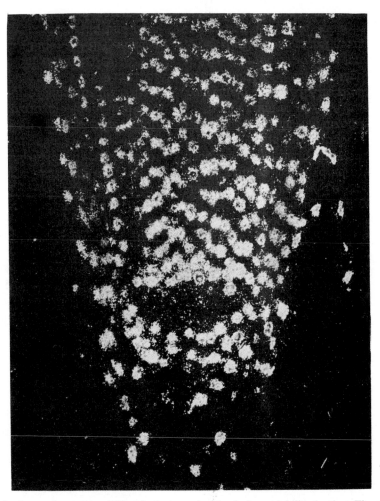

Fig. 2.17. Autoradiograph of root tip section of *Sinapis* photographed by dark ground illumination. The silver grains (white dots) are clustered over the nuclei that have synthesized DNA during the 48 hours in which tritiated thymidine was supplied to the plant. Note the quiescent centre in which no nuclei have been labelled. (From Clowes and Juniper, 1968.)

and yet a new one was determined by the meristem.

2.12 FLOWER INITIATION AND DEVELOPMENT

Sooner or later, one or more of the vegetative apices of a plant cease to produce leaves and buds and become converted into flowering apices. This transition involves a basic change in the structure of the shoot apex.

In *Xanthium,* in which the inflorescence is a capitulum, the first detectable change is an increase in cell devision in the region of the corpus between the central zone and the rib meristem (Fig. 2.19A) and this increased cell division gradually spreads to the central zone and downwards into the peripheral zone. In other species, such as *Anagallis arvensis* and *Sinapis alba,* the first changes are detected in the peripheral zone and spread to the central zone. As a result of these changes, the shoot apical region is transformed into a structure consisting of a central pith of vacuolated cells overlain by a "mantle" of smaller, closely staining meristematic

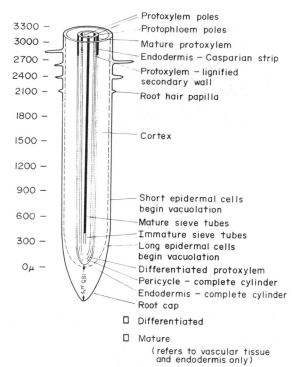

3300 –
3000 –
2700 –
2400 –
2100 –
1800 –
1500 –
1200 –
900 –
600 –
300 –
0μ –

Protoxylem poles
Protophloem poles
Mature protoxylem
Endodermis – Casparian strip
Protoxylem – lignified secondary wall
Root hair papilla
Cortex
Short epidermal cells begin vacuolation
Mature sieve tubes
Immature sieve tubes
Long epidermal cells begin vacuolation
Differentiated protoxylem
Pericycle – complete cylinder
Endodermis – complete cylinder
Root cap

☐ Differentiated

☐ Mature
    (refers to vascular tissue
    and endodermis only)

Fig. 2.18. Diagram of the root tip of *Sinapis alba* to illustrate stages of differentiation of various tissues. (From R. L. Peterson, *Can. J. Bot.* **45**, 319, 1967.)

cells (Fig. 2.19B). During these changes, the height of the apical region increases considerably in most species, but where the inflorescence is a capitulum, as in the Compositae, it may become flattened.

The subsequent pattern of development of the individual flower varies considerably from species to species. According to the classical viewpoint, the receptacle of the flower is a modified vegetative shoot and differs from the latter in that it is no longer capable of unlimited growth, and has only short "internodes". The development of the flower in species such as the periwinkle (*Vinca rosea* L.), bears out this view, since during the early stages the apex still retains essentially the same structure as the vegetative apex. The outer parts of the flower, the sepals and petals, are initiated first, followed by the stamens and carpels (Fig. 2.20).

The initiation and early development of these various flower parts are very similar to those of leaves (Fig. 2.21), although the patterns of development diverge later. During the later

development of the stamen intercalary growth in the basal region gives rise to the future filament, while the distal region gives rise to the anther.

In flowers with an apocarpus gynoecium, i.e. with free carpels, the first stage of carpel development is the appearance of a rounded primordium similar to that of the other organs. This primordium elongates and a depression appears in the tip. As a result of further unequal growth each

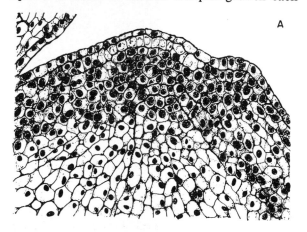

A

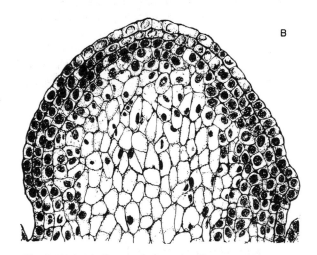

B

Fig. 2.19. A. Median section through the shoot apical meristem of *Xanthium strumarium,* during the transition from the vegetative to the reproductive state. Active division has commenced in the region between the central mother zone and the rib meristem and the zonation characteristic of the vegetative apex (Fig. 2.5) has been partly obliterated. B. Early reproductive apex of *Xanthium,* showing the meristematic "mantle" overlying the enlarged central rib meristem. (From F. B. Salisbury, *The Flowering Process,* Pergamon Press, Oxford, 1963.)

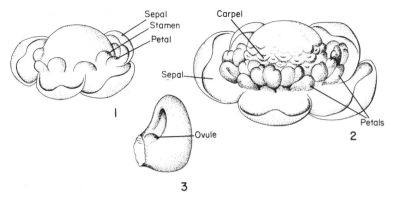

Fig. 2.20. Development of the flower in *Ranunculus trilobus.* 1 and 2. Two stages in the development of the entire flower. 3. Developing carpel. (Adapted from Payer, *Traité d'organogénie comparée de la fleur,* Paris, 1857. Reprinted from A. Fahn, *Plant Anatomy,* Pergamon Press, 1967.)

carpel adopts a horseshoe form (Fig. 2.20); these structures grow upwards and their margins meet and fuse. The process of fusion involves the interpenetration of the cells of the two adpressed surfaces, and in some plants there is cell division in various places, so that the original boundaries are obscured. This type of fusion also occurs in the development of the syncarpous ovary of many species in which two or more initially separate carpels fuse at their margins (Fig. 2.22). The fusion of carpels may extend up into the style and stigma, or the latter parts of each original carpel may remain separate. The fusion of flower parts which were separate at an earlier stage of development is referred to as *ontogenetic* or *postgenital* fusion.

Fusion of carpels or of other organs, including sepals and petals, can also occur by zonal growth. Thus, in a flower such as a primrose (*Primula vulgaris*) with a gamosepalous calyx and a gamopetalous corolla, these parts appear first as isolated lobes (representing separate sepal and petal primordia), but later growth activity extends to the areas between them and forms a complete ring of growing tissue, as a result of which a cylinder of tissue is formed on the upper edge of which are borne the original sepal and petal primordia. In this process there has been no fusion of previously separate surfaces, and structures formed by such zonal growth are said to be *congenitally fused.* The reader is referred to other texts for details of the development of the pollen grains and ovules (see for example, the book by Cutter (1978) referred to below).

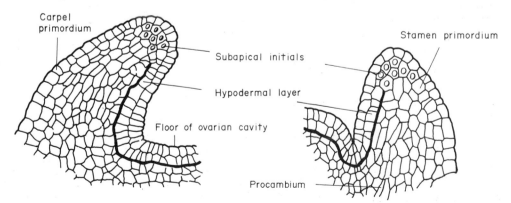

Fig. 2.21. Primordium of the carpel (A) and stamen (B) of *Frasera carolinensis,* showing growth by subapical initials. (From A. S. Foster and E. M. Gifford, *Comparative Morphology of Vascular Plants,* W. H. Freeman & Co., San Francisco and London, 1959.)

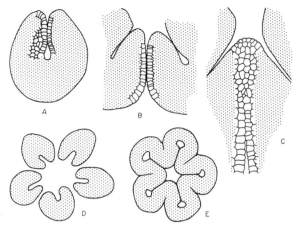

Fig. 2.22. Carpel primordia and fusion of adjacent carpels in *Aquilegia truncata* var. formosa. A, transection of carpel showing initiation (two hypodermal cells with nuclei) of ovule distal to margin; B-C, stages in the ontogenetic fusion of the edges of two adjacent carpels; B, carpel edges in contact with distinct epidermal layers; C, edges of carpels fusing, boundary of epidermal layers becoming indistinct; D, transection showing typical horseshoe shape of the free carpel primordia; E, transection showing postgenital fusion of carpel primordia. (From A. S. Foster and E. M. Gifford, 1959, after Tepfer.)

2.13 MEASUREMENT OF GROWTH

So far we have been concerned largely with qualitative and descriptive aspects of plant growth. It is also important to study growth quantitatively, however, and for this purpose we need methods for measuring growth.

As we have already seen, in the final analysis, we might say that growth involves an increase in the amount of living material. However, it is not always easy to measure this increase in living material without destroying the organism in the process. Moreover, if we simply include the protoplast material (cytoplasm and nucleus), we shall leave out increase in such materials as cell walls which form an integral part of the plant body.

It is possible to adopt a different approach to the problem. Since growth essentially involves increase in cell number, we may use this criterion as a measure of growth. Thus, we can measure the growth of a colony of unicellular organisms by counting the increase in the number of individual cells.

In multicellular organisms, such as higher plants, growth still involves large increases in cell number, but it clearly is inconvenient, if not impossible, to measure such increases. However, this increase in cell number, which is accompanied by cell growth, leads to an increase in *size* and in the case of a root or an unbranched shoot it may be convenient to measure simply the increase in length or height over a given interval of time. This method is not usually appropriate for a complex root or shoot system, however. Since, if we are studying the growth of a whole plant, we are concerned with the increase in total new tissue formed, it is frequently most appropriate to study changes in the *dry weight* of the plant, which will reflect the actual amount of new organic material synthesized by the plant. However, even a change in dry weight is not always a satisfactory measure of growth, since plant tissues may increase in dry weight due to the accumulation of reserve materials, such as starch and fat, although they may not be growing. Conversely, a germinating seed may show an overall loss in dry weight, due to the utilization of reserves in respiration, although there is no doubt that it is growing.

2.14 COLONY GROWTH IN MICRO-ORGANISMS

Before considering the growth curves of higher plants, it is useful to study the growth of a colony of unicellular organisms, such as bacteria or yeast, which multiply by division or "budding", or of a multicellular organism, such as duckweed (*Lemna*), which similarly multiplies by a form of budding.

Consider the growth of a colony of bacteria maintained under constant nutritional and environmental conditions so that there is a constant rate of cell division. Assume also that the cells divide synchronously, i.e. all cells in the colony divide simultaneously. (Synchronous division can be achieved with cultures of certain organisms.) If the initial number of cells in the colony is n_0, and *n* the number of cells after a given number of divisions, then,

at the end of the 1st generation $n = n_0 \times 2$,

at the end of the 2nd generation $n = n_0 \times 2 \times 2$,

at the end of the *x*th generation $n = n_0 \times 2^x$.

This latter relation indicates that the number of cells in the colony is increasing by geometric progression or "exponentially", i.e. at an ever-increasing rate, and if we plot n *against the number of generations, we obtain a curve of the form seen in Fig. 2.23A.*

We can rewrite the equation

$n = n_0 \times 2^x$ as:

$$\log n = \log n_0 + x \log 2. \tag{1}$$

It will be seen that we have an equation expressing the relation between the number of cells in the colony *n*, and *x*, the number of generations which have occurred, but normally we require the relation between *n* and *t*, the time. Now, if *t* is the time taken for *x* generations, and the time of one generation (i.e. time between two successive divisions) is *g*, then $x = t/g$.

Substituting in equation (1), we get

$$\log n = \log n_0 + t/g \log 2.$$

Now, $\log 2/g$ is a constant (*k*).

Therefore, $\log n = \log n_0 + kt.$ (2)

Now equation (2) is a linear equation of the form *y = a + bx,* where $\log n_0$ corresponds to *a,* and *k* corresponds to *b.* Hence if we plot the *log* of the number of cells present in the colony after different times, against *t,* we should obtain a straight line. This relation is, in fact, found to hold in practice for various organisms growing under constant conditions, whether they multiply by fission, as in bacteria, or by budding, as in yeast and duckweed (*Lemna*) (Fig. 2.23B). A colony growing in this manner is said to be increasing "logarithmically" or "exponentially".

If we consider the type of curve shown in Fig. 2.23A, giving the increase in the number of *Lemna* "fronds" in the colony, then the *growth rate* of the colony at any time is given by the increase (*dn*) in the number of cells over a short interval of time, *dt,* or we can say, growth rate = *dn/dt.*

The value *dn/dt* represents the *slope* of the curve at any given time, *t,* and it will be seen from

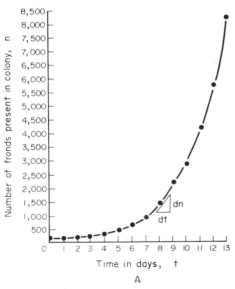

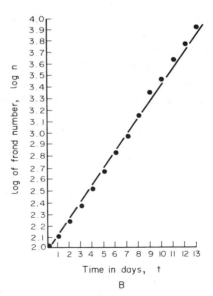

Fig. 2.23. A. Growth curve for a colony of duckweed (*Lemna*) budding synchronously at a constant rate. The initial number of fronds is assumed to be 10. B. Growth of a colony of *Lemna* in culture. Linear relationship between logarithm of frond number (log *n*) and time. (Data from E. Ashby and T. A. Oxley, *Ann. Bot.* **49**, 309, 1935.)

the graph that the value of the slope increases progressively with time. If all cells are dividing at the same rate, *r,* clearly at any time, *t,* the rate of growth of the colony is proportional to the number of cells present, i.e. *dn/dt* ∝ *n.* Thus, although the rate of cell division (*r*) remains constant, the *absolute growth-rate* of the colony as a whole does not, since as time goes on the number of cells present in the colony increases. The value of *r* is clearly given by *dn/dt* (the increase in cell number over a given short interval of time), divided by the number of cells so dividing, i.e.

$$r = \frac{dn}{dt} \cdot \frac{1}{n}$$

and this value is known as the *relative growth rate* of the colony. Thus, for a colony showing this type of growth, the absolute growth rate increases with time, but the relative growth rate remains constant.

It has been shown above (equation (2)) that

$\log n = \log n_0 + kt.$

This equation can also be written in the form

$n = n_0 e^{kt}$ (3)

where e = the base of natural logarithms (2·7182).

In practice, unrestricted growth of a colony can never proceed indefinitely and some limiting factor, such as deficiency of nutrients, must always lead to a decline in growth rate sooner or later. Under cultural conditions in a flask or tube, for example, the food supply will ultimately be exhausted and growth will finally cease. Instead of the typical "exponential" growth curve for cell number, we obtain a "sigmoid" type of curve (Fig. 2.24 *left*), in which the growth rate increases up to the point of inflection and then declines gradually to zero. When log *n* is plotted against time, growth follows a straight line initially, but later declines (Fig. 2.24 *right*). In addition to the exhaustion of some food factor, growth in colonies may also be limited by some toxic substance which is formed during growth. The production of such "staling factors" often occurs in cultures of bacteria, fungi, *Chlorella,* etc.

2.15 GROWTH OF MULTICELLULAR ORGANISMS

2.15.1. The exponential phase

The sigmoid type of growth curve observed for colonies of unicellular organisms is characteristic

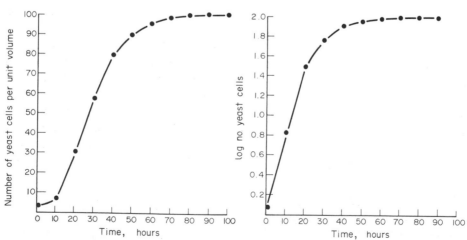

Fig. 2.24. Growth of a colony of yeast growing in a constant volume of culture solution. *Left:* Sigmoid growth curve obtained when number of cells (*n*) is plotted against time. *Right:* Plot of logarithm of cells (log *n*) against time. (Data from O. W. Richards, *Ann. Bot.* **42,** 271, 1928.)

also of the growth of individual multicellular plants. This is true not only for the whole plant (Fig. 2.26), but also for individual organs, such as

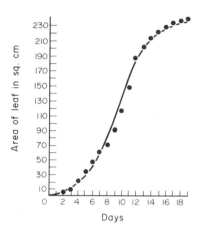

Fig. 2.25. Sigmoid growth curve of leaf of cucumber (*Cucumis sativa*). (From F. G. Gregory, *Ann. Bot.* **35**, 93, 1921.)

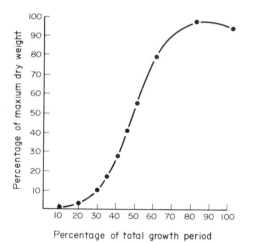

Fig. 2.26. Increase in dry weight of barley plants during growing season. The points on the curves were derived from smoothed curves drawn from the transformed original data. (From F. G. Gregory, *Ann. Bot.* **40**, 1, 1926.)

leaves (Fig. 2.25) or internodes. Initially the organism is increasing in size (or weight) by geometrical progression or exponentially. V. H. Blackman (1919) showed that during this initial phase the growth of seedlings follows a "Compound Interest Law" fairly closely and is given by the equation

$$W = W_0 e^{rt} \qquad (4)$$

where W = weight of plant after time t,

W_0 = initial weight of plant,

r = percentage (or proportional) rate of increase,*

e = exponential coefficient ($= 2 \cdot 7182$).

The equation is clearly exactly comparable with equation (3) above for colony growth and may be derived in a precisely similar manner. From equation (4) we may write

$$\log W = \log W_0 + rt \log e.$$

This equation is again of the form $y = a + bx$. This means that we should obtain a straight line when we plot log weight against t (at least for the initial phase of growth, which we are now considering) and this has, in fact, been demonstrated in a number of cases.

From equation (4), it is clear that the final weight attained will depend upon (1) the initial weight, (2) the rate of "interest" (r), (3) the time. The rate of "interest" represents the efficiency of the plant as a producer of new material and was called by Blackman the *efficiency index* of dry weight production. A small difference in the efficiency index between two plants will soon make a marked difference in the total yield, and the difference will increase with the lengthening of the period of growth.

It should be noted that the efficiency index is merely a different method of expressing the *relative growth rate* ($dW/W.dt$) as described for colony growth. Whereas the efficiency index (or relative-growth rate) remains constant through the exponential growth phase, the *absolute* increments per unit time increase progressively. The absolute growth increment over a time interval dt is clearly

$$W \times \frac{r}{100} . dt.$$

*The same symbol, *r,* has been used here as for the rate of cell division in a colony of bacteria, described above, since in both cases *r* represents the relative growth rate, whether measured by rate of cell division or increase in dry weight.

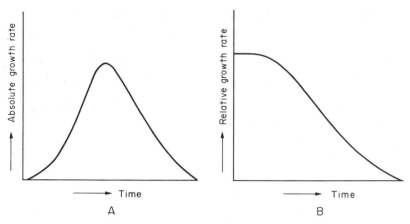

Fig. 2.27. Changes in (A) absolute growth rate dW/dt and (B) relative growth rate $(dW/W.dt)$, where the dry weight changes follow the type of curve shown in Fig. 1.10.

Thus, the absolute growth rate at any given time is proportional to the size of the plant at that time.

The physiological basis of this latter conclusion is easily understood, for when photosynthesis has become active in a young seedling, the power of the plant to synthesize new material (and hence increase dry weight) is clearly dependent upon its leaf area. Therefore, as the plant grows and increases its leaf-area, the rate at which new material is assimilated will increase proportionately.

2.15.2. Later phases of growth

Just as the growth rate (dn/dt) of a bacterial colony ultimately falls off with time due to the exhaustion of nutrients or the accumulation of toxic products, so the growth rate of a multicellular organism decreases gradually, resulting in a sigmoid growth curve. The absolute growth rate (dW/dt) is clearly given by the slope of the growth curve at any time t and if we plot the changes in this growth rate with time we get a curve of the type shown in Fig. 2.27A. We see that the growth rate attains a maximum (corresponding to the point of inflection of the S-shaped growth curve) and then falls away to zero. If, on the other hand, we plot the relative growth rate $(dW/W.dt)$ against time, we frequently get a curve of the type in Fig. 2.27B. It is seen that the relative growth rate (RGR) remains nearly constant at first but later begins to decline. The changes in RGR with time vary a great deal from one species to another and with the conditions under which the plants are growing. Sometimes it is found that the RGR declines steadily from the commencement of growth, so that a true exponential phase does not occur.

The reason for the fall in the RGR is not fully understood and various hypotheses have been suggested. The deficiency for some nutritive factor is clearly not the cause, as it is in colonies of unicellular organisms under artificial conditions. It has been suggested, however, that the reason for the departure from "exponential" growth in a normal plant is that part of the plant material formed during growth gives rise to mechanical, vascular and other tissues which do not directly contribute to further synthesis of new material. The leaves which are the organs most directly concerned in the synthesis of new material thus constitute a diminishing fraction of the total plant weight, i.e. the ratio

$$\frac{\text{Total leaf dry weight } (L)}{\text{Total plant dry weight } (W),}$$

known as the *leaf-weight ratio,* gradually falls. Now, the relative growth rate (R) is given by

$$R = \frac{dw}{dt} \cdot \frac{1}{W}.$$

Multiplying numerator and denominator by L we get

$$R = \frac{dw}{L} \cdot \frac{L}{W} \cdot \frac{1}{dt}.$$

Now $dw/L.dt$ is the rate of increase in dry weight per unit dry weight of leaf, known as the *net assimilation rate* (E_w) and is a measure of the photosynthetic activity, minus losses due to respiration; that is, E_w is a measure of the net efficiency of the plant in the production of dry matter.

It will be seen that we have the following simple relationship between the three parameters:

$$R = E_w \times \frac{L}{W}.$$

If we assume that E_w does not change appreciably with the age of the plant, then the fall in R with age must primarily be due to the fall in L/W, for the reasons already indicated. However, both E_w and L/W are frequently found to decline during the growth period and thus to contribute to the decline in R.

In the above discussion, we expressed the net assimilation rate (E_w) in terms of the rate of increase in total plant dry weight per unit leaf dry weight, but it is more usual to express photosynthetic efficiency as the rate per unit *area* of leaf. However, the net assimilation rate can easily be expressed on a leaf area basis, by making use of the following relationship:

$$E_w = E_A \times \frac{L_A}{L_w}$$

where E_A is the rate of increase in dry weight per unit leaf area and L_A/L_w is the ratio of leaf area to dry weight, known as the *specific leaf area*.

Thus, equation (5) above can be rewritten as

$$R = E_A \times \frac{L_A}{L_w} \times \frac{L_w}{W}$$

The foregoing simple mathematical relationships have been used to analyse the growth of crops. Thus, the determination of relative growth rates for different crop plants gives us a useful basis for comparing their growth rates. Similarly by determining net assimilation rates and leaf-weight ratios we can obtain some indication of how differences in R arise.

FURTHER READING

General

Causton, D. R. *A Biologist's Mathematics,* Edward Arnold, 1977.
Clowes, F. A. L. *Apical Meristems,* Blackwell Scientific Publications, Oxford 1961.
Cutter, E. G. *Plant Anatomy: Experiment and Interpretation* Part 2, 2nd Ed., Edward Arnold, London 1978.
Evans, G. C. *The Quantitative Analysis of Plant Growth,* Univ. of California Press, Berkeley, 1972.
Sinnott, E. W. *Plant Morphogenesis,* McGraw Hill Book Co., New York, 1960.
Wardlaw, C. W. *Morphogenesis in Plants,* Methuen & Co. Ltd., London, 1968.

More Advanced Reading

Cutter E. G. Recent experimental studies of the shoot apex and shoot morphogenesis. *Bot. Rev.* **31,** 7, 1965.
Dormer, K. J. *Shoot Organisation in Vascular Plants,* Chapman & Hall, London, 1972.
Halperin, W. Organogenesis at the shoot apex, *Ann. Rev. Plant Physiol.* **29,** 239, 1978.
Milthorpe, F. L. and J. Moorby. *An Introduction to Crop Physiology,* 2nd Ed., Cambridge University Press, London, 1979.
Shininger, T. L. The control of vascular development, *Ann. Rev. Plant Physiol.* **30,** 313, 1979.
Torrey, J. G. and D. T. Clarkson. *The Development and Function of Roots,* Academic Press, London, 1975.
Wardlaw, C. W. The organization of the shoot apex; *Encycl. Plant Physiol.* **15**(1), 966, 1965.
Wardlaw, C. W. *Organization and Evolution in Plants,* Longmans, Green & Co. Ltd., London, 1965.

SECTION II

**Internal Controls in
Plant Development**

Introduction

The orderly series of changes so characteristic of development clearly require control systems to ensure that there is co-ordination of growth and differentiation in both *space* and *time*. Thus, the development of a leaf primordium is accompanied by the differentiation of vascular tissue in the neighbouring stem tissue (p. 30). Similarly, the growth of an embryo following fertilization is normally accompanied by growth of the surrounding tissues of the ovule. These are examples of the phenomenon of *correlation,* which we shall consider in more detail in Chapter 5.

Correlation of growth in different regions is also seen at the whole-plant level. The indeterminate pattern of growth of higher plants by apical meristems leads to the steady accretion of mature tissue in the older parts of the plant body, while meristematic activity is maintained at the shoot and root apices. However, potential meristematic regions still remain in the older parts, such as the cambium, axillary buds, developing seeds and fruits, and storage organs. Hence there is a need for control systems to co-ordinate growth in the various regions of the plant, which we shall refer to as *spatial* co-ordination.

There is also a need for *temporal* co-ordination of development, which involves an orderly sequence of changes at all levels of organization. Thus, the whole plant passes through a succession of phases during its life cycle, viz. germination and vegetative growth, flowering and fruiting, ripening and senescence and dormancy. This orderly sequence of changes must also involve control systems which ensure co-ordination in *time*.

A control system may operate spontaneously within the plant itself, i.e. its action may by *autonomic* in origin, or it may be activated or modulated by *environmental* factors. Since environmental conditions do not normally vary appreciably around the different parts of the shoot system, spatial co-ordination within the plant is, in general, achieved by autonomic control systems, whereas environmental factors are frequently important in temporal co-ordination.

Autonomic control is expressed at both the *intracellular* and the *intercellular* levels of organization. Control of development at the intracellular level is reflected in the appearance and disappearance of activities of different enzymes during various phases of cell growth and differentiation, and this is achieved in several ways, including regulation of nucleic acid and enzyme synthesis and the activation and inactivation of preformed enzymes. Thus, intracellular control of developmental processes quite clearly involves regulation of gene activity. But intercellular control mechanisms must also, of course, ultimately be determined by the genotype.

3

Plant Growth Hormones and Their Metabolism

3.1 INTRODUCTION

The growth of a plant is a dynamic and complex, yet strictly controlled, process. This means that growth in different parts of the plant must be integrated and co-ordinated and we shall meet a considerable number of examples of such *growth correlations* in later chapters. The co-ordination of growth between different parts of the plant must clearly involve some control mechanism. Moreover, we have seen, in earlier chapters, that the development of organs, such as leaves or stems, involves an orderly sequence of phases of cell division and cell extension, so that there is also co-ordination of growth in *time*. As a result of intensive studies extending over many years, it is now known that hormones play a vital role in the control of growth not only within the plant as a whole, but apparently also within individual organs. It is now realized that there are at least three major classes of growth-promoting hormones— *auxins, gibberellins* and *cytokinins*. In addition, other classes of plant hormones exist, particularly the "growth inhibitors" such as *abscisic acid* (ABA), but also including a gas, *ethylene,* which is apparently involved in many growth phenomena.

Growth hormones are translocated within the plant, and influence the growth and differentiation of the tissues and organs with which they come into contact. This leads us to a consideration of the nature and role of growth hormones. The word "hormone" was first used by animal physiologists, to refer to a substance which is synthesized in a particular secretory gland and which is transferred in the blood or lymph to another part of the body where extremely small amounts of it influence a specific physiological process. However, plant hormones differ in certain respects from the classical concept of hormones which was originally based upon the discovery of these substances in animals. In an animal a hormone is a substance produced in one particular organ such as a gland, and which is secreted to produce its typical and usually specific effect at a site distant from its point of origin. In the case of plant hormones we cannot always differentiate so clearly between the site of hormone synthesis and its place of action, although there is much evidence, referred to in Chapters 5 and 7, that plant hormones do usually have effects at sites distant from their place of production. Another difference between animal and plant hormones is that whereas the effects of most animal hormones are rather specific, a plant hormone can elicit a wide range of responses depending upon the type of organ or tissue in which it is acting. For reasons such as these, plant growth hormones have frequently been referred to by other names, such as "growth regulators" or "growth substances". In general, though, the term "growth hormone" appears to be more appropriate, despite the difficulties discussed above, since it does intimate that these substances are active in extremely small quantities, and that in many instances they exert control over processes in tissues different from those in which they are synthesized.

Each of the chemically different categories of growth hormone has characteristic influences on growth and differentiation in plant cells and tissues. Auxins were the first plant growth hormones to be discovered, and consequently we shall now review very briefly the history of the discovery of auxins and their chemistry, followed by similar considerations of gibberellins, cytokinins, ethylene, and finally abscisic acid.

3.2 AUXINS

The basis of our modern knowledge of auxins lies in the work of Charles Darwin, published almost a century ago in a book entitled *The Power of Movement in Plants.* Darwin investigated the phenomenon of *phototropism,* the bending of plant organs in response to unilateral illumination.

Darwin experimented with seedlings of the ornamental canary grass (*Phalaris canariensis*). The coleoptile of grass seedlings proved a very convenient subject for the study of phototropism by Darwin and many other later workers. However, it was Darwin who first realized that the tip of the coleoptile *perceives* the unilateral light stimulus, but that the curvature *response* occurs lower down (Fig. 3.1). Darwin concluded that, "when seedlings are freely exposed to a lateral light some influence is transmitted from the upper to the lower part, causing the latter to bend". It was left to later researchers following Darwin to find out the nature of the "influence".

Various workers, particularly Boysen-Jensen and Paal, conducted experiments in the second two decades of this century which demonstrated that the growth-promoting influence transmitted from a coleoptile tip was of a purely chemical nature (Fig. 3.1). Paal was led to suggest that this chemical, under conditions of darkness or uniform illumination, continually moves down the coleoptile on all sides and acts as a *correlative growth promoter.*

The first successful isolation of the chemical messenger from coleoptile tips was carried out in 1926 by a Dutchman, F. W. Went, who thus extended the work of Boysen-Jensen and Paal. Went found that if he placed an excised oat (*Avena*) coleoptile tip upon a small block of agar gel, then the agar block acquired growth-promoting properties, in that if the block was separated from the tip and placed on one side of a coleoptile stump, then curvature of the stump resulted. An agar block which had not been in contact with a tip had no such effect. The conclusion was, therefore, that the chemical messenger had diffused from the tip into the agar block. Subsequent placing of the agar block on to a coleoptile stump allowed diffusion of the messenger out of the block and into the stump. However, this was not only the first separation of a growth hormone from a plant but it also afforded Went a technique whereby he was able to make quantitative measurement of the growth hormone. The method of measurement that Went devised is a biological assay (*bioassay*) based on the curvature of a coleoptile stump in

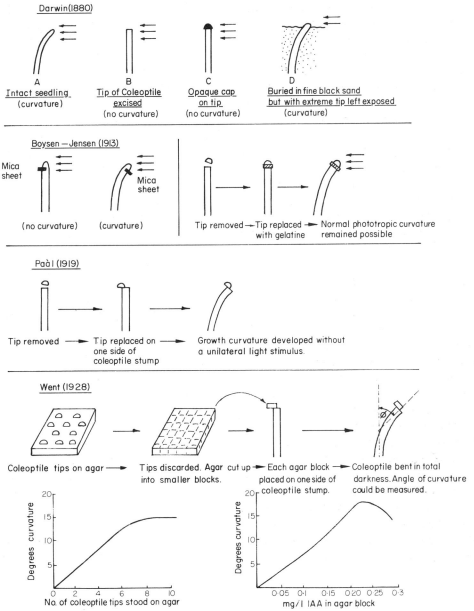

Fig. 3.1. A summary of important experiments which established the existence of auxin in plants. All experiments performed with coleoptiles of grass seedlings. Triple arrows indicate the direction of incident unilateral light. In the case of Went's experiments, dose-response curves are shown for the Went *Avena* curvature test; one for response against number of coleoptile tips diffused on agar, and the other for response against concentration of indole-3-acetic acid (IAA) in the agar block.

response to an asymmetric placing of an agar block containing the growth hormone (Fig. 3.1).

The name *auxin* (from the Greek *auxein,* to grow) was given to the growth hormone produced by the tip of a coleoptile. However, it is now known that auxins are present in *all* higher plants and not only in grass seedlings. Auxins are, as we shall see later, of fundamental importance in the physiology of growth and differentiation. They appear to be synthesized mainly in meristematic tissues such as those of young developing leaves, flowers and fruits.

3.2.1 The isolation and chemical characterization of auxin

Early attempts to isolate and chemically identify auxin suffered from the difficulty that it is present in plant tissues in extremely small quantities. It was found to be impossible at that time to obtain sufficient auxin from plants to prepare pure crystalline auxin for analysis. In fact the first crystalline auxin was obtained from human urine. Some confusion arose as a result of early attempts to determine the chemical nature of this auxin, but in 1934 it was shown to be indole-3-acetic acid (often abbreviated to IAA).

Indole-3-acetic acid (IAA)

Since the initial discovery of IAA as an auxin, it has been found that this substance occurs in most plant species, and it is now believed to be the principal auxin in higher plants.

Much research is still conducted with the aim of elucidating the true nature of endogenous (i.e. internally synthesized) auxin in plants, and this involves the use of a number of modern chemical and physical techniques, as well as biological methods. As we saw earlier, the first isolation of auxin from plants was achieved by allowing diffusion of the hormone to take place from the plant tissue into a suitable inert medium such as agar gel. This method is still utilized and auxin obtained in this way is referred to as *diffusible auxin*. More commonly, plant tissues are extracted by organic solvents, such as di-ethyl ether or methanol. There are often quantitative and qualitative differences between auxins obtained by diffusion and extraction, even from the same tissue.

Chromatographic separation of the constituents of a plant extract results in the original extract being split up into a number of separate fractions. Each of these fractions is then examined to see whether or not it contains any auxin. To do this

some sort of bioassay is usually used, such as the Went *Avena* test described earlier. There are numerous other types of biological assay for auxin that have been developed over the years, all of them making use of some measurable aspect of the effects of auxin in plants. One of the most commonly used tests is the coleoptile straight-growth test, in which sections of uniform length (say 5 mm) are cut from young *Avena* coleoptiles and placed in the extract to be tested for growth-promoting activity. However, although biological assays are indispensable in such studies, chemical identifications of the auxin must eventually be carried out. Even today this identification presents great difficulties, mainly because of the difficulty in obtaining sufficient auxin for analysis. Nevertheless, chemical characterization of auxins in plants is now possible, largely through the development and application of sensitive physico-chemical techniques such as ultra-violet and infra-red spectroscopy, and more recently mass-spectrometry has been applied to the problem.

The results of such studies have, in general, confirmed that IAA occurs as an auxin in plants, although it has become apparent that there are other substances in plants which also possess auxin activity. In many cases these other substances are indoles, closely related chemically to IAA; for example, the compound indole-3-acetonitrile (IAN) is known to be present in a number of plant

Indole-3-acetonitrile (IAN)

species, particularly members of the Cruciferae, and pea plants appear to contain 4-chloro-indole-3-acetic acid as a natural auxin. There are also compounds, such as phenylacetic acid, present in plants which are not indoles and yet possess auxin activity, but both the chemical natures and the physiological significance of most of these non-indole endogenous auxins remain obscure at the present time. It is because of the existence of more

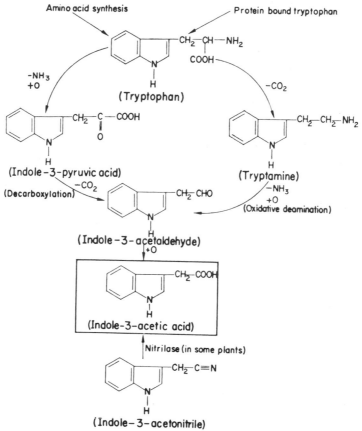

Fig. 3.2. Possible biosynthetic pathways for indole-3-acetic acid (IAA) in plants. (The pathway through tryptamine operates in some plant species, for example in oat, tobacco, tomato and barley, but not in others.)

than one type of chemical which show auxin activity, that we often speak of auxins, rather than the singular, auxin.

3.2.2 Indolic auxin metabolism

Investigations of the metabolism of endogenous auxin have concentrated on the origin and fate of IAA in plant tissues. The reasons for this are first, the belief that IAA is the principal auxin, and secondly the concern to understand normal plant growth regulatory mechanisms. Since it is known that the *concentration* of auxin available to tissues can have a determining effect on growth and differentiation, then the factors which limit the con-

centration of IAA in plant tissues have received most attention. These factors include:

1. IAA synthesis.
2. IAA destruction.
3. IAA inactivation by processes other than destruction of the molecule.
4. Regulation of IAA transport between tissues and organs (see Chapter 5).

Auxin biosynthesis. As soon as IAA was identified as an endogenous auxin, it was suggested that it is formed from the amino acid tryptophan, a compound with an indole nucleus which is universally present in plant tissues, both in the free state and incorporated into protein. Indeed, it has been demonstrated many times that higher plants, or enzyme preparations from them, are able to bring about the conversion of exogenous tryptophan to IAA. Greater amounts of IAA are formed from added tryptophan in non-sterile plants or enzyme preparations than under sterile conditions, and in the late 1960s it was suggested that *all* the IAA present in plants is synthesized by epiphytic bacteria. Subsequent research has, however, revealed that completely sterile plant tissues and enzyme preparations are able to convert tryptophan to IAA. There seems little doubt, therefore, that IAA is synthesized by higher plants, utilizing a biosynthetic pathway involving indole-3-pyruvic acid (IPyA) and indole-3-acetaldehyde (IAAld) as intermediates (Fig. 3.2). In certain species, such as oat, tobacco, tomato and barley, there is some evidence that IAA can also be synthesized from tryptophan via tryptamine and IAAld (Fig. 3.2), but other species (e.g. pea, bean, cabbage, squash) do not contain tryptamine. A third pathway for IAA synthesis exists in members of the Cruciferae, in which tryptamine may be converted to indoleacetaldoxime and thence to IAN either directly or via the thioglucoside, glucobrassicin, followed by the conversion of IAN to IAA by the enzyme nitrilase. The relative importance of these alternative pathways for IAA synthesis is not known.

Apart from the indole compounds shown in the schematic representation of IAA synthesis in Fig. 3.2, a number of other indoles are known to occur naturally in plants. It is possible that any one or all of these indoles could serve as precursors of IAA, but clearly we still lack good information on the biosynthetic pathways for IAA actually occurring *in vivo*.

IAA destruction. It is well established that IAA is in one way or another fairly readily inactivated by most plant tissues. It appears, therefore, that the concentration of IAA in plants is regulated not only by its rate of synthesis, but also by inactivation mechanisms. The indications are that IAA catabolism as well as biosynthesis may follow more than one pathway.

Photo-oxidation of IAA. IAA in aqueous solution will soon decompose if left exposed to light. This photo-oxidation of IAA is greatly accelerated by the presence of many natural and synthetic pigments. Thus, it is possible that plant pigments absorb light energy which energizes the oxidation of IAA *in vivo*. If this is so, then the most likely pigments involved are riboflavin and violaxanthin, for these are commonly present in plants in relatively large amounts and they absorb light in the blue regions of the spectrum which have been found to be the most active in inducing photo-oxidation of IAA.

Breakdown products of *in vitro* photo-oxidation of IAA include 3-methylene-2-oxindole and indolealdehyde, together with other unidentified compounds formed by cleavage of the indole ring (Fig. 3.3).

Nothing is known of the chemistry of IAA photo-oxidation within the plant, but indolealdehyde and methylene-oxindole do occur naturally in many plants, and may perhaps represent a product of *in vivo* photo-oxidation. Some research has indicated that when methylene-oxindole is applied to plants, it may show growth-regulating properties, either inhibiting or ac-

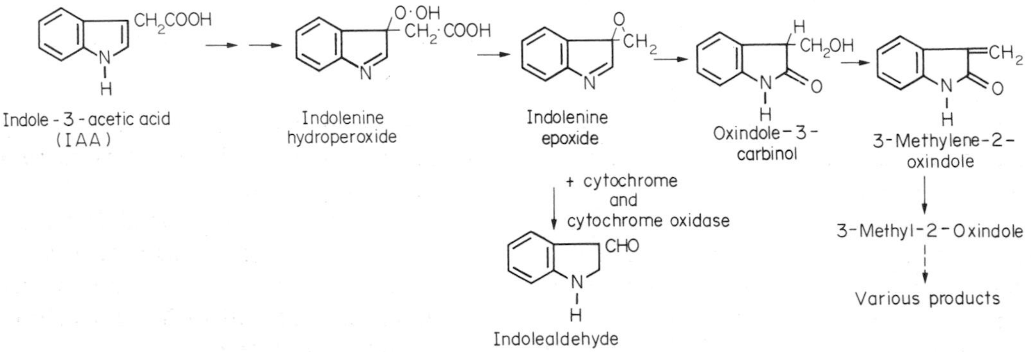

Fig. 3.3. Pathways for oxidative metabolism of IAA.

celerating the rate of growth. However, other more recent work has suggested that endogenous methylene-oxindole does not function as a growth regulator *in vivo*.

Enzymic oxidation of IAA. Many plants contain an enzyme, or enzyme system, known as "*IAA-oxidase*", which catalyses the breakdown of IAA, with the release of CO_2 and consumption of O_2. IAA-oxidase preparations from different plant species often have different properties, but they all show some similarities to the enzyme peroxidase. Full peroxidation of IAA (i.e. by H_2O_2 in the absence of O_2) does not occur, but oxygen is always required in addition to H_2O_2. Addition of H_2O_2 to some IAA-oxidase preparations does enhance the rate of IAA degradation, but in other cases it does not; probably because crude plant enzyme preparations can possess the ability to generate H_2O_2 which would be available for peroxidase action. IAA-oxidase from higher plants requires manganese as a co-factor, and its activity is increased by monophenols and reduced by *ortho-* and *para*-dihydric phenols and polyphenols.

The pathway of IAA breakdown (Fig. 3.3) by IAA-oxidase is ill-understood. A principal product *in vitro* is 3-methylene-2-oxindole, and this is probably further metabolized to 3-methyl-2-oxindole. IAA-oxidase preparations that also contain cytochrome oxidase metabolize IAA to yield indole-3-aldehyde as the main product, and this latter compound is relatively stable compared with the oxindoles. The physiological significance of these observations is not at all clear at present. Furthermore, the relationships between IAA-oxidase and naturally occurring phenolics are still rather obscure, probably because many of the experimental studies have been made with relatively crude enzyme preparations. It is quite likely, nevertheless, that naturally occurring phenolic modifiers, or regulators, of IAA-oxidase activity do function in plants.

Interest in the enzymic oxidative destruction of IAA has been maintained by indications that such a process might be important in regulating auxin levels in plant tissues. Thus, there have been a number of reports that (1) IAA-oxidase activity rises with age of tissues, (2) there is a negative correlation between growth rate and IAA-oxidase content of various organs, and (3) that root tissues contain both very low IAA concentrations and very high IAA-oxidase activity. Nevertheless, there is not, as yet, conclusive proof that such correlations are physiologically important.

Inactivation of IAA by other Processes. Apart from degradation by photo-oxidation or enzyme activity, IAA can be inactivated in plant tissues by the formation of hormonally-inert complexes of various types. Thus, IAA is readily esterified by plant enzymes to yield its ethyl ester (indole ethyl acetate).

Indole ethyl acetate

Similarly, enzymic formation of conjugates such as indole acetyl-aspartic acid have been reported many times.

Indole acetyl-aspartic acid

The nature of the esterifying or conjugating enzymes involved in these reactions remain very uncertain at present.

Conjugates are also formed between IAA and various sugars and sugar-alcohols, yielding compounds such as indoleacetylarabinose, indoleacetylglucose and IAA-myoinositol, and between IAA and proteins.

3.3 GIBBERELLINS

3.3.1 The isolation and chemical characterization of gibberellins

The discovery of a group of plant growth-hormones now known as the gibberellins dates from the 1920s, when Kurosawa, a Japanese research worker, was investigating the "bakanae" ("foolish seedling") disease of rice caused by the fungus *Gibberella fujikuroi* (also known as *Fusarium moniliforme*). A characteristic symptom shown by rice plants infected by this fungus is an excessive elongation of stems and leaves, resulting in abnormally tall plants which usually fall over due to the spindly stem structure—hence, "foolish seedling". Kurosawa and fellow Japanese workers found that if they grew the fungus in a culture medium, and subsequently filtered off the fungus, then the culture filtrate, which was completely free of the fungus itself, would induce the same abnormal growth symptoms when applied to rice plants. It was clear that *Gibberella fujikuroi* secreted some substance into infected plants, or into the nutrient medium when grown in culture, which was stimulatory to stem and leaf elongation. In

1939 a small quantity of highly active crystalline material was isolated from such culture filtrates, and was given the name "gibberellin A". The chemical composition and structure of this material was not unequivocally worked out by the Japanese. It was not until 1954 that further progress was made, when British chemists isolated and chemically characterized a pure compound from culture filtrates of *Gibberella fujikuroi*. They called this new substance *gibberellic acid* and found that it has the following structure:

Gibberellic acid when applied to many species of intact growing plants induced abnormally great extension of stems and leaves, but the response was found to be greatest when *genetic dwarfs* of various plant species were treated. Such treated dwarf plants assumed the appearance of the normal tall plants, from which the dwarfs had originally arisen by mutation (Fig. 3.4).

Fig. 3.4. Effect of gibberellic acid (GA_3) on shoot growth in dwarf pea plants (*Pisum sativum* c.v. Meteor). Plant at extreme right was not treated with GA_3, but the other plants received increasing doses of GA_3 from right to left. (Original print provided by Professor P. W. Brian.)

It might appear strange that a substance obtained from a fungus should produce essentially normal responses in higher plants. However, it is now known that gibberellic acid and substances very similar in both chemical structure and biological activity occur in healthy (i.e. non-infected) plants of all species. In fact, a number of these compounds have been isolated from higher plants, so that at the present state of knowledge there are some 58 chemically characterized compounds which produce effects similar to those elicited by gibberellic acid. These compounds are known collectively as the gibberellins, designated gibberellins A_1—A_{58} (or GA_1—GA_{58}). Gibberellic acid is numbered GA_3. Some of the known gibberellins have been isolated from culture filtrates of *Gibberella fujikuroi,* and others from various organs of higher plants. All have the same basic molecular structure as that possessed by gibberellic acid, differing from one another mainly in the types, numbers and positions of substituents (particularly hydroxyl groups) on the ring system, and the degree of saturation of the "A" ring (see Fig. 4.5). It is highly likely that further additions will in future years be added to the list of known gibberellins.

The pathways of synthesis of a number of gibberellins in plants are shown in Fig. 3.5. However, the essential major sequences may be summarized as follows:

Mevalonic Acid→*ent*-Kaurene→GA_{12}-aldehyde→ Gibberellins

Gibberellins are diterpenoid acids biogenetically derived from the tetracyclic diterpenoid hydrocarbon *ent*-kaurene. Rather fewer than half of the known gibberellins retain the full 20 carbon atoms of this precursor. Those gibberellins that have 20 carbon atoms are termed C-20 gibberellins and have the *ent*-gibberellane carbon skeleton shown below. The other (C-19) gibberellins lose carbon atom number 20 during their biosynthesis (see p. 80 for the carbon numbering system of gibberellins) and have the basic *ent*-20-*nor*-gibberellane carbon skeleton:

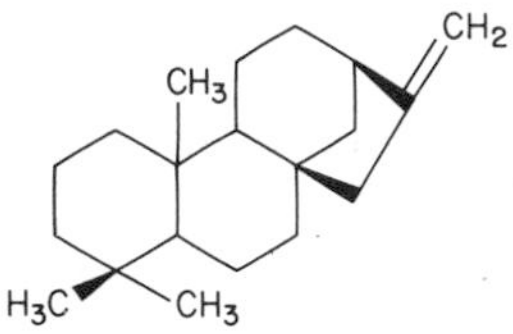

ent-Kaurene

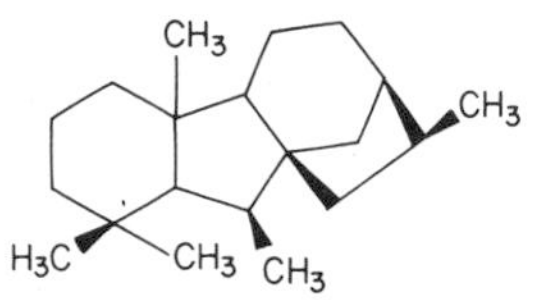

ent-Gibberellane (C-20 Gibberellins)

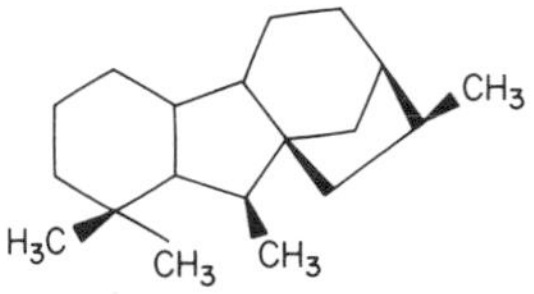

ent-20-*nor*-Gibberellane (C-19 Gibberellins)

Some naturally occurring compounds which lack the gibberellane skeleton of the gibberellins, but nevertheless possess some gibberellin-like biological activity include gibberethione (a sulphur-containing derivative of GA_3) from immature seeds of *Pharbitis nil*; antheridiogen (structurally rearranged GA_4) which is the antheridium-inducing factor in the fern, *Anemia phyllitidis*; helminthosporal isolated from the fungus *Helminthosporium sativum*; and phaseolic acid (a keto-carboxylic acid) obtained from seeds of bean plants.

The discovery that the gibberellins, or at least some of them, are natural growth hormones in higher plants, necessitated a complete reconsideration of views of the hormonal control of plant growth and differentiation. One could no longer think of development of cells and tissues as being influenced by only one growth hormone, that is auxin, but consideration had to be made of the effects of gibberellins and later, other hormones (e.g. cytokinins) on the processes affected by auxins. Indeed, as we shall see later, auxins, gibberellins and cytokinins *interact* in their influences on plant growth and differentiation, and it is highly likely that abscisic acid and ethylene also interact with other growth hormones.

3.3.2 Gibberellin metabolism

Gibberellins are chemically diterpenes, which are themselves members of a vast group of naturally occurring compounds in plants called terpenoids. Considerable knowledge of terpenoid biochemistry exists, and consequently rapid progress has been made in elucidating the outlines of gibberellin biosynthesis.

Studies of gibberellin metabolism have been much more readily investigated in the fungus *Gibberella fujikoroi* than in higher plants. *G. fujikuroi* produces gibberellins as secondary metabolites, for there is no evidence for gibberellins acting as hormones in fungi. The fungus does not by any means synthesize all the gibberellins that occur in higher plants. In recent years use of suitable cell-free preparations from seeds of higher plants (such as *Marah macrocarpus,* synonym *Echinocystis macrocarpus;* and *Cucurbita maxima*) has accelerated research into patterns of gibberellin metabolism in plants.

Gibberellin biosynthesis. All terpenoids are basically built up from "isoprene units", which are five-carbon (5-C) compounds. The linking together of two isoprene units yields a

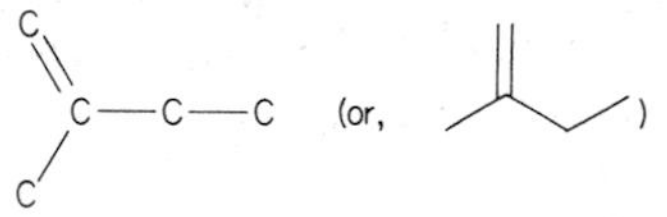

Isoprene unit (5-C)

monoterpene (C-10), of three a sesquiterperne (C-15), of four a diterpene (C-30), and of six a triterpene (C-30).

There are so many different gibberellins that it is clear there is not a single biosynthetic pathway in fungi and higher plants. The routes of formation of the different gibberellins are still not fully known, but are better understood for *G. fujikuroi* than for higher plants. The steps to GA_{12}-aldehyde (the first intermediate with the *ent*-gibberellane skeleton) are common to the fungus

and all higher plants, but thereafter pathways diverge (Fig. 3.5) giving rise to the "early 3ß-hydroxylation pathway" and the "non-3ß hydroxylation pathway". The former pathway starts with the 3ß-hydroxylation of GA_{12}-aldehyde, which is then converted to GA_4. This conversion involves (i) the oxidation of the 7ß aldehyde group to a 7ß carboxyl group and (ii) the loss of the 10α methyl group, and the formation of a lactone ring between C-19 and C-10.

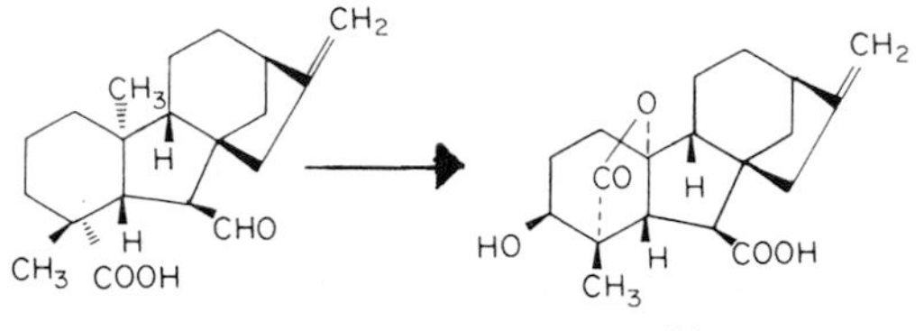

GA_{12}- aldehyde GA_4

From GA_4 the pathway divides several ways; three of these are hydroxylations to give GA_{16}, GA_{47} and GA_1, a major metabolite, respectively; the fourth involves the introduction of a Δ^1-double bond to form GA_7, another major metabolite. The latter is then 13-hydroxylated to yield GA_3.

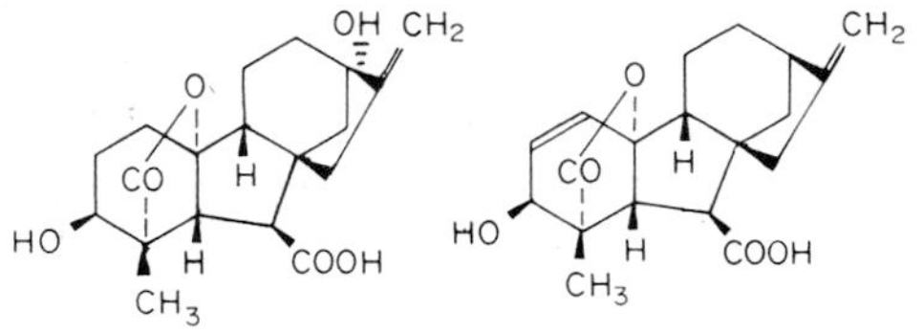

GA_1 GA_7

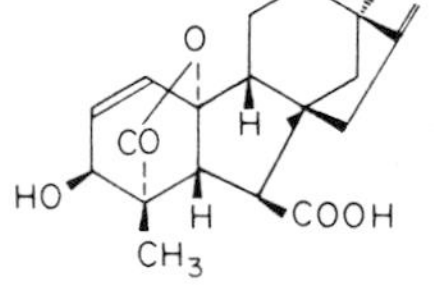

GA_3

The "non-3ß-hydroxylation pathway" yields GA_{12} and GA_9 as the main products. The first step is the oxidation of the 7α aldehyde group of GA_{12}-aldehyde to a carboxyl group, giving rise to GA_{12}, from which is formed GA_9 by the removal of the 10^{α} methyl group and the formation of a lactone ring between C-19 and C-10.

GA$_{12}$- aldehyde   GA$_{12}$ GA$_9$

Evidence is accumulating that *plastids* contain gibberellins and are capable of carrying out at least some, and probably all, of the steps involved in gibberellin biosynthesis. Also, some research has indicated that phytochrome can in some way be concerned in the regulation of gibberellin synthesis in and release from plastids. Etioplasts are able to take in mevalonic acid from the cytoplasm and convert it to gibberellins. Illumination of the plant triggers development of etioplasts to chloroplasts and when this occurs the chloroplast envelope becomes impermeable to mevalonic acid. Thus, chloroplasts synthesize their own mevalonic acid, some of which is utilized in the synthesis of gibberellins. Whether any cell organelles other than plastids have the capability of synthesizing gibberellins is not known.

It is of interest to note here that the growth inhibitor abscisic acid (ABA) is a sesquiterpenoid, therefore sharing with gibberellins a partly common pathway of biosynthesis from mevalonic acid, and ABA is also synthesized in etioplasts and chloroplasts.

A number of synthetic *growth retardants* have been discovered in recent years (e.g. below):

(2-Chloroethyl) trimethylammonium chloride
('CCC'; 'Cycocel'; 'Chlormaquat')

2,4-Dichlorobenzyltributyl phosphonium chloride
('Phosfon-D')

Ammonium (5-hydroxycarvacryl) trimethylchloride piperidine carboxylate
('AMO-1618')

Succinic acid-2,2-dimethyl hydrazide
('SADH'; 'B-995'; 'B-9'; 'Alar')

Some of these are proving of considerable importance in agriculture (see p. 146). Several of these growth retardants have been shown to act by inhibiting gibberellin biosynthesis in the plants to which they are applied. For example, the growth retardant "Amo-1618" has been shown to inhibit the biosynthesis of gibberellin in homogenates of the endosperm of the wild cucumber (*Echinocytis macrocarpa*). It appears that Amo-1618 inhibits the cyclization of geranylgeranyl pyrophosphate to *ent*-Kaurene (Fig. 3.5). The growth retardant "Cycocel", or "CCC" appears to act similarly.

Gibberellin Deactivation. Very little is known of the eventual fate of gibberellins in plant tissues. There is evidence that gibberellins retain their physiological activity for some considerable time in plants. This is in marked contrast to the rapid inactivation of applied natural auxins (such as IAA) in plant tissues, but it is a common observation that even large doses of gibberellins are not particularly injurious to plants whereas auxins can

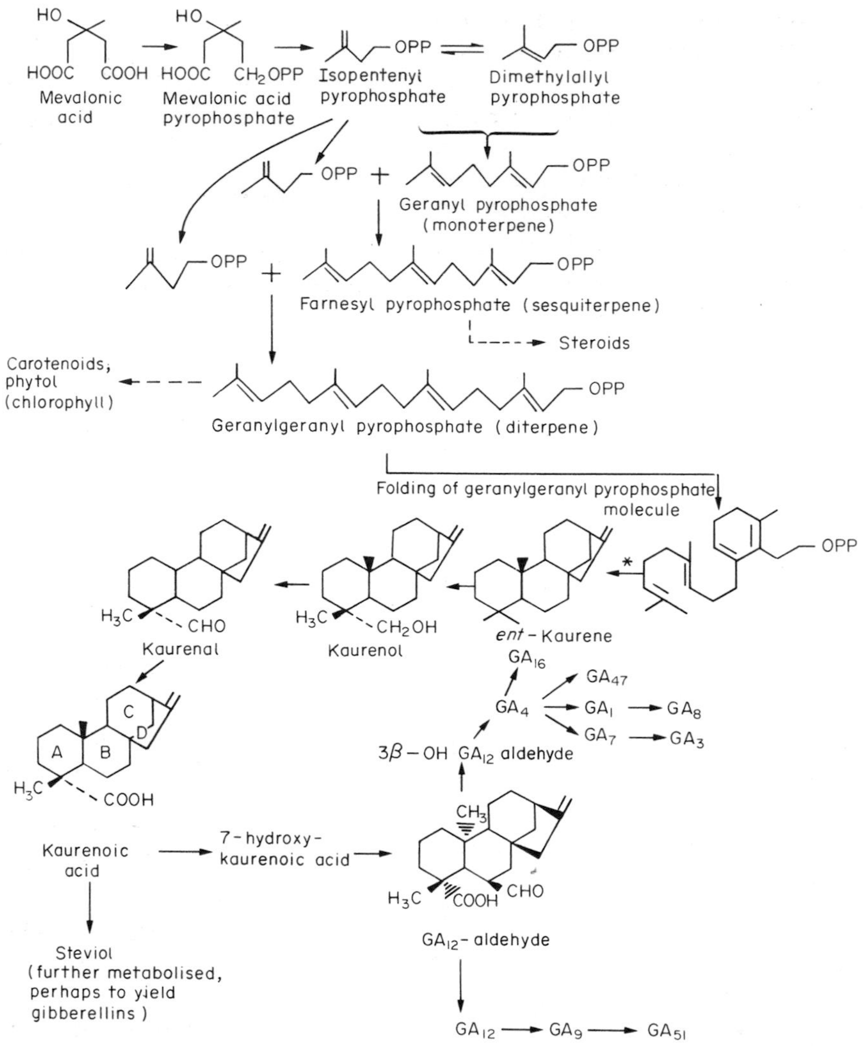

Fig. 3.5. Possible biosynthetic pathways for some of the known gibberellins from mevalonate. Others of the approximately sixty known gibberellins are formed through various transformations and interconversions, the details of which are not all yet known. The cyclization of geranylgeranyl pyrophosphate (*) is blocked by growth retardants such as CCC, Phosfon D and AMO-1618.

be, perhaps due to an effect on the production of ethylene (p. 119). Thus, it would seem more important that plants should have the capacity for speedy inactivation of auxin when its concentration exceeds a certain value.

However, there is now no doubt that plants metabolize gibberellins to a considerable extent. Inactivation proceeds by either modification of the gibberellin carbon skeleton (initially at least through specific substitution) or by conjugation with glucose. An important gibberellin inactivation process in higher plants is 2 ß-hydroxylation of the molecule (i.e. addition of -OH in the ß configuration to carbon number 2 on the "A" ring):

2,β-hydroxylated gibberellin

This 2 ß-hydroxylation occurs widely in higher plants, does not appear to be reversible and results in complete loss of or at least greatly reduced biological activity of the gibberellin. Some examples of known 2 ß-hydroxylations of gibberellins in plants are: $GA_1 \to GA_8$, $GA_{30} \to GA_{29}$ and $GA_9 \to GA_{51}$. In each case the product is an inactive gibberellin, or one of very low activity.

It is of interest that the fungus G. fujikuroi does not 2 ß-hydroxylate gibberellins, but can hydroxylate some with the 2α - configuration:

2, α-hydroxylated gibberellin

Examples of 2α-hydroxylation by the fungus are: $GA_9 \to GA_{40}$ and $GA_4 \to GA_{47}$. Both GA_{40} and GA_{47} are active gibberellins.

Conjugation of gibberellins with glucose occurs either through a substituent hydroxyl group (gibberellin glucosyl ethers), e.g.:

GA_8–glucosyl ether (GA_8–glucoside)

or by the sugar linking to the 7-carboxyl group of the gibberellin to form a glucosyl ester, e.g.:

GA_1–glucosyl ester

The highest concentrations of these conjugates have been found in mature seeds, and it has been suggested that they may be "storage" forms of gibberellins which could be hydrolysed to release free gibberellins at the time of germination. However, much work is still needed to establish whether this is so. One difficulty is that it has been found that a number of inactive 2 ß-hydroxylated gibberellins (e.g. GA_8, GA_{26}, GA_{27} and GA_{29}) all occur as glucosyl ethers in seeds. Since, as previously mentioned, 2 ß-hydroxylation is probably irreversible, these conjugates would not give rise to active gibberellins when hydrolysed. Also, studies to date have not detected enzymes in plants capable of hydrolysing glucosyl ethers of gibberellins. Nevertheless, glucosyl esters of active forms of gibberellins are enzymically hydrolysed at germination to release active free gibberellins, so that these conjugates do have the necessary properties to serve as storage forms of gibberellins in seeds.

Gibberellic acid in solution can be decomposed by acid hydrolysis, particularly at higher temperatures, to yield compounds such as gibberellenic acid, allogibberic acid and gibberic acid. The latter compound does not retain any hormonal activity, but gibberellenic and allogibberic acids can still elicit some of the physiological effects of gibberellins.

3.4 CYTOKININS

The discovery of this group of plant growth hormones came from work concerned with the *in vitro* culture of young plant embryos and tissue explants. Studies by Haberlandt in the first two decades of this century demonstrated the existence in plant tissues of a diffusible factor which stimulated parenchymatous cells in potato tubers to revert to a meristematic state. That is, cell division could be induced by the factor.

Many workers, particularly Skoog and Steward in the U.S.A., have made a close study of the growth requirements of callus-cultures (a mass of undifferentiated and usually rapidly dividing cells (p. 154)) prepared from the parenchymatous pith cells of tobacco, and from carrot roots. It is primarily as a result of their work during the 1950s that we became aware of the existence of cytokinins—plant growth hormones originally regarded as being particularly important in the processes of cell division and differentiation, but which more recently have been found to be implicated in various other physiological processes such as senescence and apical dominance.

Skoog used a tissue culture technique in which an isolated piece of tobacco pith is placed on the surface of agar gel into which had been incorporated various nutritive substances and other, hormonal, factors. The exact composition of the agar medium was varied, and the effects on the growth and differentiation of the pith cells noted. For growth to occur it was found that it was necessary to add not only nutrients to the agar, but also hormonal substances such as auxin. However, when auxin (IAA) was applied alone with the nutrients very little growth of the pith explant occurred, and that which did consisted predominantly of cell enlargement; very few cell divisions occurred, and no differentiation of cells took place. If, however, the purine base *adenine* was incorporated into the agar medium along with IAA, then the parenchymatous cells were induced

to divide and a large callus mass was created. Adenine added without auxin, however, did not cause cell division in the pith tissue. There was, therefore, an *interaction* between adenine and auxin, resulting in the triggering off of cell division. Adenine is a purine derivative (6-aminopurine) and is a naturally occurring component of nucleic acids.

Purine
(with numbering system)

Adenine
(6-aminopurine)

Later, another substance with similar, but more potent, effects as adenine was prepared from degraded deoxyribonucleic acid (DNA). This substance was found to be 6-(furfurylamino) purine, basically similar in structure, therefore, to adenine. Due to its property of actively promoting cell division (in conjunction with auxin) it was given the name of *kinetin*. Indole-3-acetic acid and kinetin were found to have interacting effects upon cell division and differentiation in tobacco pith cultures (Fig. 6.2), similar to those shown by IAA and adenine.

Kinetin
(6-furfurylaminopurine)

Benzyladenine (BA)

Tetrahydropyranylbenzyladenine (PBA)

Kinetin is a synthetic cytokinin, which does not occur naturally in plants. A number of other synthetic cytokinins have been subsequently discovered, among the most active of which are the compounds benzyladenine (BA) and tetra-hydropyranylbenzyladenine (PBA) (see above).

3.4.1 Endogenous cytokinins

Although kinetin, BA and PBA have never been shown to be present normally in plants, substances which produce similar physiological and morphological effects have been found in various organs of many plant species, particularly in "nurse tissues" such as coconut milk (a liquid endosperm), in immature caryopses of *Zea mays,* and in immature fruits of *Aesculus hippocastanum* (horse chestnut), banana and apple. These naturally occurring substances, together with other synthetically prepared compounds which have effects on growth similar to those of kinetin, have been given the generic name of *cytokinins*. A cytokinin, therefore, is a substance which, in combination with auxin, stimulates cell division in plants and which interacts with auxin in determining the direction which differentiation of cells takes.

Most available evidence suggests that naturally occurring cytokinins are purine, particularly adenine, derivatives. In 1964 the New Zealander, Letham, isolated a cytokinin from sweet corn kernels and identified it as 6-(4-hydroxy-3-methyl but-2-enyl) amino purine. For convenience, Letham has called this substance *Zeatin.*

Since the first isolation and characterization of zeatin, a quite extensive range of naturally occurring cytokinins have been identified from various plant sources. All the known naturally occurring cytokinins are adenine derivatives (i.e. they are 6-substituted amino purines). Some of the naturally occurring cytokinins are illustrated below. Zeatin is the most active of the known natural cytokinins. The cytokinin (*o*-hydroxybenzyl) adenosine, isolated from poplar leaves, is of interest since the base is identical with that of the synthetic cytokinin benzyl adenine, apart from the additional hydroxyl group.

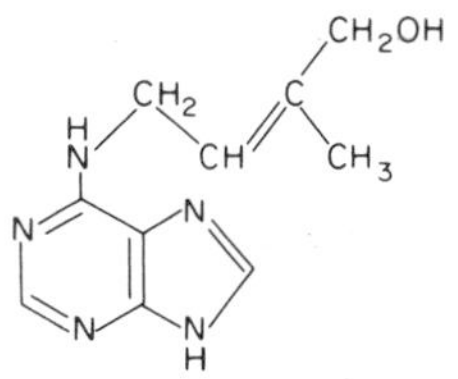

Zeatin

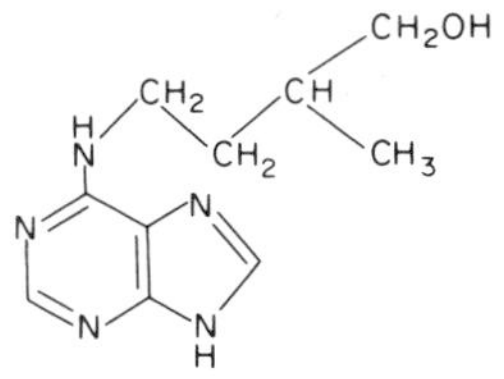

Dihydrozeatin

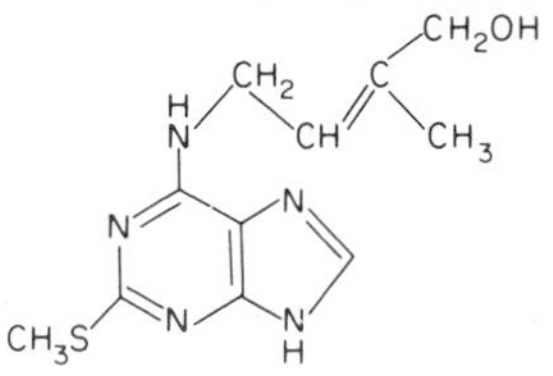

Methylthiozeatin

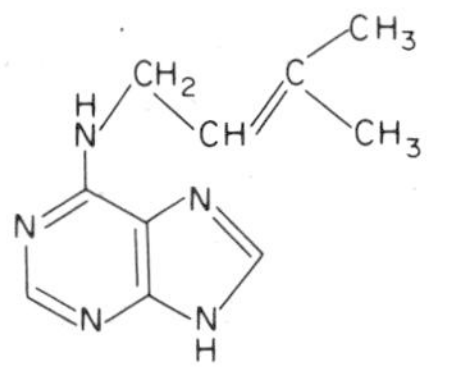

Dimethylallyladenine
(also known as
isopentenyladenine)

Like other purines, the natural cytokinins readily form the riboside (purine base + ribose sugar) and ribotide (base + ribose + phosphate group) derivatives, and may also, as we shall consider later, be contained in ribose nucleic acids. Examples of a naturally occurring cytokinin riboside and a cytokinin ribotide are shown below:

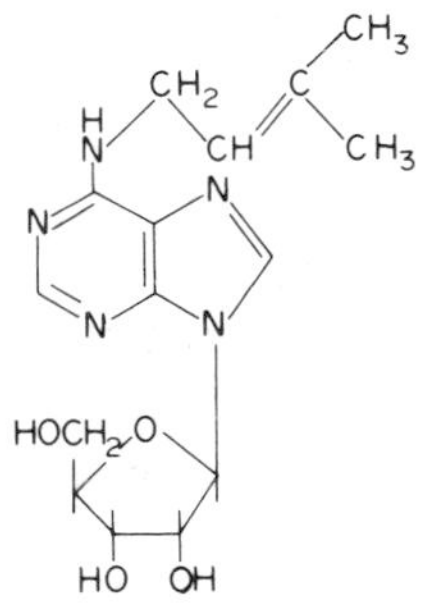

Dimethylallyladenine riboside
(= isopentenyladenosine, IPA)

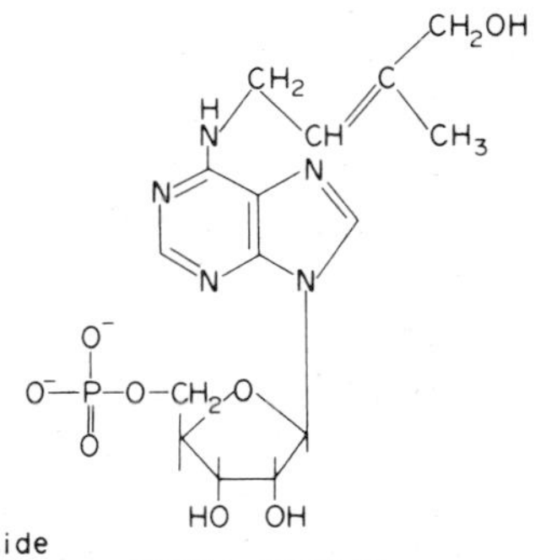

Zeatin ribotide

3.4.2 Cytokinin metabolism

Biosynthesis of Cytokinins. Free cytokinins in plant tissues could arise in one or both of two ways; (a) Through substitution of the characteristic side chains onto carbon 6 of adenine, adenine riboside, or adenine ribotide.

Adenine and its nucleoside and nucleotide forms are of course ubiquitous in plant cells. The side chain of almost all natural cytokinins contains five carbon atoms and this suggests that it is derived from the isoprenoid biosynthetic pathway. (b) By hydrolysis of certain tRNA species that contain cytokinin groups (i.e. some types of tRNA have adenine incorporated to which is attached an isopentenyl side chain at carbon 6).

Although tRNA does appear to be a potential precursor for free cytokinins, there is considerable evidence against it serving such a function in most plants. For example, some aseptically-cultured plant tissues that are cytokinin-dependent (i.e. exogenous cytokinin must be added to the medium for growth to occur) nevertheless contain cytokinin-containing tRNA—hence the tRNA does not appear able to act as a precursor of required free cytokinin. Also, the levels of free cytokinins in pea root tips have been found to be twenty-seven times greater than the cytokinins contained in tRNA. Furthermore, if free cytokinins are derived solely by hydrolysis of tRNA, various types of free cytokinins would be expected to be present which are in fact not found.

Thus, it is widely believed that free cytokinins are synthesized in plants by mechanisms independent of the degradation of tRNA. The most obvious possibility is that there occurs transfer of an isopentenyl group from IPP (isopentenyl-pyrophosphate) to carbon 6 of adenine or its riboside or ribotide. Experiments conducted to explore this have unfortunately not yet yielded conclusive information. Feeding tissues with [^{14}C]-mevalonic acid (which would be metabolized to [^{14}C]-IPP) has not been shown to result in the appearance of significant quantities of ^{14}C-labelled free cytokinin, which would be expected if [^{14}C]-IPP were being utilized in side-chain formation and attachment to adenine. On the other hand, there is good evidence that [^{14}C]-adenine is converted by plant tissues to ^{14}C-labelled free cytokinins. Thus, crown gall tissue cultures of *Vinca rosea* form [^{14}C]-zeatin ribotide from [^{14}C]-adenine, and later on [^{14}C]-zeatin riboside, [^{14}C]-zeatin and [^{14}C]-zeatin glucosides all appear. Thus, the experiments with *V. rosea* cultures indicate that cytokinin biosynthesis occurs primarily at the nucleotide level.

Apart from direct substitution of adenine ribotide with an isoprene-derived side chain, at least some of the naturally occurring cytokinins are formed by conversion from one or other cytokinins. A possible mechanism of metabolic interconversion of natural cytokinins in plants is illustrated in Fig. 3.6.

Deactivation of Cytokinins. Studies of the metabolism of externally applied radioactive cytokinins have revealed that their deactivation in plants occurs by both conjugation with glucose and alanine, and by degradation. Leaves frequently show particularly high cytokinin glycosylating activity.

Conjugates of particular interest in the context of deactivation are the glucosides of cytokinins. Use of radiolabelled cytokinins has revealed that cytokinins become conjugated with glucose at the 3-, 7- or 9-position of the purine ring:

Cytokinin 3 - Glucoside

Cytokinin 7 - glucoside

Cytokinin 9 - glucoside

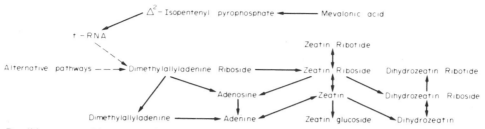

Fig. 3.6. Possible pattern of interconversions of some of the naturally occurring cytokinins in plants (*note:* dimethylallyl = isopentenyl). Dashed lines indicate that only indirect evidence or theoretical grounds exist for such conversions. (Adapted from R. H. Hall, *Ann. Rev. Plant Physiol.* **24**, 415-44, 1973.)

The cytokinin 7- and 9-glucosides are very much less active than are the non-conjugated free cytokinins from which they are derived. The 3-glucosides, though, seem to retain almost all the activity of the free cytokinins. The 7-glucosides appear to be particularly stable metabolites in plant tissues. It has been suggested that these cytokinin glucosides may be "storage" forms of cytokinins, but a great deal of research is necessary to establish the physiological significance of glucose conjugates of cytokinins. In the case of zeatin and dihydrozeatin, both metabolites and naturally occurring compounds have been found with glucose attached to the side chain -OH group. These conjugates are as active as the parent compounds, possibly as the result of facile cleavage of the glycoside linkage. The 9-alanyl derivatives of both zeatin and benzyladenine have been found as metabolites, and these derivatives exhibit negligible biological activity.

Metabolic degradation of cytokinins in plants has been found to involve an enzyme system which oxidises the side chain double bond, eventually producing adenine. This so-called "cytokinin oxidase" enzyme may well be important in controlling cytokinin levels in plant tissues since it inactivates the compounds. The adenine moiety can then be subjected to progressive oxidation to form urea *via* several intermediates (Fig. 3.7).

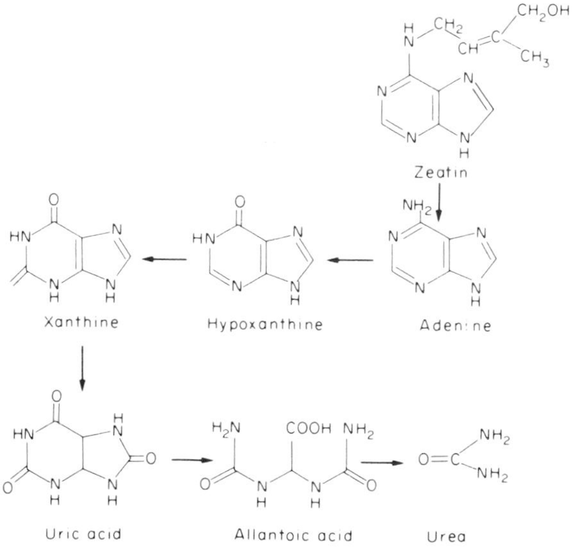

3.5 ETHYLENE

Ethylene may appear a curious substance to consider as a hormone. It is a very simple organic molecule, contrasting with the more chemically complex gibberellins, cytokinins, auxins and

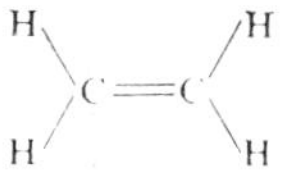

Ethylene

abscisic acid. Moreover, ethylene exists as a gas at normal temperatures. Thus, ethylene, if a plant hormone, is a gaseous hormone. It can be, and has been, argued that there are theoretical advantages to the plant in having a gaseous, diffusible growth regulator in addition to other hormones which necessarily move through living cells to reach their site of action. Treatment of plants with very low concentrations of exogenous ethylene has many profound effects on their physiological and

Fig. 3.7. Catabolism of the naturally occurring cytokinin zeatin. (Adapted from D. Schlee, H. Reinbothe and K. Mothes, *Z. Pflanzenphysiol.* **54**, 223-36, 1966.)

metabolic activities. However, evidence is also accumulating that endogenously synthesized ethylene is involved in the normal control of many aspects of plant growth, differentiation and responses to the environment.

It has been known for many years that developing fruits evolve ethylene, and that the time of maximum ethylene production by a ripening fruit coincides with the time of the *respiratory climacteric* (the latter term is applied to the large increase in respiration rate which occurs during the ripening period of many fruits, prior to a fall off in respiration as the fruit enters a senescent decline). It has been found that exposure of fruits to ethylene results in a hastened and enhanced respiratory climacteric with earlier ripening. In fact, ethylene is so effective in stimulating respiration that it does so even in fruits, such as oranges and lemons, which do not naturally experience a respiratory climacteric. The speeding of fruit ripening by ethylene has proved of great commercial value in the citrus industry.

In recent years, however, it has become apparent that we must consider that ethylene has much wider physiological significance than its effects in fruit ripening alone would indicate. In general, it would appear that many effects previously considered to be induced directly by auxin may be mediated by an intervening step in which auxin leads to an increase in the formation of ethylene, following which ethylene induces the actual response.

In those situations where auxin-induced ethylene is responsible for effects of auxin, it is possible to regard ethylene as an intermediate hormone in much the same way that cyclic AMP serves as an intermediate in the action of many animal hormones. Not all auxin effects however, appear to involve ethylene as an intermediate. Neither can all effects of ethylene be elicited by auxins. Thus, ethylene cannot substitute for auxin in (a) the stimulation of cell elongation in the vast majority of plant species, (b) in promotion of growth in tissue cultures and, (c) in inhibition of senescence, ripening and abscission. Similarly, auxin usually cannot substitute for ethylene in the promotion of processes such as leaf senescence, abscission and seed germination.

Ethylene can be accurately and sensitively measured relatively easily by use of gas-liquid chromatography, and it has been found that rates of ethylene biosynthesis in plant tissues vary widely from organ to organ and the stages of their development. In vegetative plants, the highest rates of ethylene production occur in most actively growing regions of stems and leaves, particularly in meristematic tissues. For example, in meristematic parts of the apical region of the stem of pea plants, ethylene production has been measured as $0.43\,\mu l\,hr^{-1}\,kg^{-1}$ fresh weight of tissue, whereas in older internodes only $0.04\;\mu l\;hr^{-1}$ ethylene was produced per kg fresh weight. In general, parts of the plant rich in endogenous auxin also produce greatest quantities of ethylene, although senescent tissues usually produce relatively large amounts of ethylene and yet are low in auxin content (Chapters 5 and 12). Amongst the highest levels of ethylene production in plants are those recorded for ripening fruits of some species: passion fruits, for example, can produce $500\;\mu l\;kg^{-1}\;hr^{-1}$.

3.5.1 Ethylene metabolism

Biosynthesis of Ethylene. Although ethylene synthesis in fungi appears to take place mainly by the dehydration of ethyl alcohol:

$$CH_3—CH_2OH \longrightarrow CH_2{=}CH_2 + H_2O$$

this pathway does not appear to operate in higher plants. Several substances have been suggested as precursors of ethylene in higher plants, but research over the past decade has, in the main, supported the view that the amino acid L-methionine is the principal, or perhaps sole, natural precursor. The biochemical pathway for conversion of methionine to ethylene in plants is shown in Fig. 3.8. The first reaction involves activation of methionine by ATP and is catalysed by the enzyme methionine adenosyltransferase, with the formation of S-adenosylmethionine (SAM). Conversion of methionine to SAM is not stimulated by auxin. The second reaction is the conversion of SAM to 1-aminocyclopropane-1-carboxylic acid (ACC) through the action of ACC

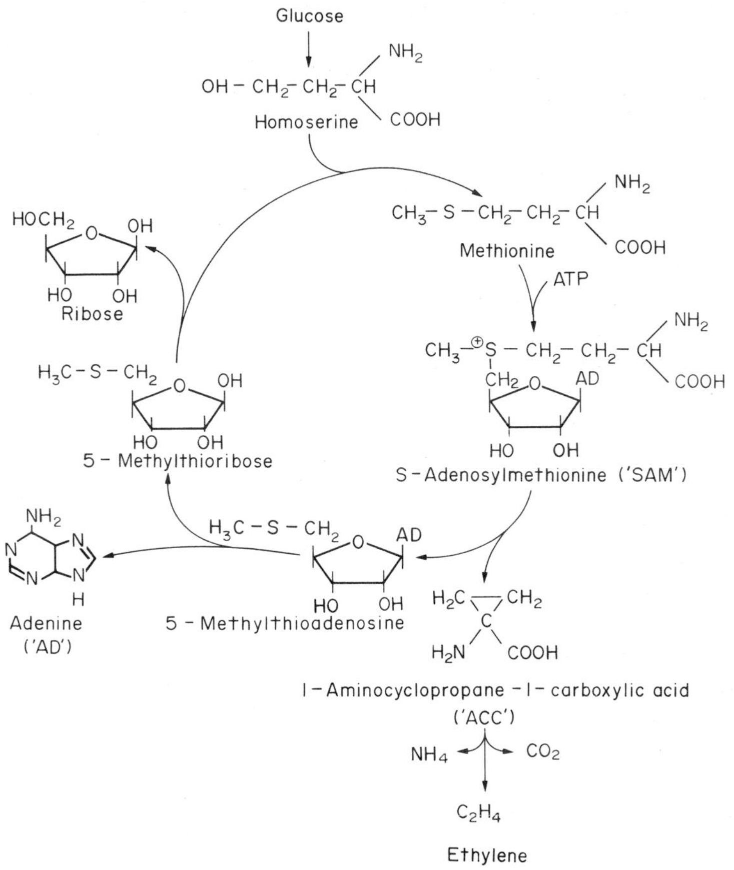

Fig. 3.8. Pathway of ethylene synthesis in plant tissues and the associated proposed cycle for methione metabolism. Note that the immediate precursor of ethylene is 1-aminocyclopropane-1-carboxylic acid (ACC) and that ACC is formed from S-adenosylmethionine (SAM). SAM is formed through activation of methionine by ATP in the presence of methionine adenosyltransferase. The conversion SAM→ACC is promoted by auxins through stimulation of synthesis of ACC synthetase. ACC→C_2H_4 is strongly oxygen-dependent.

synthetase. It appears that it is the rate of production of ACC from SAM which constitutes the rate limiting process in the biosynthesis of ethylene in plants, and that it is here that auxins exert their promoting effects on ethylene biosynthesis (see p. 119 and Fig. 5.13). Thus, auxins have been found to stimulate the synthesis or activation of ACC synthetase with a consequent greater production of ACC, the immediate precursor of ethylene. The conversion of ACC to ethylene is also enzyme-mediated, but does not appear to be auxin-sensitive.

Some studies on ethylene production in senescing *Ipomoea tricolar* petals (see Chapter 12) have suggested the possibility that the rate of ethylene production in these tissues can be regulated by availability of homocysteine to a cellular compartment in which methionine is first formed from S-methylmethionine (SMM) and then is consumed in ethylene production according to the following scheme (this illustrates that SMM can act as a methyl (—CH_3) donor, and that during senescence of the petals the methyl group is transferred to homocysteine, to give two molecules of methionine. One of the two-product methionine molecules is remethylated to SMM, and the other contributes to a rise in free methionine level and hence perhaps to increased ethylene formation):

(SMM)   (Methionine)   (Homocysteine)   (Methionine)   (Ethylene)

Deactivation of Ethylene. Until recently, it was not thought that higher plants possessed a

The probable pathway of ethylene catabolism in plants is therefore as follows:

Ethylene   Ethylene Oxide   Ethylene Glycol (1,2-Ethanediol)

mechanism for degrading ethylene. This did not seem incongruous because the ease with which the gas diffuses out of plant tissue (see p. 68) appears to provide a means for controlling endogenous concentrations. In other words, it seemed that the process of control of ethylene concentrations in plants by emanation was analogous in effect to the process of degradation for other growth regulators.

However, it has very recently been found that plants can in fact metabolize ethylene. Products detected are; ethylene oxide, ethylene glycol (1,2-ethanediol) and glucose conjugates of ethylene glycol. The oxidation of ethylene appears to take place at a copper-containing receptor site, for Cu chelators (e.g. EDTA), cobalt and silver ions all block ethylene oxidation. It has been suggested that Co^{2+} and Ag^+ may displace copper from the ethylene oxidation site. Further evidence for the involvement of copper in ethylene oxidation is provided by the fact that in nonbiological aqueous solutions copper forms a complex with C_2H_4 that results in the evolution of ethylene oxide.

The further metabolic fate in plants of ethylene glycol and its glucose esters is not yet known, but by analogy with what occurs in microorganisms it is likely that ethylene glycol is converted to acetate *via* acetaldehyde.

Thus, there is now good evidence that not all ethylene synthesized in plants is lost to the atmosphere, and this discovery has some important implications. Thus, most assessments of rates of ethylene biosynthesis have relied on the assumption that rates of emanation are proportional to rates of production (see above, p. 68). Clearly, at any rate in those plants which can metabolize ethylene, this assumption, and conclusion drawn from making it, are invalid.

3.6 ABSCISIC ACID

Studies on abscission and on dormancy in buds and seeds (Chapters 11 and 12) during the 1950s and early 1960s indicated the possible existence of a hormonal plant growth inhibitor. The chemical structure of the inhibitor present in fruits and leaves of cotton plants was eventually elucidated in 1965, and in the same year the same compound was isolated and identified from buds of dormant *Acer pseudoplatanus* trees.

The substance isolated from *Acer pseudoplatanus* and cotton was named abscisic acid (ABA), and proved to be a sesquiterpenoid (p. 60) of the following structure:

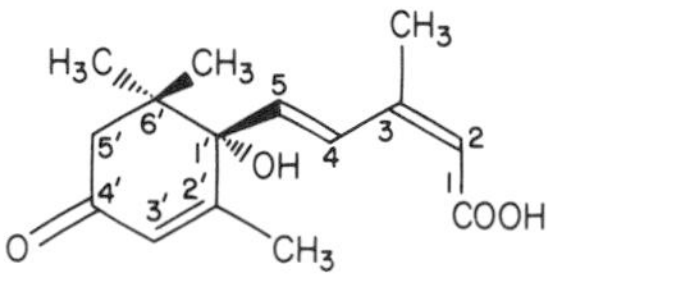

(+)-2-*cis*-Abscisic acid (ABA) Numbering system for carbon atoms shown

The molecule contains an asymmetric carbon atom (1′) and therefore exhibits optical isomerism. However, only the (+) enantiomorph occurs naturally in plant tissues. Abscisic acid also shows geometric isomerism. Steric considerations demand that the side chain always be *trans* around carbon 5 of the side chain, but the molecule can be either *cis*- or *trans*-around carbon 2 of the side chain. Most of the ABA present in plant extracts is in fact (+)-2-*cis* ABA though small amounts of (+)-2-*trans* ABA may also be present:

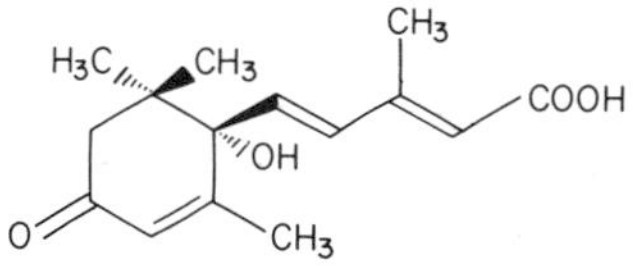

2-*trans*-abscisic acid

By convention, the naturally occurring (+)-2-*cis* form is simply referred to as abscisic acid, or ABA.

Abscisic acid has been isolated from numbers of species of angiosperms, gymnosperms, ferns and mosses, but it does not appear to occur in liverworts. However, the compound *lunularic acid* has been identified in at least eight species of liverwort and some algae, and may play roles in the physiology of liverworts similar to those served by ABA in more highly evolved plants, for it is a very potent growth inhibitor and also appears to be involved with the mechanisms of dormancy and gemma growth in certain liverworts.

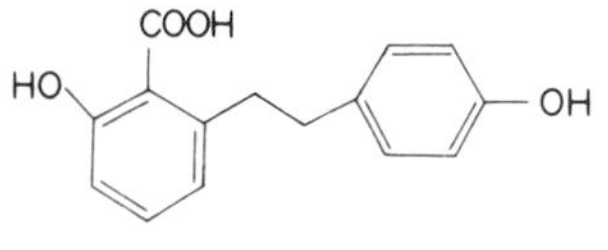

Lunularic acid

It has recently been found that *Cercospora rosicola,* a fungal pathogen that causes a rose leaf spot disease, synthesises and secretes large amounts of ABA into a culture medium. This discovery may well open the way to detailed studies of the chemistry of ABA biosynthesis.

3.6.1 Abscisic acid metabolism

Biosynthesis of ABA. Because ABA is a sesquiterpenoid, it seems likely that the compound is synthesized in plants by operation of the common pathway for terpenoid biosynthesis as far as farnesyl pyrophosphate (Fig. 3.4). The feeding of plant tissues with radioactive mevalonate has been shown to result in the formation of radioactive ABA, and overall experimental evidence strongly suggests that ABA can be synthesized directly by the reactions of the isoprenoid biosynthetic pathway. However, it has also been found that certain naturally occurring carotenoids, such as

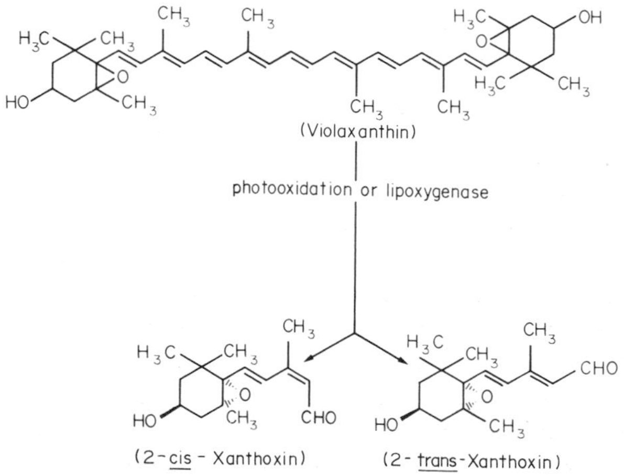

Fig. 3.9. Carotenoids such as violaxanthin can be broken down in the presence of strong light to yield growth-inhibitory products such as xanthoxin which are structurally similar to abscisic acid (ABA). It is possible that some of the ABA in plants arises by the further metabolism of carotenoid breakdown products, although most naturally occurring ABA is considered to be synthesized directly by the reactions of the terpenoid biosynthetic pathway.

violaxanthin, can be photo-oxidized to yield growth-inhibitory products that are very similar in structure to ABA (Fig. 3.9), and that these breakdown products may be further metabolised to ABA. Energization of the photo-oxidation of carotenoids to ABA-like compounds requires high light intensities, and for this and other reasons it is very unlikely that physiologically important ABA is formed in this way from carotenoids. For example, ABA occurs in etiolated plants and, furthermore, rapid rises in ABA levels occur in response to water-stress even in dim light or darkness. The principle means of ABA synthesis in plants therefore appears to be by direct synthesis from mevalonate via farnesyl pyrophosphate, although enzyme-mediated oxidation (by lipoxygenase) of violaxanthin to xanthoxin may also play some part in the biosynthesis of ABA in certain tissues.

The sites of synthesis of ABA within plants have not yet been investigated extensively, but indirect evidence suggests that most or perhaps all ABA is formed in mature green leaves and in fruits. ABA may be translocated from leaves to other regions such as the shoot apex and there inhibit growth and perhaps induce the formation of resting buds (see Chapter 11). There is also experimental evidence which indicates that plastids, particularly chloroplasts, may serve as centres of ABA synthesis.

Deactivation of ABA. As endogenous ABA levels fluctuate in relation to changes in growth rate, water potential and season, then not only synthesis but also inactivation of the molecule must take place in plant tissues. Relatively little is known of the factors concerned in the regulation of ABA inactivation, but it has been found that applied ^{14}C-ABA is rapidly conjugated in plants to form the glucose ester of ABA. This ester seems to be quite stable in plants, and possesses hormonal activity similar to that of ABA. Degradation of ABA also occurs, however, the early stages of which involve hydroxylation and oxidation of methyl substituents of the ring. Work conducted so far on the metabolism of applied 2-[14]C-(+)-ABA has indicated that in tomato exogenous ABA is rapidly converted into its glucose ester and phaseic acid, but in *Phascolus vulgaris* the major metabolites have been identified as

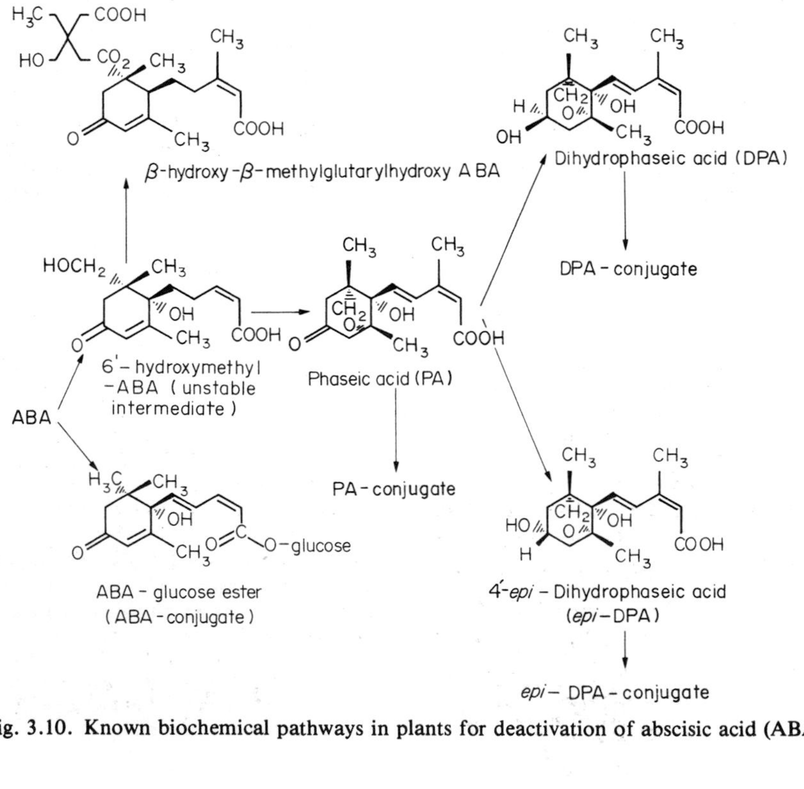

Fig. 3.10. Known biochemical pathways in plants for deactivation of abscisic acid (ABA).

phaseic acid, dihydrophaseic acid, and 4'-epi-dihydrophaseic acid, and the substance ß-hydroxy-1-methylglutarylhydroxy ABA has been recently identified as a metabolite of ABA in seeds of *Robinia pseudacacia* (Fig. 3.10). It is not yet known with certainty whether this pattern of breakdown is the same for endogenous ABA but it probably is, for PA, DPA and *epi*-DPA all occur naturally in plants.

There is some evidence that phaseic acid may have regulatory functions in the physiology of plants, for example in the inhibition of photosynthesis in plants subjected to water stress.

FURTHER READING

General

Abeles, F. B. *Ethylene in Plant Biology,* Academic Press, New York and London, 1973.
Audus, L. J. *Plant Growth Substances,* 3rd ed., Vol. I. L. Hill Ltd., London, 1972.
Paleg, L. G. and G. A. West. The gibberellins. In: F. C. Steward (ed.), *Plant Physiology—a Treatise,* Academic Press, New York, pp. 146-80, 1972.
Thimann, K. V. The natural plant hormones. In: F. C. Steward (ed.), *Plant Physiology—a Treatise,* Academic Press, New York, pp. 3-332, 1972.
Wilkins, M. B. (Ed.). *Physiology of Plant Growth and Development,* McGraw-Hill, London, 1969.

More Advanced Reading

Abeles, F. B. Biosynthesis and mechanism of action of ethylene, *Ann. Rev. Plant Physiol.* **23,** 259-92, 1972.
Blomstrom, D. C. and E. M. Beyer, Jr. Plants metabolise ethylene to ethylene glycol, *Nature* **283,** 66-68, 1980.
Dodds, J. H., S. K. Musa, P. Jerie and M. A. Hall. Metabolism of ethylene to ethylene oxide by cell-free preparations from *Vicia faba* L., *Plant Science Letters* 17, 109-114, 1979.
Galston, A. W. and W. S. Hillman. The degradation of auxin, *Encyl. Plant Physiol.* **14,** 674.

Gordon, S. A. The biogenesis of auxin, *Encycl. Plant Physiol.* **14,** 620.
Graebe, J. E. and H. J. Ropers. Gibberellins. In: *Phytohormones and Related Compounds—A Comprehensive Treatise,* Vol. I (eds. D. S. Letham, P. B. Goodwin and T. J. V. Higgins), Elsevier-North Holland Biomedical Press Amsterdam, pp. 107-204, 1978.
Hall, R. H. Cytokinins as a probe of developmental processes, *Ann. Rev. Plant Physiol.* 24, 415-44, 1973.
Hedden, P., MacMillan, J. and B. O. Phinney. The metabolism of gibberellins, *Ann. Rev. Plant Physiol.* 29, 149-192, 1979.
Hillman, J. R. (Ed.). *Isolation of Plant Growth Substances,* Cambridge University Press, 1978.
Kefeli, V. I. and Ch. Sh. Kadyrov. Natural growth inhibitors. Their chemical and physiological properties, *Ann. Rev. Plant Physiol.* **22,** 185-96, 1971.
Lang, A. Gibberellins: Structure and metabolism, *Ann. Rev. Plant Physiol.* **21,** 537-70, 1970.
Letham, D. S. Cytokinins. In: *Phytohormones and Related Compounds—A Comprehensive Treatise,* Vol. I (eds. D. S. Letham, P. B. Goodwin and T. J. V. Higgins), Elsevier—North Holland Biomedical Press, Amsterdam, pp. 205-263, 1978.
Lieberman, M. Biosynthesis and action of ethylene, *Ann. Rev. Plant Physiol.* **30,** 533-591, 1979.
Milborrow, B. V. The chemistry and physiology of abscisic acid, *Ann. Rev. Plant Physiol.* **25,** 259-307, 1974.
Milborrow, B. V. Abscisic acid. In: *Phytohormones and Related Compounds—A Comprehensive Treatise,* Vol. I (eds. D. S. Letham, P. B. Goodwin and T. J. V. Higgins), Elsevier-North Holland Biomedical Press, Amsterdam, pp. 295-347, 1978.
Pilet, P. E. (Ed.). *Proc. 9th Int. Conf. Plant Growth Regulating Substances, Plant Growth Regulation,* Springer-Verlag, Berlin-Heidelberg, 1977.
Schneider, E. A. and F. Wightman. Metabolism of auxin in higher plants, *Ann. Rev. Plant Physiol.* **25,** 487-513, 1974.
Schneider, E. A. and F. Wightman. Auxins. In: *Phytohormones and Related Compounds—A Comprehensive Treatise,* Vol. I (eds. D. S. Letham, P. B. Goodwin and T. J. V. Higgins), Elsevier-North Holland Biomedical Press, Amsterdam, pp. 29-105, 1978.
Shantz, E. M. The chemistry of naturally-occurring growth-regulating substances, *Ann. Rev. Plant Physiol.* **17,** 409, 1966.
Skoog, F. and D. J. Armstrong. Cytokinins, *Ann. Rev. Plant Physiol.* **21,** 359-84, 1970.
Skoog, F. and R. Y. Schmitz. Cytokinins. In: F. C. Steward (ed.), *Plant Physiology—a Treatise,* Academic Press, New York, pp. 181-212, 1972.
Wightman, F. and G. Setterfield (Eds.). *Biochemistry and Physiology of Plant Growth Substances,* The Runge Press, Ottawa, 1969.
Zeevaart, J. A. D. Chemical and biological aspects of abscisic acid, *ACS Symposium Series, No. 111, Plant Growth Substances* (ed. N. Bhushan Mandava), pp. 99-114, 1979.

4

Mechanisms of Action of Plant Growth Hormones

4.1 INTRODUCTION

Each of the known categories of plant growth hormone has the capacity to influence more than one type of developmental process. Furthermore, the physiological effects of these hormones overlap to a considerable extent. The multiple and interacting influences of plant growth hormones on plant development will be appreciated from reading subsequent chapters in this book. However, in this chapter we consider the possible ways in which hormonal control is exerted at the subcellular level. This is a very fundamental problem in developmental plant physiology and has consequently been the subject of a great deal of research for many years. The influence of growth hormones on all aspects of plant development is so pervasive that it is clear that an understanding of the subcellular action of these substances is essential to solving the general problems of development.

What then is meant by "mechanism of action"? It is an unfortunate fact that this term has come to mean rather different things to different plant physiologists. Thus, there has often been a failure to distinguish between two aspects of the action of a hormone:

(1) The direct and specific molecular interaction between hormone and its receptor site, which represents the initiating, or controlling, step in hormone action, and

(2) The succeeding series of events set in motion by hormone-receptor interaction. These events include biochemical changes that often culminate in physiological and morphological responses to the hormone.

It has been proposed by several workers that only the reaction between the hormone and its receptor be referred to as its "mechanism of action", and that the processes triggered or accelerated by this initial reaction should be regarded as the "mode of action" of the hormone. The distinction may seem a fine one, but it certainly helps to keep in focus that what one is mainly interested in so far as the subcellular action of a hormone is concerned is the nature of its *initial controlling reaction* which sets in motion those steps that culminate in the response. In other words, one is primarily concerned with direct rather than indirect effects of growth hormones in finding out how they affect cells.

As we consider in more detail later on (Chapter 13) it is only very rarely that hormones appear to determine the type of response shown by a cell or tissue. The same hormone can elicit different responses in different cells and it therefore seems that the nature of the response is usually predetermined in the cell; i.e. the developmental history of a cell programmes it to respond in a particular manner to a given hormonal signal. For example, a coleoptile cell may respond to auxin by undergoing cell enlargement growth whereas a cambium cell responds by dividing mitotically. Although there is as yet no proof, it is generally considered most likely that there is only one mechanism of action for each type of growth hormone regardless of the type of response (mode of action) elicited. Thus, the aim of research in this field is first to identify the initial hormone-cell reaction (presumably hormone-receptor association) and

then to discover how this induces the varied and profound modifications of patterns of cell metabolism.

It might reasonably be thought therefore that all research into the mechanism of hormone action would have been directly concerned with the reaction between hormone and the cell. In fact though, probably most published research data relate to secondary effects of hormones—i.e. aspects of their "mode of action" in eliciting a response. There are various reasons for this, but not least is that not until something has been found out about a phenomenon can one judge its relevance to the overall scheme of things. In other words it is easy to be wise after the event, and furthermore all the work done on secondary effects of hormone action has provided considerable information about cellular processes in general and given leads as to where to look for the primary processes controlled by hormones.

For various historical and practical experimental reasons, reseach into the mechanism of action of plant growth hormones has tended to concentrate upon particular physiological responses. As we consider below, for example, studies of the mechanism of auxin action have been largely concerned with its role as a regulator of cell extension growth, whereas equivalent research on gibberellins has concentrated on their effects on enzyme synthesis and secretion in aleurone cells of germinating cereal seeds. Furthermore, a number of diverse approaches have been taken over the years in attempting to elucidate the mechanisms of actions of plant hormones. Not all of these have been fruitful to date, but studies such as those of molecular structure in relation to hormone activity have yielded information that is now becoming valuable in the relatively recent attempts being made to isolate and study directly hormone binding sites from plant cells.

We will now consider what is known of the mechanism of action of each of the categories of plant growth hormone, drawing appropriately upon information derived from the various experimental approaches that have been applied to the problem.

4.2 RELATIONSHIPS BETWEEN MOLECULAR STRUCTURE AND HORMONAL ACTIVITY

The identification of indole-3-acetic acid (IAA) in 1934 as a naturally occurring plant growth hormone was followed by investigations to determine just what was "special" in the chemical structure of IAA to impart such profound influences on growth and developmental processes. It was hoped that an understanding of this would provide a lead into elucidating the mechanism of action of auxins within the cells. Similar considerations have more recently been applied to the gibberellins, cytokinins, ethylene and abscisic acid following their discovery.

4.2.1 Structure-activity relationships of auxins

The methods of studying this problem were initially largely empirical, in that many compounds were tested in suitable bioassays to find whether or not they possessed any "auxin activity". Some of these substances did in fact produce effects similar to those which IAA itself elicited, even when they were supplied at very low concentrations. Over the years a very large number of such compounds have been found, all of which have been synthesized in laboratories and are, therefore, called *synthetic auxins*. These synthetic auxins do not fall into any one particular class of compound, but despite the diversity of structure shown by the synthetic auxins, very serious efforts have been and still are being devoted to pinpointing exactly what attributes a molecule must possess for it to have activity as an auxin. It is hoped that elucidation of the molecular requirements for auxin activity will help in an understanding of the mechanism by which auxin operates in plant cells.

The first synthetic auxins found were compounds closely related to IAA (in having the indole ring) such as α -(indole-3) propionic acid, α-(indole-3)-butyric acid and ß-(indole-3)-pyruvic acid (Fig. 4.1) (which is now known also to occur naturally). However, many other synthetic auxins more markedly different in structure from IAA

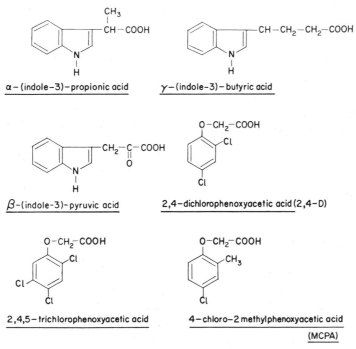

Fig. 4.1. The chemical configurations of some natural and synthetic auxins.

were subsequently discovered, and some of the more active of these, such as 2,4-dichlorophenoxyacetic acid (2,4-D), 2,4,5-trichlorophenoxyacetic acid (2,4,5-T) and 4-chloro-2-methylphenoxyacetic acid (MCPA) (Fig. 4.1), are not indole compounds.

As a result of comparing the structures of all the then available synthetic and natural auxins, in 1938 a list was drawn up of the general structural requirements of a molecule for it to behave as an auxin. Thus, it was said that an active molecule must possess: (i) a ring system with at least one double bond present; (ii) a side chain containing a carboxyl group (or group easily converted into a carboxyl-group); (iii) at least one carbon atom between the ring and carboxyl group in the side chain; (iv) a particular spatial relationship between the ring system and the carboxyl group. Later on, it was thought that another requirement of molecules which have activity as auxins is that they must have the ability to form a covalent bond at a position on the ring system *ortho* to the side chain which terminates in a carboxyl group. An examination of the structures of the synthetic auxins shown in Fig. 4.1 will reveal that they all comply with these general requirements, but there are other compounds now known which possess auxin activity and yet do not fully comply with the above list of structural requirements. For example, certain benzoic acid derivatives are active auxins (Fig. 4.3), and yet have no side chain. On the other hand, the activity of certain thiocarbamates indicates that not even the unsaturated ring is essential, although it is necessary that these latter compounds should have a planar structure. Furthermore, instead of the ability to form a covalent bond at the *ortho* position, it is now known that the requirement is that there should be a fractional positive charge at a specific point on the ring (p. 80).

On the assumption that a carboxyl-terminated side chain, and a "free" *ortho* position on the ring system are essential for activity, it was proposed that the basic reaction of an auxin within the cell involves two parts of the molecule, the carboxyl group of the side chain and an *ortho* position of the ring system. This led to what is called the

"*two-point attachment theory*" for auxin action. The research workers who put forward this theory proposed that there is covalent bond (i.e. chemical bond) formation at these two points between the auxin molecule and some constituent, possibly a protein, of the cell. The principles of the two-point attachment theory and a demonstration of the apparent validity of some of the listed structural requirements for auxin activity are clearly illustrated by a comparison of some of a series of chlorinated phenoxy compounds (Fig. 4.2).

It is necessary to stress that neither the original list of requirements for auxin activity, nor the two-point attachment theory are now accepted as valid. Clearly, the activity of a number of synthetic auxins such as benzoic acid derivatives (Fig. 4.3) cannot be adequately explained on the basis of the list of requirements drawn up in 1938 and the two-point attachment theory, and several alternative hypotheses have been put forward over the years, including "three-point" and "multipoint" attachment theories. The question is still open as to whether the auxin molecule becomes attached to some receptor in the cell by covalent (chemical) bond formation, or by some form of physical association, though the latter is generally considered much more likely. One suggestion, based on studies of the physical properties of active molecules, is that van der Waals and electrostatic forces are important in auxin-receptor association. Thus, a comparison of a range of auxins revealed that molecules active as auxins contain a strong negative charge (arising from the dissociation of the carboxyl group) which is separated from a weaker positive charge on the ring by a distance of about 5.5 A (Fig. 4.4), and it has been suggested that this is the essential structural requirement for auxin activity. This hypothesis would explain the relative activities of many synthetic auxins, and the differences in the activity of closely related compounds are apparently due to the effects of substitution in the ring on the position and size of the positive charge. Neither the nature, nor the location within the cell, of the receptor molecule is yet known.

Thus, the intensive study which has been devoted to the molecular requirements for auxin

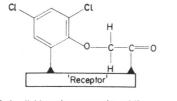

(2,4 – dichlorophenoxyacetic acid)

Active as an auxin — attached
at both receptor active positions

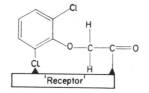

(2,6 – dichlorophenoxyacetic acid)

Inactive — due to both ortho positions
being filled by chlorine atoms.

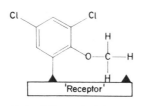

(2,4 – dichloroanisole)

Inactive — due to absence of a
terminal carboxyl group in side chain

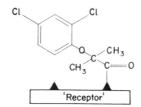

(2,4 – dichlorophenoxy iso-butyric acid)

Inactive — spatial configuration prevents
union between receptor active site and
the free ortho position

Fig. 4.2. The two-point attachment theory of auxin activity, illustrated by a comparison of 2,4-dichlorophenoxyacetic acid (2,4-D) with three inactive analogues.

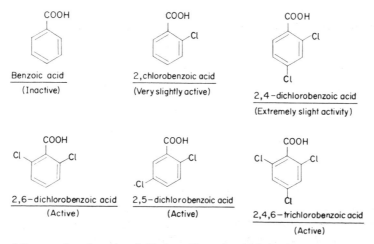

Benzoic acid

(Inactive)

2,chlorobenzoic acid

(Very slightly active)

2,4 – dichlorobenzoic acid

(Extremely slight activity)

2,6 – dichlorobenzoic acid

(Active)

2,5 – dichlorobenzoic acid

(Active)

2,4,6 – trichlorobenzoic acid

(Active)

Fig. 4.3. Activity or inactivity as auxins of a series of chlorinated benzoic acid derivatives. Note that 2,6-dichloro-benzoic acid and 2,4,6-trichlorobenzoic acid are active and also have a halogen atom at both *ortho* positions.

activity has not yet given any clear indication of the basic mechanism by which auxins produce their effects in growth and differentiation. It has, however, led to results of practical importance in the finding of a number of compounds which have proved of enormous value in agriculture and horticulture, such as selective weed-killers, fruit-setting agents and rooting-hormones (Chapter 5).

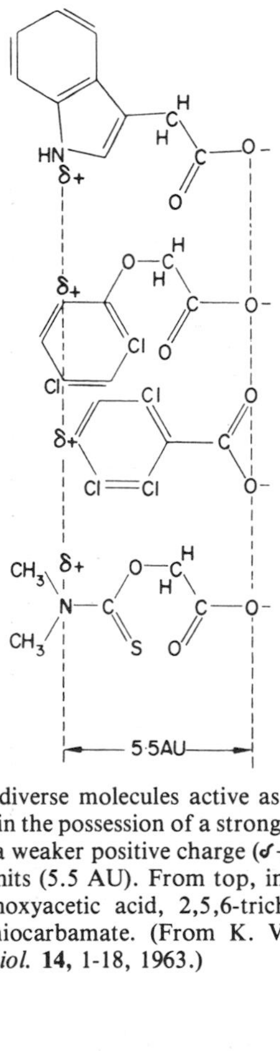

Fig. 4.4. Some diverse molecules active as auxins. They are, however similar in the possession of a strong negative charge (-) separated from a weaker positive charge (δ+) by a distance of 5.5 Angstrom units (5.5 AU). From top, indole-3-acetic acid, 2,4-dichlorophenoxyacetic acid, 2,5,6-trichlorobenzoic acid, carboxymethylthiocarbamate. (From K. V. Thimann, *Ann. Rev. Plant Physiol.* **14**, 1-18, 1963.)

4.2.2 Structure-activity relationships of gibberellins

As stated in Chapter 3, some fifty-eight chemically characterized gibberellins are currently known. All have been obtained from natural sources, either from the fungus *Gibberella fujikuroi* or from various species of higher plants.

The structures of gibberellins A_1 to A_{29} inclusive are illustrated systematically in Fig. 4.5.

They (and also GA_{30} — GA_{58} which are not shown) are all similar in the possession of the same basic carbon skeleton, the numbering system of which is as follows:

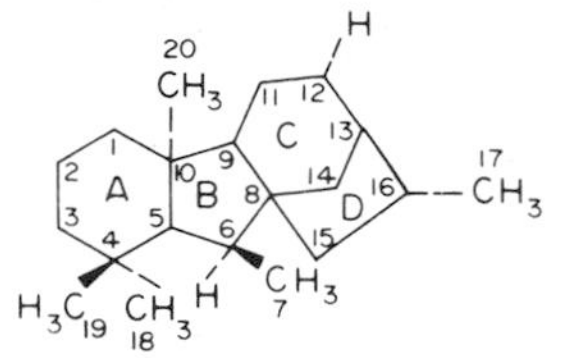

The *ent*-gibberellane carbon skeleton

Structural differences between the different gibberellins lie principally in the number and distribution of hydroxyl (-OH) groups, and the degree of saturation of the "A" ring. Present knowledge allows us to assume that for a molecule to have biological activity as a gibberellin it must have the basic gibberellane carbon skeleton together with an appropriate pattern of substitution.

It should be noted that not all of the known gibberellins are equally effective in stimulating growth. In fact, their activity when tested with different species and varieties of plants can be used as a means of distinguishing between different gibberellins. A good example of this is seen in their effect on the growth of dwarf mutants of maize (*Zea mays*). In maize there are a number of mutant genes, the presence of any one of which results in a dwarf habit of growth. Certain gibberellins have been found to promote the growth of some dwarf mutants, while others are effective with other mutants. It has been suggested that the primary effect of the mutant genes in maize is on the levels of endogenous gibberellins, by interfering with different steps in the biochemical pathway leading to a gibberellin necessary for normal growth. In the case of the d-5-mutant variety of *Zea mays,* it has been found that *iso*-kaurene is formed (instead of *ent*-kaurene, see Fig. 3.5) and that the plant is unable to metabolize this to gibberellins.

Although gibberellins vary in their activities, all active gibberellins have a carboxyl group attached

A-Ring	Substituents		C- and D- Rings 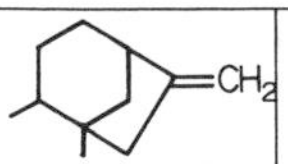	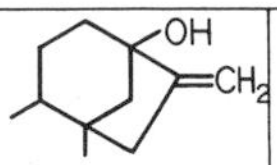	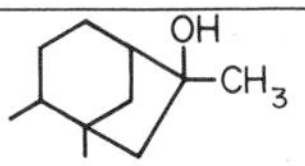	
	Carbon position	Group	=CH₂	OH =CH₂	OH -CH₃	
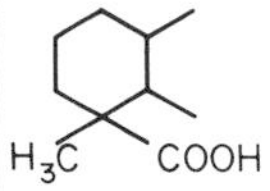 H₃C COOH	10	— CH₃	A_{12}			C — 20 Gibberellins
	10	— CHO	A_{24}	A_{19}		
	10	—COOH	A_{25}	A_{17}		
	{10, 3}	{—CH₃, —OH}	A_{14}	A_{18}		
	{10, 3}	{—CHO, —OH}		A_{23}		
	{10, 3}	{—COOH, —OH}	A_{13}	A_{28}		
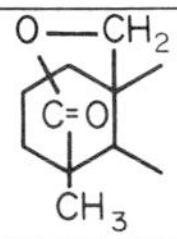 O—CH₂ C=O CH₃	—	—	A_{15}			
	3,2	—OH	A_{27}			
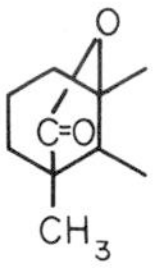 (C=O, CH₃)	—	—	A_9	A_{20}	A_{10}	C — 19 Gibberellins
	3	—OH	A_4	A_1	A_2	
	2	—OH		A_{29}		
	3,1	—OH	A_{16}			
	3,2	—OH		A_8		
	3,2 / 12	—OH / =O	A_{26}			
(C=O, COOH)	—	—		A_{21}		
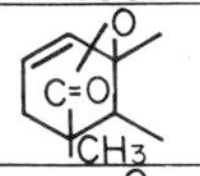 (C=O, CH₃)	3	—OH	A_7	A_3		
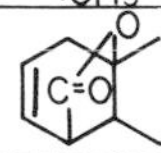 (C=O)	4	—CH₃		A_5		
	4	—CH₂OH		A_{22}		
(O, C=O, CH₃)	—	—		A_6		
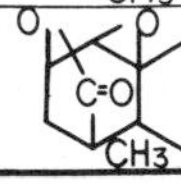 (O, C=O, CH₃)	—	—	A_1			

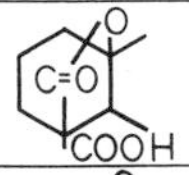

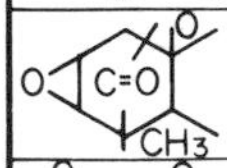

Fig. 4.5. A summary showing the range of chemical structures and chemical relationships in the first twenty-nine gibberellins to be isolated and characterized. (Adapted from L. J. Audus, *Plant Growth Substances,* Vol. 1, *Chemistry and Physiology,* 3rd ed., Leonard Hill, London, 1972.)

to carbon 7. Furthermore, the C_{19}-gibberellins (i.e. those based on the *ent-*20-nor-gibberellane structure—see Chapter 3) are more active than the C_{20}-gibberellins. Within the C_{19}-gibberellins 3ß-hydroxylation, 3ß,13-dihydroxylation, and 1,2-unsaturation all confer higher activity in general. Among the C_{20}-gibberellins greatest activity is shown by those that have the δ-lactone across the A-ring or an aldehydic group (-CHO) at the C-20 position. Also, as we considered in Chapter 3, one of the most striking features of gibberellin structure is the loss of activity which is associated with 2ß-hydroxylation. On the other hand a naturally occurring diterpenoid in plants called *steviol,* which does not have the gibberellane carbon skeleton, has been found to have some slight growth-promoting properties similar to those of gibberellins. However, this is probably due to conversion of steviol by plant enzymes to an active form, rather than to its possessing hormonal activity itself.

Steviol (*ent*-13-hydroxykaurenoic acid)

4.2.3 Structure-activity relationships of cytokinins.

As we considered in Chapter 3, the important known natural cytokinins are all substituted adenine compounds that possess side chains of five carbon atoms attached to carbon-6 of the adenine moiety. Systematic studies of many 6-substituted purines have revealed that where the side chain does *not* contain a ring system, the optimum number of side chain carbon atoms is five. Increasing or decreasing the size of the aliphatic side chain reduces physiological activity of the cytokinin, but does not necessarily abolish it completely. Thus, lengthening the side chain to contain as many as ten carbon atoms (e.g. 6-decylaminopurine) does not completely remove activity. Further features of the side-chain requirements are that it must be non-polar and the presence of a double bond in the aliphatic side chain usually increases cytokinin activity.

The range of molecules which, though not naturally occurring as cytokinins in plants, do possess cytokinin activity is very large indeed. We have previously mentioned synthetic cytokinins such as kinetin and benzyladenine (p. 64) that contain ring systems in their side chains. Some of these synthetic cytokinins are at least as, or even more, active than naturally occurring cytokinins. Benzyladenine, for example, is more active than even zeatin (the most potent of the natural cytokinins) in some types of bioassay.

Generally speaking, any modifications made to the adenine ring result in reduction in activity as a cytokinin. Thus, riboside or ribotide derivatives of cytokinins (p. 65) are less active than the free cytokinins, and various other types of attachment or substitution in the adenine ring always reduce, if not actually remove completely, hormonal activity. However, some 1-substituted adenine compounds have been found to have cytokinin activity, but it is possible that this results from enzymatic conversion to the 6-substituted active forms in plant tissues.

Exceptions to the general rule that cytokinins are 6-substituted adenine compounds are seen in a number of phenylureas and the related biurets. Although the phenylureas are very much less active than the subsequently discovered purine cytokinins, several hundred compounds of this

Diphenylurea   Chlorophenylphenylurea

Fluorophenylbiuret

type do possess cytokinin activity, among the most active of which is chlorophenylphenylurea. The minimum requirements for activity in the phenylureas are the —NH—CO—NH—bridge and a planar phenyl ring. At first sight it is difficult to recognize what structural features these compounds have in common with adenine derivatives, but one is the —N—C—N— linkage, of which adenine derivatives have four and the ureas one. The six-membered pyrimidine ring of adenine may be analogous to the phenyl ring of the ureas and active biurets (e.g. fluorophenylbiuret, above) and the amino nitrogen of the purine analogous to a *meta* substituent (e.g. Cl) in phenylureas.

An interesting discovery to come from studies of cytokinin analogues has been the finding of some compounds, the pyrazolo [4,3-*d*] pyrimidines, that act as competitive anatagonists of cytokinins. One of the most potent of these is:

3-methyl-7-n-pentylaminopyrazole [4, 3-d] pyrimidine

These competitive antagonists of cytokinins (and other cytokinin antagonists that act in a non-competitive manner) are likely to prove useful tools in studying the mechanism of action of cytokinins.

It can be seen, therefore, that studies of the structure-activity relationships of cytokinins have revealed a situation essentially the same as that derived from similar work on auxins, in that a quite bewildering array of compounds are seen to have the capacity to influence growth and differentiation in the manner of cytokinins. This has, so far, prevented any meaningful conclusions being drawn as to a possible receptor site for the initial action of cytokinins.

4.2.4 Structure-activity studies on ethylene analogues

Ethylene, $CH_2 = CH_2$, is a small unsaturated hydrocarbon, and the physiological activities of a series of ethylene analogues have been compared with those of ethylene itself. None of the analogues possesses activity as great as that of ethylene, whether the comparison is based on concentrations in parts per million (ppm) or molarity. On a ppm basis, for example, propylene, $CH_3—CH = CH_2$, has activity only about 1/100th that of ethylene, $CH = CH$, less at 1/2800th (except in the case of the induction of flowering in pineapple, where for unknown reasons acetylene can be as active as ethylene); and allene, $CH_2 = C = CH_2$, even less at 1/29000th ethylene activity.

Carbon monoxide, $C = O$, has effects on plants similar to those elicited by ethylene, but with only approximately 1/2700th the potency of ethylene itself. Carbon dioxide may act as an antagonist to ethylene action in plants, and this is possibly because CO_2 is a close structural analogue of allene and carbon monoxide but nevertheless lacks certain molecular characteristics which are essential for ethylene action. Thus, CO_2 may compete with ethylene for the active receptor sites for ethylene action in plants, and the physiological response of a tissue to a given concentration of ethylene is determined by, among other factors, the prevailing concentration of CO_2 in the tissue.

The characteristics that a molecule must show to allow it to function, however weakly, as a substitute for ethylene were summarized as follows by Burg and Burg in 1967:

(1) The molecule must be unsaturated. A double bond confers more activity than a triple bond, and single-bond compounds are inactive.
(2) Activity decreases with increasing chain length.
(3) Substitutions that lower the bond order of the unsaturated position by causing electron delocalization reduce biological activity, although steric factors are also important. Thus, the nature of the substituent can affect activity by influencing both electron density in the double bond and overall size and shape of the molecule.
(4) The unsaturated position must be adjacent to a terminal carbon atom.
(5) The terminal carbon atom must not be positively charged.

These structural features, derived from studies of the varying degrees of physiological activities shown by ethylene and ethylene analogues, together with the known antagonism to ethylene action shown by carbon dioxide, do not help us to decide the nature of the initial receptor site for ethylene, nor the type of bonding involved in linking hormone and receptor, except that ionic and hydrogen bonding seem unlikely. However, various other experiments have yielded results that indicate that ethylene may be bound to its site of action by means of weak van der Waals forces, rather than by covalent or co-ordination bonding.

The existence of ethylene analogues which possess varying degrees of biological activity similar to that of ethylene itself has very recently proved of value in studies of an isolated putative ethylene receptor from plant tissues. Thus, by comparing the binding capacity of the receptor for ethylene with that for less active analogues, it has proved possible to evaluate the specificity and functional significance of the ethylene binding site (see Bengochea *et al.,* 1980).

4.2.5 Abscisic acid structure and physiological activity.

The natural (+), and synthetic (−), optical enantiomers of ABA have equal activity on plants, but only the 2-*cis* geometric isomer possesses hormonal activity (see p. 71 for an account of isomerism in ABA). Plant extracts have been found to contain only traces of the 2-*trans* isomer, and it is likely that even these small amounts are formed by isomeration of the natural 2-*cis* isomer during extraction and purification procedures. It therefore appears that only the 2-*cis* form of ABA will fit an unknown receptor site in cells. Studies of ABA analogues have not, so far, been successful in further defining molecular requirements for ABA-like physiological activity, except that the ring double bond appears essential for activity.

4.3 DIRECT EVIDENCE FOR PLANT HORMONE RECEPTORS

The previous section of this chapter showed how activity in each class of plant hormone is determined by structural and often stereospecific properties of the molecules. It is therefore reasonable to consider it probable that there exists receptor molecules in cells able to recognize subtle differences between active and inactive analogues of plant growth hormones. Many scientists think that plant hormone receptors are probably proteins, because of this capacity for recognition of

appropriate molecular structure of hormones. Such a view is supported by research in animal endocrinology, where several specific proteinaceous receptors for steroid hormones have been isolated and characterized. On the other hand, careful and very recent work on the isolation of a gibberellin receptor from lettuce hypocotyls strongly suggests that it is not a protein but may be a cell-wall polysaccharide (a polygalacturonide fraction).

However, none of the attempts made to date have succeeded in unequivocally identifying a plant hormone receptor. Most of such work has concentrated upon the possibility of isolating protein receptors. Various approaches to the problem have been made, mainly using radioisotopically labelled hormones of high specific activity. One method has been to supply the radioactive hormone to plant tissues, and to subsequently homogenize and fractionate the cell constituents by centrifugation or gel filtration, in the hope that radioactivity will be found to be associated with a particular cell fraction. Another has been to apply the labelled hormone to various cell fractions, such as nuclei, isolated chromatin material and membrane fractions, with measurement of the affinity for binding between these and the hormone. Thirdly, attempts have been made to localize radiolabelled hormones within cells by autoradiographic methods. Reasons for the failure to isolate a proteinaceous receptor for any of the plant growth hormones are not at all certain. It should be mentioned that a number of pieces of research have yielded results suggesting that receptor proteins for cytokinins and gibberellins are present in plant cells, but their existence is by no means proven. Some very recent and promising research has led to the isolation of a cell-free preparation from plants that binds ethylene actively and specifically, and appears to have some of the properties one would expect of a receptor-site for this hormone (see Bengochea *et al.,* 1980 and Hall *et al.,* 1980). This putative ethylene binding site is apparently proteinaceous and probably associated with subcellular membranes such as those of Golgi bodies or endoplasmic reticulum. Lending support to the view that the isolated ethylene-binding fraction may indeed function as an ethylene-receptor *in vivo* are observations that physiologically active structural analogues of ethylene (such as propylene, vinyl chloride, carbon monoxide and acetylene, etc.) all competitively inhibit ethylene binding to the putative receptor to extents that would be expected on the basis of their relative biological activities. Finally in considering the ethylene receptor site, it should be mentioned that there is some evidence that ethylene binding to its receptor in plant cells leads not only to its hormonal effects taking place, but also to its metabolism to ethylene oxide and ethylene glycol (see Chapter 3). In other words, that the action of ethylene results in its inactivation.

As a cautionary note it must be mentioned that it has been pointed out by Kende and Gardner (1976) that some features of the action of plant hormones indicate that they may be significantly different from animal steroidal hormones with respect to their association with receptor sites, and that it may therefore be incorrect to assume the existence of proteinaceous hormone receptors in plant cells. The dose-response curves for plant hormone action characteristically show that their effects vary over a very wide range of concentration (typically over some four or five orders of magnitude) which contrasts with the much more restricted effective concentration range for animal hormones. Comparisons of hormone dose-response curves also suggest that whereas the binding of an animal steroid hormone to its receptor fits Michaelis-Menten-type saturation kinetics, the binding of plant hormones do not, which in turn poses the possibility that plant hormone receptors are not specific proteins. If the receptors are not proteins, then one can only guess as to their possible nature.

However, some experiments referred to below have indicated that the primary point of plant hormone action lies in cell membranes, so that it appears to be a property of certain cell membranes to serve as specific receptors for growth hormones. It is of course possible that protein receptors may be located in or associated with cell membranes. Also, if plant cells contain multiple receptor sites (maybe proteins) for a particular hormone, or

group of hormones such as the gibberellins, then this could possibly explain some of the difficulties enumerated by Kende and Gardner. There is, therefore, no reason to suppose at present that the current research activity aimed at isolating and identifying hormone receptors from plant cells will not eventually be as successful as have been the equivalent studies on receptors for steroidal hormones in animal cells.

4.4 MECHANISM OF ACTION OF PLANT HORMONES IN CELL EXTENSION GROWTH

As we saw in the first chapter of this book, the major component of overall growth in plants is a consequence of vacuolation of cells derived from the meristems. The most obvious manifestation of this is seen during the elongation growth of stems, coleoptiles, petioles and roots. Thus, control of plant growth by growth hormones to a very significant extent results from effects of these substances on cell extension processes. Because of this, and the experimental convenience offered by elongating structures such as coleoptiles and stems, a considerable proportion of research into the mechanism of action of plant hormones has been devoted to the phenomena of straight extension growth.

As would be expected, many experimental methods have been used in studying the action of hormones on extension growth. However, probably without exception a feature of all such work has been to remove the endogenous source of the hormone under study (e.g. by excising segments of stems or coleoptiles) and then to apply exogenous hormone. What happens with such a procedure is that removing the natural hormone results in a change in the rate of extension growth (either falling or increasing, depending on the type of hormone), but addition of exogenous hormone partially or completely restores the original rate of growth. The effect under study therefore is a *quantitative* one. This means that there is no need to invoke ideas of hormonal induction of new types of metabolic activity such as changes in patterns of protein synthesis, though of course the hormone may well and usually does affect overall rates of protein synthesis in cells that respond by growing at a different rate.

A general and fundamental feature of plant cells is that they are bounded by cell walls, and that consequently for a cell to enlarge the cell wall must increase in total area. The general process of plant cell enlargement can be expressed by the following equation:

$$\dot{v} = \phi. L (\Delta \pi - Y)/L + \phi$$

where $\dot{v}$ is the rate of cell enlargement (i.e. relative change in cell volume; dv/dt), ϕ is the wall extensibility (i.e. its yielding compliance), L is the water conductivity (i.e. the permeability of the cell to water), $\Delta \pi$ is the difference in osmotic potentials of the cell and its bathing medium, and Y is the yield turgor (i.e. the value of turgor pressure that must be exceeded before any cell expansion can occur). If, as is normally apparently the case, L is large relative to ϕ, then the L terms cancel out and the equation simplifies to:

$$\dot{v} = \phi (\Delta \pi - Y)$$

It can be seen from the first of the above two equations that there are only four possible ways by which a growth promoting hormone could increase the rate of cell enlargement: (i) by causing a decrease in the osmotic potential (π) of the cell (i.e. an increase in cell solute concentration), (ii) by decreasing yield turgor (Y), (iii) by increasing wall extensibility (ϕ), and (iv) by an increase in water conductivity (L) should this be low enough to be limiting.

There is no evidence that auxins exert their effects through decreasing the cell osmotic potential (though there is a possibility that part of the growth effect of gibberellins may occur through effects on π). There is a limited amount of evidence that auxin can increase the water conductivity of stem cells, but there is a great deal of information that demonstrates that hormonal effects on plant cell enlargement growth are mainly, or solely, achieved through effects upon wall extensibility.

4.4.1 Cell wall properties in relation to plant growth hormone action

During cell enlargement, due to vacuolation, irreversible plastic stretching of the cell wall takes place. It is, therefore, tempting to consider that cell vacuolation is a consequence of a softening of the cell wall, for this would inevitably lead to an influx of water into the protoplast for the reasons given in Chapter 1 (p. 4). Many experiments have revealed that auxin increases the plasticity of the cell walls. This can be shown by increased plastic deformation of plant organs treated with auxin following the application of a mechanical force (Fig. 4.6). The possible physiological significance of auxin effects on cell wall plasticity is increased by observations that there is a positive correlation between the effects of different auxin concentrations on promotion of elongation growth, and on cell wall plasticity (Fig. 4.7). The effects of other types of plant hormones on cell wall extensibility have not received by any means as much study as those of auxins. Promotion of elongation growth by gibberellins has in most investigated cases been found to be associated with an increase in wall plasticity, but there is also a certain amount of evidence that gibberellins may also promote cell growth by causing a decrease in the osmotic potential of the cell sap. Where ethylene induces cell enlargement, it is usually isodiametric expansion that occurs rather than cell extension—in other words in the presence of ethylene the cells do not elongate and consequently the stem, or other organ, becomes thicker and not as long as it would do otherwise. How ethylene induces this transverse expansion of cells is not yet clear, though changes in cellulose microfibril orientation and increased cellulase activity in the wall appear to be involved.

During cell enlargement, the cell wall not only stretches, but it also increases in thickness by the deposition of new cell wall material (p. 5). This cell wall growth is stimulated by auxin, and can occur even when cell enlargement is completely suppressed by various means (e.g. by surrounding the tissue with a hypertonic solution of mannitol).

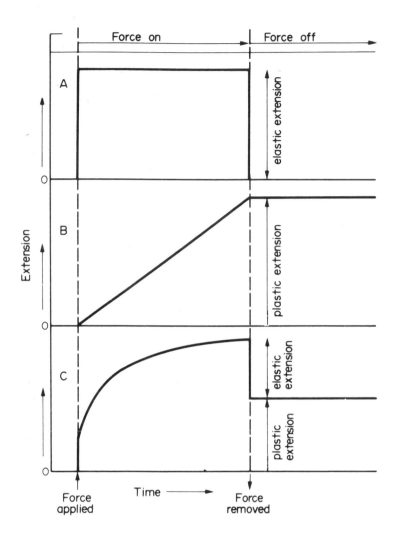

Fig. 4.6. Relationships between extension and time for three contrasting types of materials.

A. An elastic material such as rubber (the molecules of which are bonded together by extensive crosslinks) shows nearly instantaneous extension followed by no further increase with time, but the extension is fully reversible upon removal of the imposed force which caused the extension.

B. Where only few crosslinks or entanglements are present between the molecules, as in shortchain plastics, then irreversible extension occurs by *viscous flow* which is directly proportional to time.

C. In the case of materials such as the primary walls of plant cells, which contain polymers of varying lengths and degress of crosslinking, the extension is of an intermediate type called a *viscoelastic extension,* in which a certain amount of instantaneous extension occurs initially but is followed by a period of slower and continuous extension at a rate which is nearly proportional to the logarithm of time; the time-dependent component of viscoelastic extension is known as "creep". In fact, the instantaneous part of viscoelastic extension is just the creep which occurs too rapidly to be measured, so that there is no fundamental difference between the two components. Viscoelastic extension may be either partially or completely reversible, depending on the previous history of the material under test. When subject to mechanical stress for the first time, recovery behaviour is as illustrated in C for plant cell walls—part of the viscoelastic extension can be seen to have been elastic and part plastic in nature. Treatment with auxin increases the plastic component of total extension. Subsequent extensions, provided that the maximum length reached in the first extension is not exceeded, are entirely elastic, and this change in behaviour is termed mechanical conditioning. Note, however, that the extension pattern for plant cell walls was illustrated in C applies only to *dead* tissues. The walls of *living* cells extend by a continuous *series* of such viscoelastic extensions, so that extension may proceed at a steady rate for a considerable period of time.

The type of results shown in Fig. 4.6 indicate that cell walls show *viscoelastic* extension, that is, an initial rapid extension followed by a further slower extension (termed "creep"). However, *living cells* may elongate at a *constant* rate for a considerable time and it is thus accepted that growth proceeds by a *series of viscoelastic extensions* driven by the turgor pressure of the cell sap. Moreover, whereas cell walls derived from living tissue treated with auxin give the type of response shown in Fig. 4.6, isolated cell wall material does not respond in this way. This has led to the proposal that auxin does not act directly on the cell wall but rather than it controls certain events in the protoplast which result in a change in the properties of the cell wall. What are these events which are so influenced? In order to understand this we must consider briefly the structure and physical properties of primary cell walls.

remaining 80 per cent being water. In general, one can regard the cell wall as being analogous in structure to steel-reinforced concrete or glass-fibre-reinforced plastic, the cellulose microfibrils acting as the reinforcing element, and the noncellulosic matrix serving as the stabilizing component. The cellulose molecules within the microfibrils are held together by hydrogen bonding whereas the components of the matrix—both polysaccharides and protein—appear to be connected by covalent bonds. In dicotyledons at least it appears that hydrogen bonding also occurs between microfibrils and matrix and that the matrix polysaccharide involved is a xyloglucan (a polysaccharide with a ß1 ⟶ 4 linked glucan backbone as in cellulose but also with frequent xylose side chains and occasional galactose, fucose and arabinose units attached) (Fig. 1.4). The cell wall protein is unusual in that in addition to amino acids it contains a very high proportion of the iminoacid hydroxyproline (Fig. 4.8).

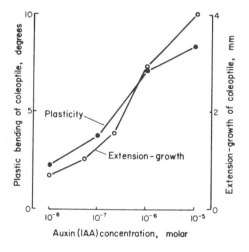

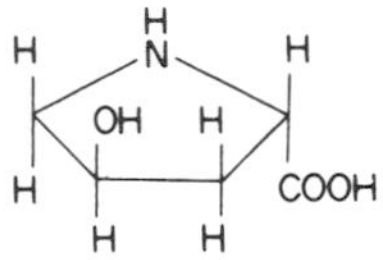

Fig. 4.7. Positive correlation between effects of auxin (IAA) on cell wall plasticity (measured by plastic bending) and on elongation in the oat coleoptile. (From J. Bonner, *Z. Schweiz. Forstv.* **30**, 141-59, 1960.)

As we saw in Chapter 1 the walls of young growing cells consist of interwoven chains of cellulose microfibrils embedded in a dense matrix of noncellulosic polysaccharides (of several different types) and protein. These components make up approximately 20 per cent by weight of the wall, the

Fig. 4.8. Structure of hydroxyproline.

Each hydroxyproline unit in the protein is glycosidically connected to an arabinose chain four units long (termed a tetraarabinoside) which is not, however, connected to the rest of the matrix (Fig. 4.9). The polysaccharides of the matrix appear to be linked to the wall protein via the serine residues of the latter.

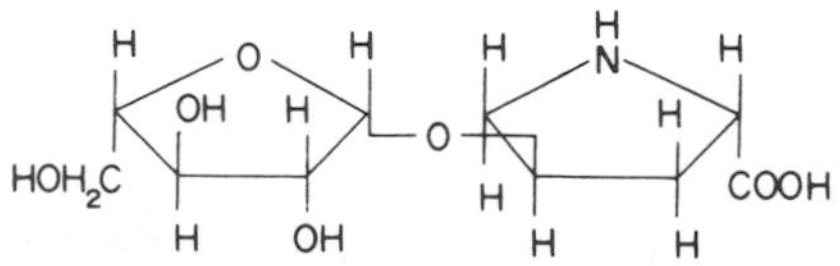

Fig. 4.9. Glycosidic linkage of hydroxyproline to arabinose. Each hydroxyproline unit in cell wall protein is connected in this manner to a tetraarabinose molecule (i.e. four joined arabinose units).

As in the man-made structures mentioned above, so in the cell wall the mechanical properties are the resultant of interactions within and between the microfibrillar and matrix components.

4.4.2 The kinetics of hormone-promoted cell enlargement growth

Although it is well established that auxins, and perhaps also other growth hormones such as gibberellins and ethylene, can cause changes in the mechanical properties of plant cell walls, the question is how do these effects occur? The great majority of experiments on changes in the cell wall associated with cell growth have been concerned with auxins, and consequently most of the following discussion deals with auxin-promoted growth.

Now it is known that although cell extension requires continued protein and RNA synthesis (Fig. 4.10) and respiration, nevertheless if auxin is applied to stem or coleoptile tissue the growth rate

increases after a "lag" of a matter of a few minutes (Fig. 4.11) which makes it unlikely that growth is accelerated by changes in the rates of transcription or translation, but rather that auxin is affecting some "pre-formed" system. It is also well established that marked changes occur in the

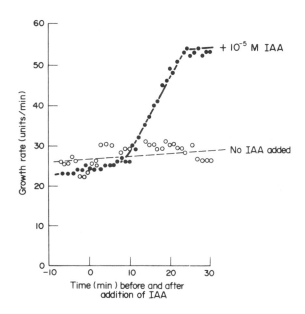

Fig. 4.11. The rate of response of excised pea-stem segments to 10^{-5} M IAA. The elongation growth rate of two batches of segments are shown, and it can be seen that treated segments commenced elongating at a more rapid rate after a lag period of about 10 minutes after application of the auxin. (Adapted from Pauline Penny, *New Zealand J. Bot.* **7**, 29-301, 1969.)

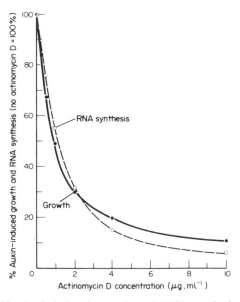

Fig. 4.10. Auxin-induced growth and RNA synthesis are similarly inhibited by various concentrations of actinomycin D. Soybean hypocotyl segments were pretreated for 4 hours with the indicated concentrations of actinomycin D prior to the addition of 5×10^{-5} M 2;4-dichlorophenoxyacetic acid (2,4-D, an auxin). RNA synthesis was meaured by incorporation of (^{14}C)-ADP over the 4-hour growth period. (Adapted from J. L. Key *et al., Ann. N. Y. Acad. Sci.* **144**, Art. I, 49-62, 1967.)

polysaccharides of the wall in response to auxin treatment and indeed, as shown above, it is difficult to conceive how the mechanical properties of the cell wall can be altered without changes of this nature. It seems that since such changes involve the making and/or breaking of covalent bonds then it is likely that enzymes are involved. In the past many workers have attempted to define what changes occur in the polysaccharides of the wall and whether these changes are correlated with variations in enzyme activities within the wall.

These efforts have not on the whole been conspicuously successful, principally because the exact structure of cell wall matrix polysaccharides, and how they were interconnected was unknown.

This made interpretation of such work very difficult indeed.

Early work emphasized the possibility of direct effects of auxin on the cell wall itself, particularly upon ionic bridging between its polyuronide components ("pectic substances") (Fig. 4.12), but such ideas have been generally superseded by subsequent work. Nevertheless, very recent studies have indicated that polyuronides, perhaps cell wall associated, may be acting as receptor molecules for gibberellins in the promotion of elongation growth in lettuce hypocotyls (see p. 84).

Recent work has given us a much clearer picture of cell wall structure and armed with this knowledge and some new experimental techniques a number of new hypotheses have been advanced. These hypotheses have been greatly influenced by another recent finding, namely that incubation of coleoptile or etiolated stem tissue at low pH (around 3·0) results in extension growth (termed the "acid growth effect") which in the short term at least is similar to that shown by exposure to auxin but without a significant "lag" phase. This led to further work which showed that auxin appears to promote the secretion of H^+ ions (protons) by stem or coleoptile tissue resulting in a lowering of the pH of the wall. The kinetics of the acid-growth effect are very similar to those of

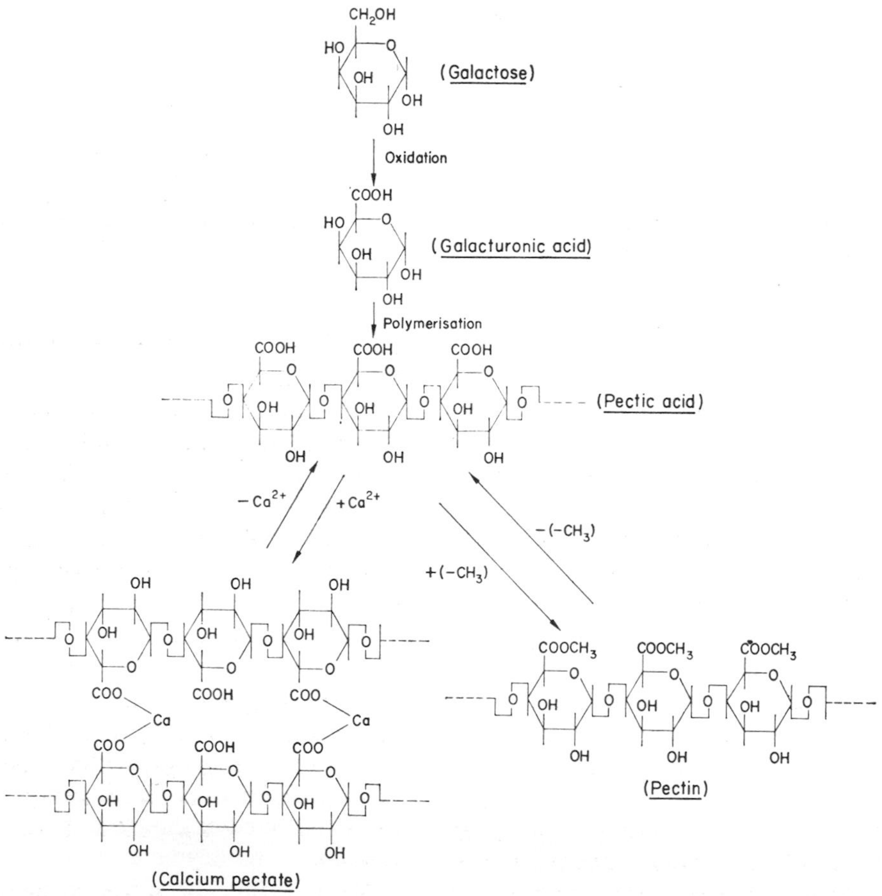

Fig. 4.12. Formation and interconversion of some pectic substances (polygalacturonides) in plant cell walls. Note the ionic bridging through calcium ions (Ca^{2+}) that can occur between adjacent molecules of pectic acid (but not of pectin) to form calcium pectate.

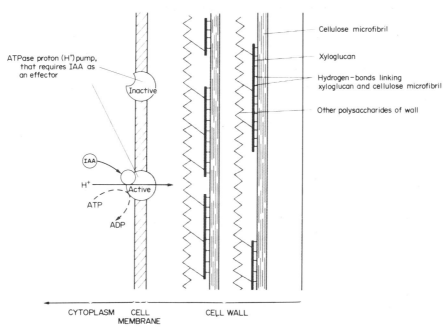

ATPase proton (H⁺) pump, that requires IAA as an effector

Inactive

IAA

H⁺

Active

ATP

ADP

Cellulose microfibril

Xyloglucan

Hydrogen-bonds linking xyloglucan and cellulose microfibril

Other polysaccharides of wall

CYTOPLASM   CELL   CELL WALL
            MEMBRANE

Fig. 4.13. Diagram to illustrate the hypothetical scheme of auxin (IAA) action as an effector of a membrane-bound proton (H⁺) pump. The pump is envisaged as containing ATPase, and being active in pumping H⁺ ions from the cytoplasm into the cell wall only when IAA is present. The secreted H⁺ ions could cause breakage of hydrogen bonds that link together xyloglucan polymers and cellulose microfibrils. When the wall is under tension (cell turgor pressure) this would result in a creep of the xyloglucan along the microfibril and thus give wall extension and cell enlargement. (Adapted from P. J. Davies, *The Botanical Review,* **39,** 139-71, 1973.)

growth stimulation shown by auxin-treated tissue. These findings would certainly account for the observed requirement for respiration mentioned above since ion "pumps" and charge separation are features of this process. The hypothetical auxin-sensitive proton-pump is generally assumed to be located in the plasma membrane (Fig. 4.13).

If we accept that this phenomenon represents the mode of action of auxin in increasing growth rate then three further questions present themselves: (1) How does auxin promote H⁺ ion secretion? (2) Why are RNA and protein synthesis required for extension growth? (3) How does changing the pH of the cell wall alter its properties?

The answer to the first question is quite unknown and much further work is needed before a solution is likely to present itself. As regards protein and RNA synthesis the question does not really arise if we distinguish between the *initiation* of increased growth and its continuance once in-

duced. Thus, although there is no sound evidence that the initial promotion of growth requires RNA and protein synthesis, *continuance* of *normal* cell growth requires that the wall must not only change its mechanical properties but much synthesis of new wall material must also occur—this is clearly so because cell walls do not become thinner as they extend. Thus a continuous supply of biosynthetic and hydrolytic enzymes is necessary and this probably involves both RNA and protein synthesis. Indeed it appears that the amount and/or activity of such enzymes—cellulose synthetase for example—is increased by auxin treatment, although this occurs some time after the promotion of growth; conversely, such changes are often suppressed if growth is inhibited even though auxin is present.

No firm answer is yet available to the third question but clearly H⁺ ions could act in two obvious ways, namely by breaking acid-labile bonds directly or by making conditions more favourable for various enzyme mediated modifications of the wall, e.g. by changing the pH of the wall so that it is nearer the pH optimum of some critical enzyme(s). If the process is enzyme controlled, the xyloglucan component of the matrix in dicotyledons may be the critical component. It has been shown auxin influences the "turnover" of xyloglucan more markedly than it does of other polysaccharides. That is, both breakdown and synthesis of this molecule are accelerated by auxin. Moreover, this effect occurs very rapidly in response to auxin treatment, and is also brought about by lowering the pH of the cell wall. As we showed above, the xyloglucan appears to interact with the microfibrillar component of the wall and a change in this interaction would be expected to affect wall properties. At present, much work is being done to identify the enzyme(s) responsible for this increased turnover. It should be emphasized that whereas the above hypothesis is feasible for dicotyledons, monocotyledons do not appear to possess xyloglucans and hence another mechanism is called for in their case.

At this point it should be noted that we have so far in this consideration of the mechanism of action of growth hormones (mainly auxins) in the control of cell enlargement growth, been almost exclusively concerned with short-term or relatively short-term effects. This emphasis reflects the activities of research workers over the past decade, who have recognized that if the key to the role of auxin in promoting growth is to be found, it is logical to look at processes occurring *within* the latent, lag, period (see Fig. 4.11). But, like all other developmental processes, cell extension growth involves changes in protein levels and enzyme activities. Many experiments have recorded positive correlations between growth and auxin (or other hormone)-promoted growth on the one hand, and rates of RNA synthesis on the other. A significant proportion of such experiments have involved the use of various inhibitors of protein synthesis (e.g. actinomycin D which blocks DNA-dependent RNA synthesis, and cycloheximide which acts at translational level by inhibiting assembly of proteins on ribosomes). For example, Fig. 4.10 illustrates that the capacity of exogenous auxin to enhance extension growth is inhibited by actinomycin D to the same extent that RNA synthesis is suppressed.

There is no doubt that normal, continued, extension growth does require continued protein synthesis, even if short-term rapid growth responses (those with latent times of less than approximately 10 minutes) to hormones can take place without synthesis of new RNA or protein. In the case of auxin-controlled cell extension rapid growth responses appear to involve H^+ secretion into the wall, the lowered pH either directly weakening certain intermolecular bonds or favouring the actions of certain wall-loosening enzymes.

Other types of growth hormone have not yet received such detailed study as have auxins in connection with their effects on cell enlargement growth. The latent period for promotion of cell growth by gibberellins appears to be significantly longer than is the case for auxins (30 minutes or more compared with less than 10 minutes), and there is still some uncertainty as to whether gibberellin-promotion of cell extension involves promotion of H^+ secretion by the responding cells. On the other hand ethylene affects growth rates within 5 minutes of exposure of a tissue to the gas.

The kinetics of inhibition of growth by abscisic acid (ABA) have not been by any means as carefully determined as they have for growth promotion by auxins. It is known that auxin-dependent elongation growth in coleoptiles can be inhibited within a few minutes of time of addition of ABA, and that closure of stomata in response to ABA (see Chapter 5) occurs within a minute of time of application of the hormone. Such observations of rapid effects of ABA suggest that at least some of the regulatory effects of ABA are independent of nucleic acid-directed protein synthesis. Longer-term inhibition of growth by ABA does though involve suppression of protein synthesis. Several possible ways have been suggested by which ABA could reduce protein synthesis. Findings that ABA enhances ribonuclease (RNase) activity led to suggestions that it is this effect of ABA that leads to lower RNA levels and a consequent fall in the rate of protein synthesis and growth rate. However, some subsequent work indicated that ABA can reduce total RNA synthesis in coleoptiles within 3 hours of its application, but that RNase activity does not increase until after 8 hours (Fig. 4.14), which suggests that the first effect of ABA on cell growth is not achieved through increased RNase synthesis or activity. The mechanism by which RNA and protein synthesis are reduced in some tissues by ABA is considered again later in this chapter (p. 99).

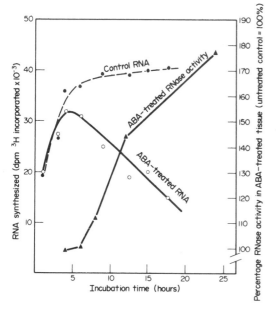

Fig. 4.14. Time-course of effects of 3.8×10^{-5} M ABA on total RNA synthesis (expressed as d.p.m. ^{3}H-labelled cytidine incorporated in RNA per gramme fresh weight of tissue) and on total RNase activity, in *Zea mays* coleoptile cells. (Adapted from J. H. M. Bex, *Planta*, **103**, 1-10, 1972.)

Thus, the mechanisms involved in plant cell enlargement growth are gradually being elucidated, and the role of auxin in the overall process is beginning to be understood. However, it should be remembered that although cell enlargement is in many respects the most characteristic result of auxin action, it is by no means the only one, nor even necessarily the first to appear. Thus, auxin can induce rapid increases in the rates of respiration and of protoplasmic streaming. Also, a number of responses to auxin do not immediately involve cell vacuolation (e.g. cambial division,

root initiation, and correlative inhibition of axillary buds). In other words, although auxin may induce a rapid increase in the rate of cell-wall loosening, this does not necessarily represent the only point of auxin action.

Finally, a further important aspect of hormonal control of plant cell enlargement growth is the effect that these substances have upon cell wall biosynthesis. Promotion of cell enlargement by a hormone such as auxin clearly involves stimulation of wall synthesis as well as a wall "loosening" effect, for the wall maintains a constant thickness during the period of cell growth. How these effects on wall biosynthesis are achieved is not yet known. However, of significance during the deposition of new cell wall material are the microtubules and there have been reports that growth hormones can affect the orientation of these. The orientation of the microtubules that lie adjacent to the plasmalemma (see Fig. 1.6) seems to determine the orientation of the cellulose microfibrils being deposited in the cell wall, and it is the orientation of microfibrils that determines the *direction* in which the walls can most readily expand.

4.5 MECHANISMS OF ACTION OF PLANT HORMONES IN CELL DIVISION

Cell division can be controlled through the actions of plant growth hormones, and we shall now consider some of the ways in which this may occur. Because mitosis is usually associated with DNA replication, the effects of hormones on DNA metabolism have received attention in relation to their effects in cell division. However, control of cell division can of course occur at a point in the cell cycle subsequent to DNA replication, and there is some evidence that at least certain instances of hormonal regulation of cell division are expressions of action on mitosis rather than on synthesis of DNA.

The first study directly aimed at exploring the effects of plant hormones on DNA synthesis and cell division were conducted on aseptic cultures of tobacco pith parenchyma by Skoog and his associates in the 1950s. They discovered that auxin was required for both DNA synthesis and mitosis to take place, but that mitosis and cytokinesis occurred only when appropriate levels of cytokinin were also present. Thus, these early studies suggested that auxin could stimulate DNA synthesis but that this was not necessarily followed by mitosis or cytokinesis. Cytokinins appeared to regulate mitosis and cytokinesis. These conclusions have been largely borne out in subsequent research by many different research workers. However, there is still no real understanding of the mechanism by which auxin can stimulate DNA synthesis, though there is some evidence that the activity of DNA polymerase may be regulated by auxin. Auxins therefore appear to play a permissive role in DNA synthesis, whereas most research has indicated that cytokinins have only a stimulatory (but not regulatory) role. On the other hand, there seems little doubt that cytokinins affect mitosis and cytokinesis, apparently through influencing the synthesis or activation of proteins specifically required for mitosis.

A very distinctive feature of cytokinins is that all known naturally-occurring examples of this class of hormone are purine derivatives. The obvious chemical relationship to the nucleic acids has from the time of their discovery tended to concentrate researchers' attentions upon the possibility that cytokinin action is in some way exerted through influencing nucleic acid metabolism and protein synthesis. No clear picture has yet emerged, however, although, like auxins and gibberellins, cytokinins quite clearly do have the capacity to stimulate RNA and protein synthesis in plant cells. Some workers have reported that all fractions of RNA (m-RNA, r-RNA and t-RNA) are increased after cytokinin treatment, but others have found that only r-RNA levels are raised.

From the mid-1960s much interest was centred upon the possibility that cytokinins may exert their hormonal effects through modification of specific transfer-RNAs, following the discovery that cytokinin groups occur in certain species of t-RNA. In both serine t-RNA and tyrosine t-RNA adenine nucleotides occur which possess side chains which are isomers of those of the most hor-

monally active cytokinins. Furthermore, in each case the substituted adenine moiety of the cytokinin was found to be located immediately adjacent to the anticodon of the t-RNA (Fig. 4.15). It was realized that a 6-substituted purine (all natural cytokinins are of this chemical nature) positioned adjacent to the anticodon loop would

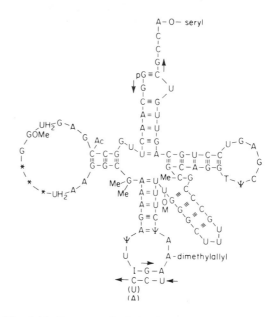

Fig. 4.15. Structure of t-RNA for serin (isolated from yeast cells), showing the location of the cytokinin dimethyl-allyladenine adjacent to the anticodon. A = adenine; C = cytosine; G = guanine; T = thymine; U = uracil. (From H. G. Zachau *et al.*, *Angew. Chem.* **78**, 392, 1966.)

permit maintenance of the correct spatial arrangement of the t-RNA, and so preclude the possibility of an incorrect triplet of nucleotides being recognized by the codon of m-RNA on the ribosome. In other words, it seems likely that the presence in t-RNA of a molecule of the cytokinin type is essential for normal codon-anticodon interaction between m-RNA and t-RNA on the ribosome. The hypothesis that cytokinins exert their regulatory effects through t-RNA functioning at translation therefore became very attractive.

However, this hypothesis has more recently received severe criticism on a number of grounds. For example, in the normal biosynthesis of t-RNA modification of component bases probably occurs after the primary structure of the polynucleotide has been established, which means that the characteristic side chain on carbon-6 of the adenine moiety of an incorporated cytokinin would be attached *after* the adenine portion is incorporated. This would make it impossible for substances such as kinetin, zeatin, etc., to be incorporated intact into t-RNA. Other evidence against cytokinins acting through their being incorporated into t-RNA includes the finding that t-RNA of *Zea mays* seed contains *cis*-zeatin, whereas the naturally occurring cytokinin in the same seed is *trans*-zeatin, which makes it difficult to believe that the cytokinin is a precursor in t-RNA synthesis. Particularly convincing evidence against the t-RNA hypothesis for cytokinin action has come frcm studies of rates of incorporation of radioisotopically labelled cytokinins into t-RNAs of cells responding to the hormone. Although the literature on the subject is rather conflicting, in general it has been found that the quantitative limits of cytokinin incorporation into t-RNA are far too low to regard the process as being the basis for the regulatory effects of cytokinins. Moreover, [^{14}C]-6-benzylamino-9-methyl-purine, although active as a cytokinin, does not become incorporated into t-RNA at all due to the masking of the carbon-9 position by a methyl group.

Experimental information on the effects on DNA synthesis and cell division of growth hormones other than auxins and cytokinins is rather sparse. There have been reports of gibberellin treatment causing elevated DNA levels and rates of cell division in certain plant organs and tissues, but the data are inconclusive in that it is impossible to decide whether the effects were direct or indirect.

Abscisic acid treatment of particular plants or tissues can result in a reduction in their rate of DNA synthesis and cell division, but it is fairly certain that these represent indirect effects of the hormone's action elsewhere in the metabolism of the cell.

Ethylene can either promote or inhibit growth of cells and these effects are often paralleled by effects on DNA synthesis and cell division. Examples of ethylene-promoted cell growth are stem-swelling, leaf epinasty, apical hook closure and seed germination. Inhibition of growth by ethylene is seen in the arrested development of some buds, leaves and apical meristems. Whenever ethylene inhibits growth there are measurable decreases in rates of cell division and DNA synthesis. However, it is not clear whether the change in DNA synthesis is the cause or an effect of ethylene inhibited growth, and detailed kinetic studies are required to resolve this question. It has been suggested that ethylene regulates cell division through an effect on the microtubules of the mitotic spindle apparatus. This proposal does have a slight amount of experimental evidence to support it, in that ethylene-induced isodiametric enlargement of cells is known to involve alterations in the arrangement of peripheral microtubules in the responding cells (see p. 8).

4.6 STUDIES OF HORMONE ACTION IN ISOLATED TISSUES AND CELL-FREE SYSTEMS

So far this consideration of the mechanism of action of plant growth hormones in development has mainly dealt with results of experiments on either whole plants or isolated organs of plants, which consist of integrated assemblages of different tissues, the cells of each of which may be expected to react differently to an applied hormone, complicating analyses of experimental results. There is therefore obvious theoretical advantage to the use of a single tissue of uniformly responding cells in studies of hormone action. By far the best example of this is the cereal, particularly barley, aleurone tissue. A further step in the direction of obtaining a more defined system to examine how hormones act is to study their effects on cell-free systems (i.e. homogenized and fractionated components from cells reacted *in vitro* with hormones). Some of the work on effects of hormones on transcriptional and translational processes is of such a nature, and this is described in section 4.6.2 below, although it should be remembered that direct studies of hormone-binding with putative receptors (see section 4.3) are also examples of cell-free experimental systems.

4.6.1 The cereal aleurone system

The aleurone is a peripheral layer of protein-rich cells lying around the endosperm in seeds of grasses and cereals. During germination and early seedling growth the aleurone tissue is very active, but then rapidly deteriorates and dies. The functions of the aleurone are to serve as a storage tissue prior to germination and as a source of a range of hydrolytic enzymes that are secreted to digest the reserves of the endosperm during germination. The aleurone therefore represents a single tissue of fairly uniform cells preprogrammed with a limited number of functions in the early life of the plant. After imbibition of water by a non-dormant seed, the aleurone cells perform their functions on receipt of appropriate hormonal signals. The principal hormone involved in barley seed aleurone is gibberellin, originating from the germinating embryo, but there also appear to be complex interactions between gibberellin and abscisic acid, and possibly with ethylene. The great majority of experiments have been conducted on barley seed aleurone, but a similar situation pertains in other cereal seeds, except that auxins and cytokinins have also been implicated in the initiation of hydrolysis of food reserves in the endosperm of wheat seeds.

The importance of gibberellins from the embryo in the activation of aleurone functions was first demonstrated in the early 1960s, when it was found that de-embryonated but imbibed barley seeds failed to develop hydrolytic enzyme activity and the endosperm reserves therefore remained undigested. Experiments, initially with embryoless half-seeds and later with isolated aleurone layers, revealed that their treatment with GA_3 caused an increase in amylolytic enzyme activity leading to release of reducing sugars from endosperm starch. Thus, gibberellin can substitute

for the embryo. The principal food reserve of the cereal endosperm is of course starch, and a major aspect of the effect of gibberellins on the aleurone cells is upon the enzyme amylase. α-amylase is not present in the dry, unimbibed, barley seed, but appears in and is secreted from aleurone layer cells in response to gibberellins. Treatment of aleurone layers with gibberellin also enhances the activity of a range of other enzymes, all of which are formed or activated in the aleurone cells but secreted to exert their hydrolytic action outside the protoplast of those cells. Some of the released enzymes act in the endosperm cells (α-amylase, ß-amylase, protease, ribonuclease, peroxidases) whereas others act on the walls of aleurone cells (ß-1→3 glucanase, ß-xylanase, ß-xylopyranosidase and α-arabinofuranosidase) which are degraded during germination. The dissolution of aleurone cell walls aids the outward movement of endosperm-digesting enzymes. The effect of gibberellin on the activity of these enzymes could occur in several ways: (i) through stimulation of their synthesis, (ii) by influencing their secretion from the aleurone cells, and (iii) by affecting enzyme-activation processes. In some cases gibberellin stimulates both the synthesis and secretion of an enzyme (e.g. α-amylase and protease) whereas in others only enzyme secretion is affected (e.g. ß-amylase, ß-1→3 glucanase, and ribonuclease).

The best studied of these enzymes of the aleurone system in relation to the mechanism of action of gibberellin is α-amylase, particularly in barley, but wheat and rice seeds behave in a similar fashion to barley. There is very good evidence that gibberellins (usually GA_3 has been used in experiments) induce *de novo* synthesis of α-amylase by aleurone cells. In 1967 Filner and Varner demonstrated that *all* the α-amylase produced in response to GA_3 treatment is synthesized *de novo* from amino acids. They obtained this proof of the effect of GA_3 on enzyme synthesis by use of *density-labelling* techniques. Barley aleurone layers were incubated with GA_3, together with either normal water (H_2O^{16}) or water containing the heavy isotope of oxygen (H_2O^{18}). The natural storage proteins in the aleurone cells were thus hydrolysed during protease action in the presence of either H_2O^{16} or H_2O^{18}, resulting in the formation of O^{16}-containing or O^{18}-labelled amino acids. The latter, heavy-isotope containing, amino acids are said to be density labelled, and proteins formed from them will also be density labelled (i.e. they will be heavier, or more dense) than proteins formed from O^{16}-containing amino acids. Following ultracentrifugation of enzyme extracts, Filner and Varner found that α-amylase formed in GA_3-treated aleurone cells incubated with H_2O^{18} was of about the theoretically expected 1 per cent greater density than α-amylase from H_2O^{16}-incubated aleurone cells (Fig. 4.16), which demonstrated that all of the induced α-amylase was newly synthesized from amino acids during incubation with H_2O^{18}. In fact, it has been subsequently shown by similar methods (though D_2O has been more commonly used than

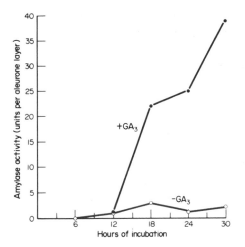

Fig. 4.17. Time course of enzyme (α-amylase) release from isolated barley aleurone layers incubated in media containing 10^{-6} M $GA_3(-GA_3)$ or no gibberellin (GA_3). Not until after 8 to 12 hours incubation do measurable quantities of enzyme appear, and the stimulatory effect of gibberellin become clear. (From K. M. Bailey, I. D. J. Phillips and D. Pitt, *J. Exp. Bot.* **27** 324-36, 1976.

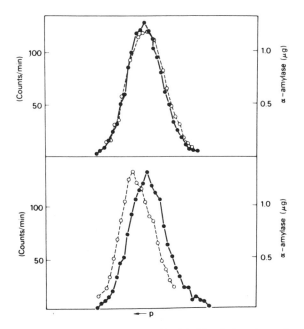

Fig. 4.16. Evidence from "density-labelling" that the entire α-amylase molecule is synthesized *de novo* from amino acids in barley aleurone cells in response to gibberellin treatment. The graphs show the distribution of α-amylase on a CsCl density gradient (ρ = density). *Above:* Coincidence of densities of α-amylase formed in presence of H_2O^{16} (O---O) and tritiated (^{3}H) marker α-amylase (•——•). *Below:* Greater density of α-amylase formed in presence of H_2O^{18} (O---O) compared with the marker α-amylase of normal density (•——•). (Adapted from P. Filner and J. Varner, *Proc. Nat. Acad. Sci.,* **58**, 1520-6, 1967.)

$H_2^{18}O$) that four α-amylase isoenzymes and enzymes such as ribonuclease and ß-1,3-glucanase are all synthesized *de novo* in response to GA_3.

Induction by GA_3 of *de novo* enzyme synthesis suggests that the mechanism of action of gibberellin may involve direct regulation of gene expression and the production of mRNAs required for translational assembly of the enzyme(s). A significant observation is that there is a lag period of at least 8 hours between application of GA_3 and measurable synthesis of α-amylase (e.g. see Fig. 4.17). Also, GA_3 must be continuously present throughout both the lag period and subsequently during the period of α-amylase synthesis (Fig. 4.18) if α-amylase synthesis is to persist. Treatment of GA_3-incubated aleurone during the lag period with actinomycin D (an inhibitor of RNA synthesis) inhibits subsequent synthesis of α-amylase, and the later the treatment with actinomycin D during the lag period the smaller becomes its effect. When actinomycin D is applied after the lag period it has only a small or perhaps no effect on α-amylase synthesis. These features suggest that for gibberellin to induce the appearance of α-amylase conditions must permit RNA synthesis during the lag period, but that

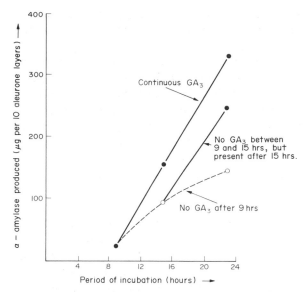

Fig. 4.18. Effect of removal and re-adding gibberellin (GA₃) upon synthesis of α-amylase by barley aleurone tissue. All aleurone layers were incubated with GA₃ up to 9 hrs (the lag period for α-amylase production). One set of aleurone layers was then maintained in GA₃ (Continuous GA₃), whereas the other two sets were washed free of GA₃ at 9 hrs. GA₃ was added back to one of the two washed sets after 15 hrs. It can be seen that gibberellin must be continuously present for maintenance of α-amylase synthesis even after the end of the lag phase. (Adapted from M. J. Chrispeels and J. E. Varner, *Plant Physiol.* **42,** 1008-16, 1967.)

synthesis of a number of other enzymes by aleurone cells in response to gibberellin similarly involves effects of gibberellin on the relevant mRNA species. What such data do not tell one, however, is whether the effect of gibberellin is direct (acting on transcriptional processes themselves) or indirect through earlier effects elsewhere in the cell. One thing seems reasonably clear. The effect of gibberellin on aleurone cells is not to cause, directly or indirectly, *selective* activation of genes for enzyme synthesis. The responses of aleurone cells to gibberellin are qualitatively predetermined—i.e. they do what they, as aleurone cells, have been programmed to do during their developmental history in the course of seed development. Gibberellin from the germinating embryo is required to initiate the whole sequence of enzyme formation, activation and secretion, but the hormone does not on its own determine the overall pattern of events.

In addition to the developmental preprogramming of aleurone cells that makes them respond to GA₃ in their own specific manner, it is also well established that the aleurone system is not a "one

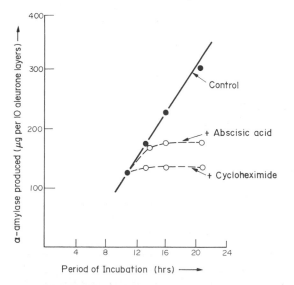

Fig. 4.19. Inhibition of α-amylase synthesis in gibberellin-incubated barley aleurone tissue by the protein synthesis inhibitor cycloheximide and by the plant hormone abscisic acid (ABA). Cycloheximide or ABA was added after 11 hours incubation. The controls were treated only with GA₃. (Adapted from M. J. Chrispeels and J. E. Varner, *Plant Physiol.* **42,** 1008-16, 1967.)

thereafter α-amylase synthesis can continue using as template RNA already formed (i.e. the α-amylase m-RNA appears to be long-lived). Inhibitors of ribosomal protein synthesis such as cycloheximide, and also abscisic acid, continue to inhibit the formation of α-amylase after the end of the lag period (Fig. 4.19). Some experiments have indicated that only approximately 1 per cent of total RNA synthesis is constituted of the gibberellin-promoted long-lived mRNA for α-amylase synthesis.

Thus, work on aleurone layers has provided good evidence for gibberellin stimulation of the synthesis of a new mRNA required for the production of α-amylase. Further evidence in support of this conclusion is provided by the results of experiments with a cell-free protein synthesizing system involving mRNA for α-amylase obtained from barley aleurone tissue (p. 100). Although less well documented, it also appears that the *de novo* hormone-one enzyme" system. We have already seen that the activities and synthesis of many enzymes other than α-amylase are influenced by application of GA_3. Furthermore, other types of growth hormone appear to be involved in α-amylase synthesis. Thus, ABA inhibits GA_3-induced α-amylase synthesis (Fig. 4.19) without significantly affecting the overall rate of protein synthesis and only slightly depressing RNA synthesis. This effect of ABA appears to be highly specific, in that it does not influence the activities of certain other enzymes even when α-amylase activity is almost completely inhibited, but no clear picture has yet emerged as to how ABA exerts its effects on aleurone cells. Cytokinins do not influence α-amylase production when applied alone to aleurone layers, but can overcome the inhibition by ABA of GA_3-induced α-amylase synthesis. Ethylene does not have any effect on the synthesis of α-amylase, but it can promote release of the enzyme from the aleurone cells and also partially relieve inhibition by ABA. Thus, the aleurone cells are sensitive to all the known plant growth hormones, which can act either alone or in combination to regulate the pace of hydrolytic enzyme synthesis, activation and secretion.

Although the main thrust of research aimed at elucidating the mechanism of hormone (particularly gibberellin) action in the aleurone system has been concerned with effects on RNA and protein synthesis, it is now known that GA_3 stimulates the level of α-amylase activity in barley aleurone cells *before* a rise occurs in the rate of RNA synthesis, which suggests that the *first* responses of aleurone cells involve the release of preformed enzyme and that only after this has occurred does the stimulatory effect of GA_3 on the synthesis of α-amylase become important. In fact, it is now generally considered that the earliest effect of gibberellins in the barley aleurone system is to influence various membrane systems already existing in the cells. The effects of gibberellin on membranes of the aleurone cells have been found to include (a) an increase in membrane synthesis (particularly rough endoplasmic reticulum), (b) stimulation of formation of hydrolytic enzyme-containing vesicles and microbodies from the en-

doplasmic reticulum, and (c) promotion of the secretion of α-amylase out through the plasma membrane.

Taken all together these various observations indicate that the synthesis of hydrolytic enzymes in aleurone cells before the addition of gibberellin may be limited not by the availability of mRNA but by the availability of appropriate membranes for the attachment of polysomes which carry hydrolase-specific mRNAs. It seems likely that at least some of the effects of gibberellins are exerted through influences on membranes, and that these effects precede stimulation of RNA and protein synthesis. For example, synthesis of lecithin, an important phospholipid of cell membranes, is increased within 2 hours of GA_3 treatment of aleurone cells. Thus, it may be that general *de novo* membrane synthesis, and especially production of rough endoplasmic reticulum, is controlled by gibberellin and is a necessary prerequisite for hydrolytic enzyme synthesis.

It is perhaps chastening to remember here that the aleurone layer system is, at least superficially, one of the simplest that has been studied with respect to the problem of the mechanism of action of hormones in plants. Yet, the complexity of the manifold responses shown have proved such that it is not yet possible to discern with clarity which, if any, of them represent the mechanism of action of gibberellin.

4.6.2 Effects of growth hormones on in vitro (cell-free) transcription and translation

As we have already seen, promotion of growth, or of enzyme synthesis in systems such as the aleurone layer, is associated with an increased rate of protein (and usually RNA) synthesis. In attempting to find out whether plant growth hormones influence gene expression or activity there have been numerous studies using cell-free (*in vitro*) assays based upon transcription of isolated chromatin and nuclei, and translation of products of transcription. Such experiments involve incubation of the isolated material with radioisotopically labelled RNA-precursors (ATP, CTP, GTP and

UTP). Newly synthesized RNA can therefore be measured in terms of incorporated radioactivity. Translation of RNA formed can be determined *in vitro* by use of cell-free protein synthesizing systems.

In the case of auxins, it seems clear that auxin treatment does increase the capacity for total RNA synthesis in isolated nuclei and in chromatin preparations. However, these effects of auxin occur only when the cells are exposed to auxin *before* isolation of the nuclei or chromatin. This suggests that the *initial* site of auxin action is located outside the nucleus and that auxin does not act directly at the transcriptional level of control. Also, there is no evidence for *selective* gene derepression *in vitro* by auxins. There is some evidence that auxin effects on transcription may be mediated *via* the release of a "transcription factor" from the plasma membrane. Incubation of plasmalemma fragments with auxin leads to rapid release of the factor which stimulates RNA polymerase *in vitro*. This observation can be incorporated into a hypothetical model for auxin action that encompasses both rapid and long-term effects on growth (see Fig. 4.20, section 4.8 below).

Experiments on the effects of auxins on *in vitro* translation have yielded much more equivocal results. Treatment of pea epicotyls with auxin increased the level of membrane-bound polysomes capable of synthesizing cellulase *in vitro,* and this effect was associated with enhanced mRNA levels. Epicotyls not treated with auxin contained no translatable cellulase mRNA whereas in treated tissue the level increased rapidly without appreciably lag. This meant that the level of cellulase mRNA was not correlated with the rate of cellulase synthesis since the latter began only after a 24 hour lag. Consequently, it has been proposed that auxin has a secondary control over translation, but there are alternative explanations (particularly the possibility that although mRNA may have been transcribed, it may not have been fully "processed" in post-transcriptional steps that are required to make it usable in translation). Thus, more research is needed to establish whether or not auxins can control translation of fully processed mRNA.

Studies of *in vitro* effects of gibberellins on nuclei and chromatin have yielded results that are very similar to those described for auxins in the previous paragraph. That is, gibberellin treatment of intact tissues can cause changes in the translational capacity of subsequently isolated nuclei or chromatin, but the effects are not seen when gibberellin is applied to the chromatin or nuclei after their isolation. The effects that gibberellins have upon the translational capacity of chromatin may involve transcription of additional template as well as changes in polymerase activity, though there is not rigorous proof of this. One of the clearest pieces of evidence that links gibberellin-stimulated *de novo* enzyme synthesis and the formation of the complementary mRNA has come from some work on barley aleurone tissue extracts. Treatment of aleurone layers with GA_3 resulted in an increase in the level of *in vitro* translatable α-amylase mRNA at a rate that was in parallel with the increased rate of enzyme synthesis. This observation has been taken by some people to have demonstrated *selective* induction of mRNA by gibberellins, but as we have indicated earlier in this chapter the types of mRNA and enzymes synthesized by aleurone cells are determined by the still obscure mechanisms of developmental programming, and gibberellin serves to initiate their preprogrammed responses—in other words the effect cannot be regarded as an example of selective gene derepression by a hormone. Furthermore, once again one must recognize that the effects of gibberellin in these experiments were not truly *in vitro,* in that the hormone had been applied to *intact* aleurone cells and the effects only subsequently determined by an *in vitro* assay. There is no reason, therefore, to believe that even non-selective effects of gibberellin on transcription represent the initial action of the hormone. Furthermore, one cannot yet discount the possibility that gibberellins have post-transcriptional effects.

Just as with auxins and gibberellins, treatment of plant organs and tissues with cytokinins results in enhancement of RNA and protein synthesis. There has been little informative work, however, on the effects of cytokinins, added either *in vivo* or *in vitro,* on transcription by isolated nuclei, chromatin or DNA. Chromatin extracted from several plant sources that had been previously treated with kinetin has been shown to have increased template activity, and some work indicated that the effect may be a result of direct kinetin-chromatin interaction rather than through effects on RNA polymerase or some other system. If there really is direct binding of cytokinin with chromatin, this would appear to provide a very promising basis for further detailed research into the possibility of direct control of transcription by cytokinins. However, such work has not been done to the present time. Much more attention has been devoted to discoveries that cytokinins occur not only as free bases but also incorporated in tRNA (see section 4.5 above) and that cytokinins bind with rather high specificity to certain proteins in isolated ribosomes. Thus, there has been a tendency to explore the possibility that cytokinins act through effects at translation rather than transcription. Reports of specific binding between cytokinins and ribosomal protein are being followed up in research, and a certain degree of purification and characterization of the protein has been carrried out. The biological significance of these cytokinin-binding proteins has not yet been established, though interest in them persists as attachment of cytokinins to ribosomes could conceivably be of importance in the control of translational activities during protein synthesis.

It is well established that ethylene does affect RNA synthesis *in vivo,* either promoting or inhibiting in line with its effect on growth rate. In abscission processes ethylene is the principal controlling hormone (see Chapter 12) and ethylene-stimulated synthesis of cellulase by cells of the abscission zone is preceded by increased formation of RNA. Very few studies have been made of the effects of ethylene on *in vitro* RNA and protein synthesis, but these have given some evidence that ethylene-treatment of tissues results in both quantitative and qualitative changes in template activity of chromatin preparations. How direct these effects of ethylene on chromatin activity are is not known.

A number of the physiological effects of abscisic acid (ABA) are opposite to those of gibberellins, and the inhibitory effect of ABA on gibberellin-stimulated synthesis of α-amylase in the barley aleurone system has been the main basis for speculation that ABA acts by specifically inhibiting DNA-dependent RNA synthesis. *In vitro* studies have, to a certain extent substantiated such an idea. ABA depresses translatable α-amylase mRNA and there is other evidence that ABA acts by regulating levels of individual mRNA species. Such effects appear to be the result of changes in transcription judging by the effects of ABA on *in vitro* transcription and polymerase activity.

4.7 HORMONES AND DIFFERENTIATION IN PLANTS

Cells which undergo vacuolation and enlargement normally cease to divide. Such cells are said to be mature, and the process of transition from the meristematic to mature state is referred to as *maturation.* Only under certain conditions, as in response to wounding or on a culture medium (see Chapter 6), are mature cells seen to resume meristematic activity.

In contrast to maturation, the term *differentiation* is applied to the processes that lead to the formation of the wide variety of types of cell and tissue found in a mature organ. Since both maturation and differentiation occur during the vacuolation and enlargement of cells, the term differentiation is frequently applied loosely to both maturation and differentiation, but this leads to a lack of precision in thinking about these matters.

The role of hormones, such as auxin, in cell vacuolation and enlargement is clear, but do hormones also control the pathways of differentiation in various cells and tissues? For example, is the development of some cambial derivatives into xylem cells and others into phloem controlled by hormones?

Now, it is characteristic of each class of plant hormone that it evokes a wide spectrum of responses in different parts of the plant and, in general, the specific pattern of differentiation which occurs in an organ appears to be determined by the "pre-programming" of the "target" cells or tissues themselves. We do not yet understand what is involved in this pre-programming of target cells, but the nature of hormone receptors formed during development of a cell may be important in determining how it will respond to hormonal signals. Thus, in many instances the specific pattern of differentiation which is evoked by the hormone is not controlled by the latter, but by the programming or "competence" of the target cells.

However, there is evidence that, in a few instances, hormones do actually determine the pattern of differentiation. For example, callus tissue derived from the pith of tobacco plants can be induced to regenerate buds or roots, depending upon the relative concentrations of auxin and cytokinin to which is is exposed (Chapter 6). The question as to the role of hormones in differentiation is discussed later (pp. 328-329) and will not therefore be considered further here, apart from emphasizing that differentiation involves changes in gene expression and activity; subjects that have received considerable attention in relation to the mechanisms and modes of action of hormones discussed in this chapter.

4.8 GENERAL CONCLUSIONS

When a hormone acts on a responsive plant tissue it brings about a change which results eventually in measurable physiological effects. There are two distinct aspects of the action of the hormone: first of all there is the molecular interaction of the hormone at its initial site of action which is referred to as its *mechanism* of action, and this initiates a subsequent sequence of reactions (referred to as the *mode* of action) leading to the physiological effect. Each type of hormone must be presumed to have its own distinctive mechanism of action, but the mode of action of a given hormone in a particular situation varies depending on factors such as the competence of the responding cell (i.e. its developmental preprogramming) and the presence of other factors including different types of hormone.

It must be apparent from reading this chapter that despite a very great deal of effort having been devoted to the problems of the mechanisms of action of plant growth hormones, and the resultant existence of a truly massive body of published experimental data, a real understanding of the subcellular action of these substances is still not available to us. Perhaps the main difficulty in elucidating the mechanism of action of plant hormones is that they have overlapping and complementary effects on the actions of one another.

It is not easy to formulate a single mechanism of action for each hormone that would apply to all situations. For example, the effects of gibberellin on hydrolytic enzyme synthesis in aleurone cells on the one hand, and the delay of leaf senescence in certain plants by the same hormone on the other, would appear to be responses so different in nature as to necessarily involve different mechanisms of action. It is nevertheless possible that a single master reaction (mechanism) is responsible for triggering the different modes of action which lead to all the varied responses to a particular hormone. If this is the case, then all the observed responses result from differing expressions of the master reaction depending on the state of genetic expression and metabolic status of the responding cell. Such a concept therefore envisages that the hormone is always perceived by the same factor (its receptor site) and that the nature of the response is not determined by the hormone. In other words, the hormone acts only as a messenger, the arrival of which activates a preset system in the cell, so that the initial reaction between hormone and receptor is amplified by a process of "cascade" regulation of metabolic reactions. This generalized view is satisfactory enough in the most common situation where plant hormones are acting on tissues whose subsequent development is clearly already determined (e.g. gibberellin on aleurone layer cells), but there are a few cases where the hormones themselves appear to determine the actual pattern of response (e.g. interactions between auxin and cytokinin in regulating the initiation of shoot buds and roots in callus cultures—see Chapter 6) and in these the hormone may perhaps be both messenger and message. A possible basis for the apparently few instances where a plant hormone determines the type of response is that the responsive cells have sites with different affinities for a given hormone, together with interactions between receptors for different hormones.

As we have seen in this chapter, plant physiologists and biochemists are actively pursuing hormone receptor sites in plant cells, and a certain guarded optimism is warranted on the basis of progress to date. Where in the cell would such receptors be situated? There are numerous indications that cell membranes of various types are important in this respect. For example, we have seen that all the plant hormones seem able to regulate plant growth through both short-term and long-term systems, the former apparently initiating the latter. The quickest (short-term) responses appear to be mediated *via* effects of the hormone on membranes and the slower (long-term) effects in-

volve modifications of gene transcription and/or translational processes. A good example of this is seen in the effect of auxin on cell enlargement growth, where the initial response occurs within 10 minutes and does not appear to require protein synthesis, whereas continued cell growth over a matter of hours depends on the maintenance of RNA and protein synthesis. The rapid growth response to auxin appears to involve proton (H^+) secretion from an ATPase "proton-pump" situated in the plasmalemma (see Fig. 4.13). But, as we have also seen, *in vitro* experiments have demonstrated that plasmalemma fragments incubated with auxin release a factor which apparently migrates to the nucleus to enhance (and perhaps qualitatively alter) RNA polymerase activity (Fig. 4.20). Such schemes, although still largely conjectural, do provide a rational framework for future research in this very challenging area of biology.

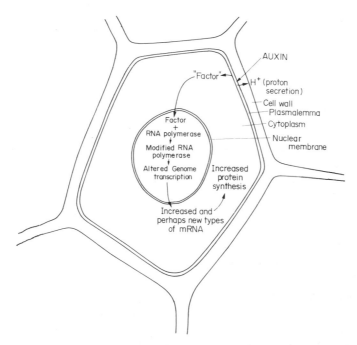

Fig. 4.20. Hypothetical scheme to account for short- and long-term stimulation of cell growth by auxin. The model proposes that auxin binds with a receptor located in the plasmalemma, resulting in rapid secretion of protons (H^+) into the cell wall which is consequently "loosened" allowing turgor-driven expansion to take place immediately. In addition, binding of auxin with a plasmalemma receptor-site (either the same site that is involved in H^+ secretion or a different one) results in rapid release of a factor that migrates to the nucleus and there modifies the activity of RNA polymerase. (Adapted from J. W. Hardin, J. H. Cherry, D. J. Moré and C. A. Lembi, *Proc. Nat. Acad. Sci. U.S.A.* **69**, 3146-50, 1972.)

FURTHER READING

General

Abeles, F. B. *Ethylene in Plant Biology,* Academic Press, New York and London, 1973.

Audus, L. J. *Plant Growth Substances,* 3rd ed., Vol. 1, L. Hill Ltd., London, 1972.

Leopold, A. C. and P. E. Kriedemann. *Plant Growth and Development,* 2nd ed., McGraw-Hill, New York, 1975.

Letham, D. S., P. B. Goodwin and T. J. V. Higgins (Eds.) *Phytohormones and Related Compounds: A Comprehensive Treatise, Volume 1: The Biochemistry of Phytohormones and Related Compounds,* Elsevier-North Holland, Amsterdam, 1978.

Moore, T. C. *Biochemistry and Physiology of Plant Hormones,* Springer-Verlag, New York/Heidelberg/Berlin, 1979.

Wilkins, M. B. (Ed.). *Physiology of Plant Growth and Development,* McGraw-Hill, London, 1969.

More Advanced Reading

Abeles, F. B. Biosynthesis and mechanism of action of ethylene, *Ann. Rev. Plant Physiol.* **23,** 259-292, 1972.

Bauer, W. D. Plant cell walls. In: *The Molecular Biology of Plant Cells* (ed. H. Smith), Blackwell Scientific, Oxford, pp. 6-23, 1977.

Bengochea, T., M. A. Acaster, J. H. Dodds, D. E. Evans, P. H. Jerie and M. A. Hall. Studies of ethylene binding by cell-free preparations from cotyledons of *Phaseolus vulgaris* L.: Effects of structural analogues of ethylene and of inhibitors, *Planta* **148,** 407-411, 1980.

Bengochea, T., J. H. Dodds, D. E. Evans, P. H. Jerie, B. Niepel A. R. Shaari and M. A. Hall. Studies on ethylene binding by cell-free preparations from cotyledons of *Phaseolus vulgaris* L.: Separation and characterisation, *Planta* **148,** 397-406, 1980.

Cherry, J. H. Hormone action. In: *The Molecular Biology of Plant Cells* (ed. H. Smith), Blackwell Scientific, Oxford, pp. 329-364, 1977.

Cleland, R. E. The control of cell enlargement, *Symposia Soc. Exper. Biol. XXXI, Integration of Activity in the Higher Plant,* pp. 101-115, 1977.

Grierson, D. The nucleus and the organisation and transcription of nuclear DNA. In: *The Molecular Biology of Plant Cells* (ed. H. Smith), Blackwell Scientific, Oxford, pp. 213-255, 1977.

Hall, R. H. Cytokinins as a probe of developmental processes, *Ann. Rev. Plant Physiol.* **24,** 415-444, 1973.

Higgins, T. J. V. and J. V. Jacobsen. Phytohormones and subcellular structural modification. In: *Phytohormones and Related Compounds: A Comprehensive Treatise,* Vol. 1 eds. D. S. Letham, P. B. Goodwin and T. J. V. Higgins), Elsevier-North Holland, Amsterdam, pp. 419-465, 1978.

Higgins, T. J. V. and J. V. Jacobsen. The influence of plant hormones on selected aspects of cellular metabolism. In: *Phytohormones and Related Compounds: A Comprehensive Treatise* (eds. D. S. Letham, P. B. Goodwin and T. J. V. Higgins), Vol. 1, Elsevier-North Holland, Amsterdam, pp. 467-514, 1978.

Jacobsen, J. V. Regulation of ribonucleic acid metabolism by plant hormones, *Ann. Rev. Plant Physiol.* **28,** 537-564, 1977.

Jacobsen, J. V. and T. J. V. Higgins. The influence of phytohormones on replication and transcription. In: *phytohormones and Related Compounds: A Comprehensive Treatise* (eds. D. S. Letham, P. B. Goodwin and T. J. V. Higgins), Vol. 1. Elsevier-North Holland, Amsterdam, pp. 515-582, 1978.

Jacobsen, J. V. and T. J. V. Higgins. Posttranscriptional, translational and posttranslational effects of plant hormones. In: *Phytohormones and Related Compounds: A Comprehensive Treatise* (eds. D. S. Letham, P. B. Goodwin and T. J. V. Higgins), Vol. 1, Elsevier-North Holland, Amsterdam, pp. 583-621, 1978.

Kende, H. and G. Gardner. Hormone binding in plants, *Ann. Rev. Plant Physiol.* **27,** 267-290, 1976.

Lieberman, M. Biosynthesis and action of ethylene, *Ann. Rev. Plant Physiol.* **30,** 533-591, 1979.

Penny, P. and D. Penny. Rapid responses to phytohormones. In: *Phytohormones and Related Compounds, A Comprehensive Treatise,* Vol. II (eds. D. S. Letham, P. B. Goodwin and T. J. V. Higgins), Elsevier-North Holland, Amsterdam, pp. 537-597, 1978.

Pilet, P. E. (Ed.). *Proc. 9th Int. Conf. Plant Growth Regulating Substances, Plant Growth Regulation,* Springer-Verlag, Berlin-Heidelberg, 1977.

Skoog, F. and R. Y. Schmitz. Cytokinins. In: F. C. Steward (ed.), *Plant Physiology—a Treatise,* Academic Press, New York, pp. 181-212, 1972.

Stoddart, J. L. Interaction of [^{3}H] Gibberellin A$_1$ with a subcellular fraction from lettuce (*Lactuca sativa* L.) hypocotyls. I. Kinetics of labelling, *Planta* (Berl.) **146,** 353-361, 1979. II. Stability and properties of the association, *Planta* (Berl.), 363-368, 1979.

Varner, J. E. and D. T-H. Ho. Hormonal control of enzyme activity in higher plants. In: *Regulation of Enzyme Synthesis and Activity in Higher Plants* (ed. H. Smith), Academic Press, London, pp. 83-92, 1977.

Walton, D. C. Biochemistry and physiology of abscisic acid, *Ann. Rev. Plant Physiol.* **31,** 453-489, 1980.

Wightman, F. and G. Setterfield (Eds.). *Biochemistry and Physiology of Plant Growth Substances,* The Runge Press, Ottawa, 1969.

Zeroni, M. and M. A. Hall. Molecular aspects of hormone treatment on tissue. In: *Encyclopedia of Plant Physiology* (*N.S.*) Vol. , (ed. J. Macmillan), Springer-Verlag, Heidelberg and Berlin, pp. 511-586, 1981.

5

Hormonal Control in the Whole Plant

5.1 INTRODUCTION

Spatial co-ordination of development in plants appears to depend upon the movement of substances between cells and tissues. Such movement may be either (1) *short range,* between adjacent or neighbouring cells or (2) *long range,* involving relatively long-distance interactions. Short-range interactions, involving the movement of protein molecules, appear to be involved in pollen/stigma recognition reactions (p.329), but whether they also are important in interactions between somatic plant cells is problematical. On the other hand, there is considerable evidence that plant growth-regulating substances (plant growth hormones) play a vital role in intercellular interactions.

Growth substances have been demonstrated to perform various important functions in growth and differentiation, particularly those (a) in which relatively long-distance correlative control is exerted by one organ or region on another, and (b) where environmental effects are apparently mediated through modulation of internal growth substance levels and distribution within the plant body. There is much more evidence for the role of hormones in the control of growth and differentiation in *existing* organs, than for their possible role in the initiation of tissues and organs. Nevertheless, there is a possibility that the major groups of growth substances do play a role in determining the sites of initiation of tissues and organs. Thus, it may be significant that the known growth substances can induce the formation of roots on shoot cuttings and the initiation of buds and roots in callus cultures (Chapter 6).

In order to fall within the traditional definition of a hormone, a substance must be released from the cells in which it is formed and produce an effect in other cells, i.e. the sites of production and action must be separate and movement of the hormone is required. Moreover, to effect control of a process the hormone must be capable of modulation in space or in time. These criteria are certainly met in the hormonal control of growth of the *Avena* coleoptile, where auxin produced in the tip stimulates cell extension in the base, and we shall meet other examples of control by auxin. However, it is more difficult to demonstrate that the other main groups of plant growth substances conform strictly to the traditional definition of a "hormone".

It is clear that the transport and distribution of growth substances within the plant must play a crucial role in their function in spatial co-ordination, and much attention has been devoted to elucidating (a) the sites of biosynthesis, (b) the patterns of hormone movement and distribution from these sites, and (c) the manner in which various environmental factors, such as light and gravity, affect hormone levels and distribution in the plant. Hence we shall first consider some of these latter topics, and then discuss the evidence for hormone control and co-ordination in various aspects of development. In later chapters we shall see that growth substances probably also play an important role in temporal co-ordination in growth responses to environmental factors, such as daylength and temperature.

5.2 THE TRANSPORT OF PLANT GROWTH HORMONES

From their sites of synthesis, growth hormones are transported to other regions of the plant, influencing the cells and tissues with which they come into contact. One would reasonably expect, therefore, that the translocation of these substances would be strictly regulated. Nevertheless, as we shall consider below, although a limited amount of evidence exists to suggest that the transport of abscisic acid, cytokinins and gibberellins may normally follow particular patterns, only auxin translocation has been unequivocally demonstrated to be polarized (i.e. auxins are usually transported along the longitudinal axis of the plant more rapidly in one direction than in the opposite direction). The polar nature of auxin transport is undoubtedly of great importance in the co-ordination of growth and differentiation in different regions of the whole plant. For this reason we will deal first with what is known of auxin transport in plants, following which the transport of the other growth hormones will be considered.

In the shoot tissues which have been studied (coleoptiles, stems, hypocotyls, petioles and flower-stalks), auxin moves more rapidly *basipetally* (i.e. from morphologically apical to more basal regions) than *acropetally* (from basal to apical regions). As we shall consider later, auxin transport in roots also appears to be polar, but there is evidence that the preferred direction of transport may be either acropetal or basipetal, depending on the region of the root.

5.2.1 Auxin transport in shoot tissues

Polar basipetal auxin transport occurs in all organs of the vegetative shoot. The majority of experiments which have shown this have been conducted with short excised segments (usually 5-10 mm long) of coleoptiles, stems, petioles, etc. In principle, the technique is to apply an auxin to one end and to follow its movement along the segment. Various methods have been adopted to determine how much auxin has been transported, and how far, in such segments, but most commonly a "donor-receiver" agar block system has been employed. In this, an agar block containing auxin (the "donor block") is placed against one cut end of a segment of tissue, and another agar block (the "receiver block") against the opposite end. Auxin molecules enter the segment from the donor block, are transported through the segment and eventually emerge into the receiver block. Once auxin starts to enter the receiver block, its concentration there rises linearly with time under carefully controlled experimental conditions. The intercept on the time axis of the straight line of increase in auxin content of the receiver block provides an estimate of the average time taken for auxin molecules to pass from one end of the segment to the other (Fig. 5.1). Since the length of the segment is known, auxin movement can be expressed in terms of velocity (distance moved in unit time).

Using the donor-receiver block method, Went, in 1928, found that auxin moved only basipetally in *Avena* coleoptile segments. Irrespective of the orientation of a segment with respect to gravity, auxin appeared only in a receiver block placed against the morphological basal end with a donor block positioned at the morphological apical end. The auxin used in Went's experiments was unknown, but it was collected from *Avena* coleoptile tips and was probably IAA. Other investigators repeated Went's experiment, and confirmed the existence of polar basipetal IAA transport in coleoptiles, stems, hypocotyls and petioles. Earlier workers, like Went, measured the quantity of IAA in receiver blocks by bioassay. More recently, the availability of radioactive aux-

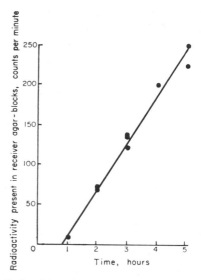

Fig. 5.1. Estimation of the velocity of basipetal polar transport of indole-3-acetic acid (IAA) in bean petiole segments. Radioactive IAA (^{14}C-IAA) at a concentration of 50 μm in agar-gel (the "donor block") was supplied to the apical end of each segment, and a blank agar "receiver block" was placed against the basal end. Radioactivity appearing in the receiver block was determined at hourly intervals. The graph line showing radioactivity present in the receiver block intercepts the time axis at 0·8 hour. The petiole segments were 5·44 mm long, which means that IAA was transported basipetally at velocity of 6·8 mm/hour. (From C. C. McCready and W. P. Jacobs, *New Phytol.* **62**, 19-34, 1963.)

ins has allowed more precise experimentation, and this has revealed that (a) auxin transport in aerial organs is not exclusively polar, for some acropetal as well as basipetal movement takes place (Fig. 5.2), (b) in addition to IAA, certain synthetic auxins, such as 2,4-dichlorophenoxyacetic acid (2,4-D), indole-3-butyric acid and naphthalene acetic acid (NAA) are transported in a polar manner, and (c) that polar auxin transport occurs only in tissues able to respire (Fig. 5.2).

The velocity of basipetal polar auxin transport has been measured in various organs by a number of workers. Values obtained for IAA polar transport all lie between 5 and 15 mm per hour. Synthetic auxins, although transported in a polar manner, apparently move more slowly. For example, 2,4-D moved basipetally in *Phaseolus vulgaris* petiole segments at a velocity of only 1 mm per hour, whereas the equivalent figure for IAA was 6 mm per hour.

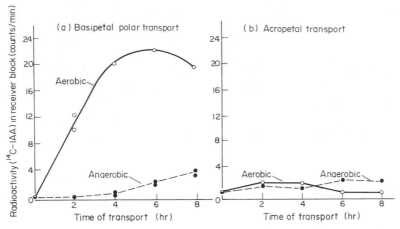

Fig. 5.2. Effect of anaerobic conditions upon polar basipetal, and nonpolar acropetal, transport of auxin in oat coleoptile segments. The agar donor-receiver block method was used with ^{14}C-IAA applied to 5 mm long segments. Lack of oxygen depressed polar basipetal transport of IAA (a), but had no significant effect on the already very low velocity of acropetal transport (b). (From M. B. Wilkins and M. Martin, *Plant Physiol.* **42**, 831-9, 1967.)

The velocity of acropetal auxin transport has not been rigorously determined for aerial organs, but it is normally very much lower than that of basipetal transport. However, the differential between the velocities of basipetal and acropetal auxin transport is influenced by a number of factors. Thus, polarity of auxin movement declines with increasing age of the transporting tissue. It is not yet clear whether this is due to a decrease in basipetal transport or an increase in acropetal transport, or both. There is, nevertheless, no doubt that maturation processes in a tissue are associated with a gradual reduction in the polarity of auxin transport. Because of this, it has been suggested that polar transport of auxin occurs only in association with cell elongation. However, careful experiments by McCready and Jacobs in 1967 showed that, in bean petiole segments, basipetal polar transport of 2,4-D was less when the segments were elongating rapidly in the presence of gibberellic acid (GA$_3$) than when their growth was inhibited by mannitol. On the other hand, earlier experiments demonstrated that GA$_3$ stimulated basipetal IAA transport in stem tissues. Thus, we know neither the significance nor the basis of reduced polarity of auxin transport in mature tissues.

Gravity appears to have some influence on basipetal polar transport of auxins, for several workers have found that when a normally erect organ is placed horizontally, or inverted, then the velocity of basipetal auxin transport is reduced. This phenomenon may be involved in geotropic responses of plant organs, although much more work needs to be done to evaluate this possibility.

Despite detailed studies of polar auxin transport, extending over 40 years, we still do not know the pathway of auxin movement. The velocity of polar auxin transport (0.5-1.5 cm per hour) is very much less than that of solute movement in the phloem (10-100 cm per hour), and the direction of solute flow in the phloem in the upper stem is acropetal rather than basipetal. For this and other reasons, it does not seem likely that auxin normally travels in the phloem. The other vascular tissue, the xylem, clearly does not usually serve as a transporting channel for auxin, for here again the flow is upwards, and dead xylem elements would be unable to provide the energy required for polar auxin transport. Early work indicated that *all* cells in coleoptile segments are capable of transporting auxin basipetally at 1 cm per hour. We cannot be certain that this is true for stems, for *Coleus* stem pith-segments failed to transport IAA at all unless a strand of vascular tissue was present. In fact, a number of workers have found over the past few years that basipetal auxin transport in stems occurs primarily or solely in tissues of the vascular strands. Although by no means certain, it does appear likely that, in stems at least, the procabium, cambium and newly formed derivatives of the cambium (particularly phloem initials) may provide the principal routes for polar auxin transport. In coleoptiles, on the other hand, several workers have reported that basipetal polar auxin transport takes place through the non-vascular parenchyma at least as readily as through the vascular tissues.

5.2.2 Auxin transport in roots

Until relatively recently, very few direct studies were made of auxin movement in roots, and perhaps because of this, a great deal of confusion existed over this matter for many years. However, experiments by several groups of workers since 1964 with root segments, using radioactive IAA and the donor-receiver agar block method, have amply confirmed that roots of a range of species show polar transport of auxin, and that the direction of movement along most or all of the root is *acropetal* (Fig. 5.3). This is the reverse of the

situation in shoot tissues, and the full physiological significance of the difference with respect to the normal regulation of root growth and geotropism remains to be evaluated. The velocity of polar acropetal auxin transport in roots has been found to be approximately 1 cm per hour, which is similar to that of basipetal polar auxin transport in shoot tissues. Despite the evidence for acropetal movement along most of the root, some recent studies have indicated that auxin transport may show basipetal polarity in the more apical parts of roots, but again the possible physiological significance of such a situation is obscure at present.

As is the case for stem tissues, available experimental evidence suggests that auxin transport in roots occurs primarily in tissues of the vascular strands, particularly the cambium and newly formed phloem.

5.2.3 The mechanism of polar transport of auxin

Polar transport of auxin in plant organs is a manifestation of the existence of polarity in each individual cell (Chapter 1). Experiments with segments of organs such as coleoptiles and stems have shown that the polarity of transport (i.e. the ratio, basipetal transport/acropetal transport) increases approximately exponentially with the length of segment studied (Fig. 5.4). This suggests that a small polarity of auxin transport occurs in each cell, and that when auxin is being transported along an organ through a series of cells there is an amplification of the individual small polarities of transport. This would be analogous to the linking up of electrical batteries in series to obtain a higher total voltage than that produced by each individual cell. Leopold and Hall in 1966 devised a mathematical model for a situation such as a file of cells each possessing a small polarity of auxin transport:

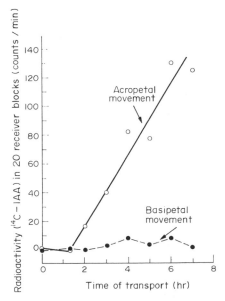

Fig. 5.3. Polar auxin transport in root segments is acropetal. The data shows transport of ^{14}C-IAA in 6 mm long segments of *Zea mays* roots, following its application in agar at a concentration of 1 μM. (From T. K. Scott and M. B. Wilkins, *Planta* **83**, 323-34, 1968.)

$$p = 1 + \frac{\log Q}{0.875\,N}$$

where p = individual cell polarity; Q = polarity quotient (experimentally measured basipetal/acropetal ratio of auxin movement through a segment); and N = number of cells along the length of the segment.

Using this formula it was calculated for *Zea mays* coleoptile segments that the observed overall polarity of auxin transport could result from each cell in the longitudinal series possessing a 1 to 10 per cent basipetal polarity (i.e. a single cell need secrete no more than 1.01 to 1.10 times as much auxin basally as apically). The data for sunflower hypocotyls shown in Fig. 5.4 can be similarly evaluated. The 10 mm long segments were 80 cortical cells long and had a polarity quotient of 149,

$$\therefore p = 1 + \frac{2.17}{0.875 \times 80} = 1.03$$

Thus, although in the 10 mm long segments 10,314 cpm of radioactive IAA were transported basipetally and only 69 cpm acropetally (giving the polarity quotient of 10,314/69 = 149), this large overall polarity can be accounted for by only a 3 per cent basipetal polarity in each cell along the hypocotyl.

The mechanism by which polar transport of auxin occurs is not yet understood. For many years it has been known that the polar transport of auxin involves metabolic processes because, (a) it takes place more rapidly than can be explained by simple physical diffusion, (b) velocities of auxin movement are greatest at temperatures favourable for the activities of most enzymes (20-30°C), (c) respiration, especially aerobic respiration, is required for the maintenance of polar transport, (d) auxin can be transported in a polar manner against its own concentration gradient, and (e) the polar transport system appears to be specific for molecules that possess biological activity similar to that of IAA.

Polarized movement of auxin occurs by cell-to-cell transport via plasma membranes and cell walls. Any movement of auxin through the plasmodesmata that connect adjacent plant cells is insignificant. Thus, in experiments where plasmodesmatal connections were ruptured by osmotic-shock treatments it has been found that there is no impairment of normal polar auxin transport in the tissues.

Until very recently, it has been generally considered that auxin is taken up passively by each cell, moves in the cytoplasm (and perhaps also through the vacuole) and is secreted across the plasma membrane at one end by a *carrier-mediated, energy-requiring* mechanism. In other words, that a localized *active-transport* system is responsible for the secretion of auxin and hence its polar transport. As with all active-transport systems in cells energy would be consumed only when the transported molecule (e.g. IAA) is being moved across the cell membrane. However, no evidence has ever been obtained for energy consumption directly attributable to the movement of auxin into or out of a cell. Because of this lack of unequivocal evidence for a truly active-transport mechanism for polar auxin transport there have

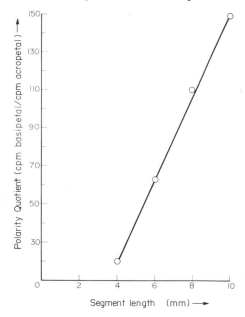

Fig. 5.4. Polarity of auxin ([^{14}C-IAA]) transport in segments of different lengths from *Helianthus annuus* hypocotyls. Radioactive IAA accumulated in agar receiver blocks was determined after 4 hour incubation with donor blocks (of [^{14}C]-IAA) at the opposite end of segments. Transport was measured in both basipetal and acropetal directions. The results are expressed as polarity quotients (basipetal transport/acropetal transport) and it can be seen that measured basipetal polarity increased with the length of segment used. (Previously unpublished data of S. P. Watkinson and I. D. J. Phillips.)

been recent proposals that auxin transport can be explained in another way; what may be termed the "*chemiosmotic*" *hypothesis* of polar IAA transport.

The chemiosmotic hypothesis of IAA transport takes into consideration the facts that; (a) IAA is a weak acid (pk = 4.7) with lipophilic properties and, (b) undissociated IAA (IAAH) is more hydrophobic than its anion (IAA⁻). This means that cell membranes are approximately 200 times more permeable to IAAH than to IAA⁻. Because of the acidic pH of plant cell walls, IAA present in them will be in the undissociated (IAAH) form, but since the cytoplasm of cells is more alkaline (pH > 4.7), IAAH that readily enters the cells is immediately dissociated to IAA⁻, in which form it

is very much less able to leave the cell again.

The chemiosmotic hypothesis of polar auxin transport proposes that there exist carriers for IAA⁻ in the plasma membrane, but that these carriers (which would facilitate outward movement of IAA⁻) are asymmetrically distributed around the cell circumference, so that each cell in a tissue that shows basipetal polar auxin transport would have a greater concentration of IAA⁻ carrier sites in the plasmalemma at its basal than at its apical end. Because the plasmalemma is very permeable to IAAH, passive equilibration of IAAH occurs all round the cell circumference. Within the cell membrane however, there will be a high concentration of IAA⁻. Since the cytoplasm is electrically negative with respect to the cell wall, there is a net

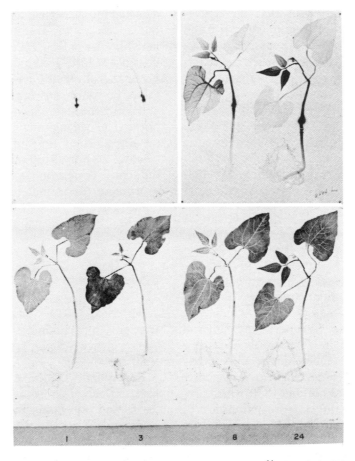

Fig. 5.5. Apparently free, non-polar, transport of gibberellin in bean plants. ^{14}C-labelled gibberellin was applied to the cotyledonary node of the plant, and its distribution with time (1-24 hrs) within the plant followed by autoradiography. (From G. Zweig *et al., Adv. Chem. Ser.* **28**, 122-34, 1961. Original print donated by Dr. Gunter Zweig.)

driving force on IAA⁻ tending to push it out of the cell. Given the rather high hydrophilic properties of IAA^- it cannot easily pass out through the cell membrane despite the electrical gradient favouring its outward movement, except at regions where there are high concentrations of its transmembrane carrier. Polarity of auxin transport will occur if the ratio of the permeability of the anion to the undissociated auxin, i.e. P_{IAAH}/P_{IAA^-} (where P = permeability coefficient) is greater at one end of the cell than the other.

Thus, the chemiosmotic hypothesis of polar auxin transport resembles the active-transport hypothesis in two respects: (i) In both hypotheses uptake of IAA (as IAAH) is regarded as a "passive" diffusion process, (ii) Both hypotheses envisage the existence of asymmetrically distributed carrier molecules for auxin in the plasmalemma. They differ in that the active-transport hypothesis envisages energy being directly required for operation of the carrier, whereas the chemiosmotic hypothesis postulates that the movement of auxin is thermodynamically "downhill" and that metabolic energy will be only required to maintain pH and electrical gradients and polar permeability. Although some experimental evidence supports the chemiosmotic hypothesis of polar auxin transport more research is required before one can fully evaluate its validity in relation to the long-held active-transport hypothesis.

5.2.4 The transport of gibberellins in plants

Far fewer studies have been made of gibberellin translocation than of auxin transport, but these have nevertheless provided fairly convincing evidence that gibberellin transport is not polar in nature, except possibly in leaf petioles. Thus, it is generally noticed that gibberellins applied to any one part of a plant elicit developmental responses in all other regions of shoot and root, which provides indirect evidence for non-polar transport of the hormone. More direct evidence that gibberellins are able to move freely in all directions within the plant has been obtained by use of radioactive gibberellins (Fig. 5.5).

Experimentation on the transport of gibberellins has not usually involved use of the agar "donor" and "receiver" blocks technique that has provided such a convenient and major tool in studies of auxin transport in segments of stems, coleoptiles, etc. The principal reason for this is that it has been found that ^{14}C- or ^{3}H-labelled gibberellins are taken up readily enough by tissue segments, but are not exported into agar receiver blocks to any significant extent. The reasons for this are not known, but it provides an interesting contrast to the facility with which ^{14}C-auxins pass out into such receiver blocks, in that it emphasizes the secretory nature of polar transport of auxins.

It is considered that the movement of gibberellins within the plant occurs through the normal general circulatory system of the phloem and xylem vascular tissues, since they have been detected in both xylem and phloem sap. However, one aspect of the transport of endogenous gibberellins for which phloem or xylem transport does not readily account, is the movement of gibberellins from their putative major regions of synthesis in young growing leaves downwards into and along the stem. Such young leaves are net importers of organic and inorganic materials, and the direction of flow of solutes in phloem and xylem is acropetal (upwards) in the apical part of the shoot. If phloem transport occurs in response to demand for assimilates by "sinks", then it is impossible to visualize basipetal gibberellin transport in the phloem of the apical region of the shoot.

5.2.5 Cytokinin translocation

No clear picture can be presented of the transport of cytokinins in plants. Available relevant evidence is slight, fragmentary and often contradictory.

Experiments that have demonstrated the role of roots in maintaining a supply of cytokinins to leaves and preventing their premature senescence (Chapter 12) are clearly indicative of an upward transport of cytokinins in the stem. Furthermore, cytokinins are known to be present in the xylem sap ascending from the root system. On the other hand, cytokinins that are synthesized in young developing fruits do not appear to be transported out at all. Similarly, numerous studies with exogenous cytokinins such as kinetin have indicated that they may remain for some considerable time in the localized region of application, even though general metabolite transport may be taking place away from the point of application of the cytokinin.

A certain amount of evidence suggests that cytokinins may be transported not as free purines, but in conjugated forms, such as ribosides or glucosides, both of which have been shown to be present in xylem and phloem sap.

5.2.6 Ethylene transport

Although as a substance of low molecular weight it might be expected that ethylene would move freely through plant tissues by normal physical diffusion, this does not appear to be the case. Thus, in broad bean for example, the "resistance" to longitudinal movement up or down the plant is such that lateral emanation effectively isolates different parts of the plant from one another. Similarly if a leaf is fed with ethylene, only a small proportion of the gas which passes into the leaf ever reaches the stem, the remainder being emanated from the petiole. Hence, ethylene does not move between different parts of the plant in physiologically significant amounts.

Despite the fact that ethylene is not translocated to any significant extent, nevertheless it appears that changes in ethylene levels in one part of a plant can influence those in another. Thus an increase in ethylene levels in the roots can induce increased levels in the shoot apex. The mechanism of this effect is not understood at present.

5.2.7 Transport of abscisic acid

Application of exogenous ABA to mature leaves, or even to roots, can result in developmental responses such as growth inhibition in all other parts, e.g. the shoot apex or the vascular cambium. Such observations indicate that, like gib-berellins, ABA is able to move freely in all directions within the plant. Also, studies with ^{14}C-labelled ABA have given no clear evidence of polarity of ABA transport in stem and coleoptile segments, but have provided evidence to suggest that in root segments ABA is transported preferentially in a basipetal manner (i.e. away from the root apex). Some research has indicated that ABA may be synthesized in the root cap, and a basipetal mode of transport in roots would therefore provide a mechanism for this ABA to reach and influence the elongating region of the root (p. 197).

5.3 GROWTH HORMONES AND SHOOT DEVELOPMENT

We shall now consider the role of growth hormones in the control of several aspects of shoot development.

5.3.1 Stem elongation growth

Although effects on elongation growth are possibly the most intensively investigated aspects of hormone action in plants, it is surprisingly difficult to draw any sort of clear picture of the overall mechanism by which stem elongation is normally regulated. Each of the five known categories of growth hormone can certainly influence stem growth. The principal problem is to visualize how they interact, for their individual effects appear to overlap, duplicate, reinforce, or antagonize one another to a bewildering extent. Furthermore, not only are we still uncertain of the sites of synthesis of some types of growth hormone (especially the cytokinins and abscisic acid), but we have only sketchy information on the factors which determine patterns of transport of hormones other than auxin. In our present state of knowledge we are therefore compelled to consider individually the evidence for the involvement in stem elongation of each class of hormone, and to attempt only briefly the synthesis of available information for all hormones towards a general understanding of the control of stem elongation growth.

5.3.2 Auxin and internode elongation

As we considered in Chapter 3, the discovery of the natural auxin, IAA, took place as a result of experiments on phototropism and elongation growth in etiolated coleoptiles. In particular, observations that removal of the apical end of a coleoptile resulted in suppression or cessation of elongation growth in the remaining coleoptile, and that IAA could substitute for the tip, led to the conclusion that the apical part of a coleoptile normally supplies auxin to the newly formed cells and

that the auxin is necessary for elongation growth of those cells. Convenient though are etiolated coleoptiles in experiments on the role of apically-synthesized auxin in the regulation of extension growth, one must remember that they are modified, tubular leaves, and that results obtained by their use should be applied only with caution to the problem of how internode elongation growth is normally controlled.

It is now generally believed that auxin which is involved in the control of internode growth is synthesized in young growing leaves (or in cotyledons in very young seedlings) from where it is translocated into and basipetally down the stem. Evidence for this view is largely circumstantial,

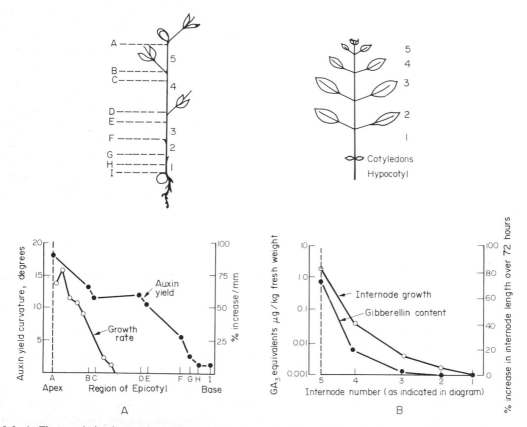

Fig. 5.6. A. The correlation between growth rate and auxin content along the length of a pea seedling stem (epicotyl). *Above:* diagram of a green 9-day-old "Alaska" pea seedling. The figures indicate the number of each internode, and the letters delimit regions assayed for auxin content. (From T. K. Scott and W. R. Briggs, *Amer. J. Bot.* **47**, 492-9, 1960.)

B. The relationship between elongation rate and gibberellin content along the length of the stem of a young sunflower (*Helianthus annuus*) plant. (From R. L. Jones and I. D. J. Phillips, *Plant Physiol.* **41**, 1381, 1966.)

but includes positive identification by mass-spectrometry of IAA in young leaves of bean plants, and measurements of quantities of unidentified auxin (presumably IAA) diffusing out from the petioles of variously aged leaves of *Coleus* plants (Fig. 5.7).

It has been found that those internode tissues which are most rapidly elongating contain the highest levels of diffusible endogenous auxin (Fig. 5.6A), which is consistent with the view that auxin is required for elongation growth in internodal regions.

Further evidence that auxin is concerned in the control of stem growth is afforded by experiments in which isolated segments of internodes are used.

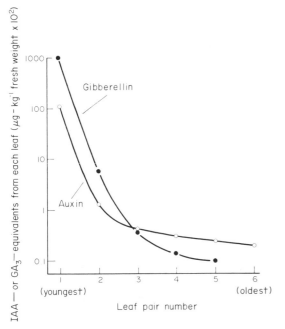

Fig. 5.7. Amounts of gibberellin and auxin transported out from leaves of various ages. Young growing leaves synthesize and export to the stem greater quantities of both types of growth hormone than do older leaves. Gibberellin data from experiments on sunflower plants (R. L. Jones and I. D. J. Phillips, *Plant Physiol.* **41**, 1381-6, 1966) and auxin data from experiments with *Coleus* plants (R. H. Wetmore and W. P. Jacobs, *Amer. J. Bot.* **40**, 272-6, 1953.)

Such segments have been deprived of their supply of auxin from the apical region of the shoot, and consequently the effects of known concentrations of auxin can be measured. Thus, if the internode

segments are floated in an appropriate solution of auxin, then they will grow more than when placed in water alone. Further, the amount of extension growth that occurs is proportional to the logarithm of the concentration of auxin in solution (Fig. 5.8A). It can be seen that stem segments show increased growth with increasing concentrations of IAA up to approximately 5×10^{-5} M. At concentrations greater than this less growth occurs, until at about 10^{-3} M IAA and above *inhibition* of excised stem segment elongation occurs, so that less growth takes place than in the control segments to which no IAA was applied. Consequently, we can say that for stem tissues there is a concentration of auxin which is *optimal* for cell elongation. Concentrations greater than this are, therefore, *supra-optimal* and lower concentrations are *sub-optimal*. In some species, such as pea, much higher concentrations of auxin are required to produce supra-optimal inhibition in light-grown (green) stem segments than in etiolated stem segments of the same species.

Since application of auxin to an intact shoot (e.g. as a spray) rarely stimulates stem elongation, it may be assumed that the stem tip normally supplies sufficient auxin to the elongating internodes to maintain an optimal auxin concentration in those tissues. The possible reasons for inhibition of growth by auxins will be discussed below (p. 119).

5.3.3 Gibberellins and internode elongation

There are several pieces of evidence which indicate that gibberellins, as well as auxins, are involved in extension growth of plant tissues. The most striking and characteristic response of a plant treated with a gibberellin such as gibberellic acid is that stem elongation is stimulated, so that the treated plant becomes taller than it would normally. This response of the stem is usually due to an increased elongation of the internodes and there is generally no increase in the number of internodes formed. The increased internode length is a consequence of increased cell extension and cell division. Thus, gibberellin treatment of intact

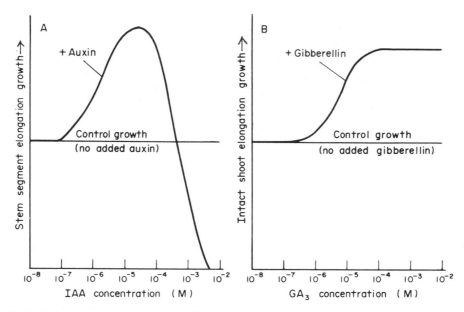

Fig. 5.8. A. Typical dose-response curve for elongation growth of stem or coleoptile segments in varying concentrations of the auxin indole-3-acetic acid (IAA). B. Typical dose-response curve for elongation of intact shoots treated with various concentrations of the gibberellin gibberellic acid (GA_3).

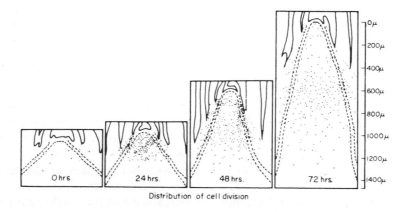

Fig. 5.9. Stimulatory effect of gibberellic acid (GA_3) upon sub-apical meristematic activity in *Samolus parviflorus,* a rosette plant. Each dot represents a sub-apical cell undergoing mitosis seen in median longitudinal sections of the stem apical region. Twenty-five micrograms of GA_3 were applied 0, 24, 48 and 72 hours previously. (Reprinted from R. M. Sachs, C. F. Bretz and A. Lang, *Amer. J. Bot.* **46**, 376-84, 1959.)

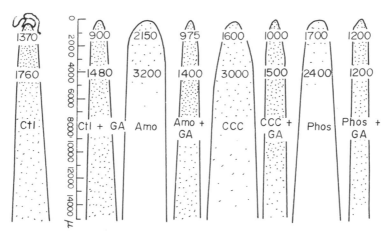

Fig. 5.10. Density and distribution of sub-apical meristematic activity in stems of *Chrysanthemum morifolium* following treatment with gibberellic acid (GA), growth retardants (Amo, CCC, and Phosfon-D), and combinations of GA with retardants. Control plants untreated ('Ctl' at extreme left). Each dot represents one mitotic figure in a 60 μm thick logitudinal slice. The numbers at 1 mm and 4 mm below the apex are the transverse diameters of the pith tissue in μm at these levels. Mitotic activity was greatly reduced by the growth retardants but transverse growth increased. GA increased mitotic activity, primarily 6 mm and more below the apex, and reduced transverse growth. The effects of the growth retardants and GA were antagonistic. (From R. M. Sachs and A. M. Kofranek, *Amer. J. Bot.* **50**, 772-9, 1963.)

rosette-habit and tall-growing (caulescent) plants can cause enhanced elongation of existing internodal cells, and also increase the number of cells present in each internode, principally as a result of an increase in mitoses in the sub-apical region of the stem (Fig. 5.9).

Growth-retardants such as Cycocel, AMO-1618 and Phosfon-D (p. 61), which inhibit endogenous gibberellin biosynthesis, all have inhibitory influences upon cell-division rates in the sub-apical meristem, and their effects can be counteracted by simultaneous application of exogenous gibberellin (Fig. 5.10).

The magnitude of the stem-elongation response to gibberellin varies from species to species and from variety to variety within a species. As mentioned in Chapter 3 (p. 58), the response is greatest in genetically dwarf plants—so much so, that following gibberellin treatment dwarf varieties often grow to the same height as that of related tall varieties (Fig. 3.4). Tall varieties of the same species respond only slightly or not at all. In contrast to the typical dose-response relationships between auxin and stem segment elongation (Fig. 5.8A), gibberellins rarely show supra-optimal inhibition of elongation, for even very high concentrations of exogenous gibberellic acid can elicit a maximum growth response (Fig. 5.8B). The reasons for this difference in pattern of response to varying auxin or gibberellin concentration are not fully understood, but supra-optimal auxin concentrations can induce increased release of ethylene (Fig. 5.13), particularly in dicotyledonous species, and it is likely that it is the additional ethylene which causes growth inhibition. Gibberellins, on the other hand, have been found to have very variable effects on ethylene production. In the majority of investigated situations, a small promotive effect of gibberellin on ethylene production has been found, but in others gibberellins had either no effect or even decreased ethylene levels.

The responsiveness of different varieties to applied gibberellin may perhaps be related to the amount of endogenous gibberellin present in the tissues. However, in some species, such as *Pisum sativum,* contradictory results have been obtained by different research groups in that some have found lower levels of gibberellins in dwarf than in tall pea varieties, whereas others have been unable to detect any quantitative difference. Consequently, one cannot yet conclude that dwarfness of a plant is always due to impaired gibberellin synthesis, and further research is being conducted to resolve this question.

It is possible to extract gibberellins from various organs, and to compare the amounts present in them; also, it is possible to collect gibberellins from organs by the agar diffusion technique previously used in studies of auxins. Such studies have revealed that the apical bud of the shoot, and young leaves, synthesize and export gibberellins to the stem (Fig. 5.7). Also, a positive correlation has been obtained in sunflower (*Helianthus annuus*) between the growth rates of internodes of different ages and the gibberellin contents of the same internodes (Fig. 5.6B). Thus, as with auxins, endogenous gibberellins are present in highest concentration in those regions of the stem which are undergoing most rapid extension growth, providing strong circumstantial evidence that gibberellins are concerned in the normal control of stem extension growth. Cytokinins may be involved in the control of cell division rates at the stem apex, but we have no evidence that internode extension is in any way under a direct influence of this class of hormone.

5.3.4 Auxin-gibberellin interactions in stem elongation

Following the discovery of gibberellins and the realization that these hormones occur naturally in higher plants (see Chapter 3), many plant physiologists were led to study the interactions of gibberellins and auxins in stem and coleoptile extension growth.

Spraying an intact plant with gibberellin was found to enhance internode extension growth, whereas it was already known that similar treatment with an auxin rarely induced greater internode elongation. Conversely, if young internodes or coleoptile segments were excised and floated in solution, then it was noted that the opposite usually occurred; auxins stimulated internode or coleoptile segment elongation but gibberellins had little or no effect. However, when gibberellin and auxin were present in solution together, then the elongation of excised segments was much greater

than when auxin alone was supplied (Fig. 5.11). In other words, for the characteristic effects of gibberellins on stem elongation to appear, auxin also has to be present. The growth-promoting action of gibberellins when applied to intact plants would, therefore, be a consequence of an interaction between supplied exogenous gibberellin and the natural endogenous auxins present. Because of such observations, it was proposed that gibberellins exert their physiological effects through some "auxin-mediated" mechanism. Thus, there is good evidence that application of gibberellins leads to increased endogenous auxin levels, by effects on either the biosynthesis or the destruction of auxin.

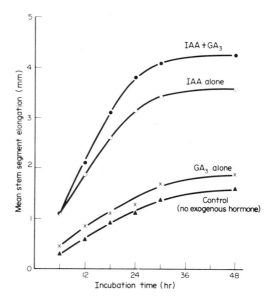

Fig. 5.11. Effects of auxin (10 μg ml⁻¹ IAA) and gibberellin (10 μg ml⁻¹ GA₃) on elongation growth in excised pea internode segments. Elongation of stem segments is much more markedly enhanced by auxin than by gibberellin, but the addition of both auxin and gibberellin results in more growth than with auxin alone. The interaction between exogenous auxin and gibberellin is sometimes additive and sometimes synergistic. (Adapted from P. W. Brian and H. G. Hemming, *Ann. Bot.*, N.S., **22**, 1-17, 1958.)

However, it is now clear that gibberellins are a class of growth hormone in their own right. If gibberellin effects were due solely to an influence on auxin activity then it would be expected that these effects would always be the same as those produced by auxins. Whilst gibberellins often are able to duplicate the known effects of auxins (e.g. induction of parthenocarpic fruits, promotion of cambial activity), there are many other examples of gibberellins having physiological effects not possessed by auxins (e.g. promotion of stem elongation in intact plants, breaking dormancy of buds or seeds, stimulation of mesophyll growth), or sometimes instances where gibberellins have the opposite effects of auxins (e.g. auxins promote but gibberellins inhibit root initiation in stem cuttings).

Nevertheless, it is apparent that interactions between auxins and gibberellins occur in many physiological responses apart from extension growth. For example, auxin and gibberellin together are often more effective in inducing the development of parthenocarpic fruits and in stimulating cambial activity (p. 122) than is either type of hormone on its own.

5.3.5 Ethylene and internode elongation

Impressive evidence has accumulated in recent years which suggests that ethylene may play some part in the control of stem growth. With most species, exposure of stems or isolated internodes to ethylene reduces cell elongation (Fig. 5.12) but

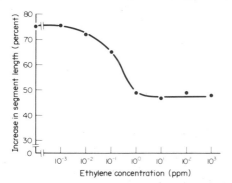

Fig. 5.12. Inhibition by ethylene of elongation growth in excised pea-stem segments. (Adapted from S. P. Burg, *Regulateurs Naturels de la Croissance Végétale*, Edition de la Rech., Sci., 1964, pp. 718-24.)

enhances isodiametric cell expansion. Such ethylene-treated internodes are shorter and thicker than untreated internodes. These effects of ethylene are similar to those which can be induced

by high (supra-optimal) concentrations of auxins. It is possible that auxins are not themselves inhibitors of stem elongation, but rather that at high concentrations they stimulate the synthesis of ethylene in plant tissues, which in turn, suppresses cell elongation (Fig. 5.13).

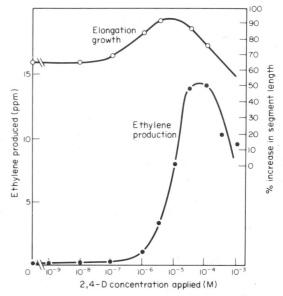

Fig. 5.13. Effects of various concentrations of the auxin, 2,4-dichlorophenoxyacetic acid (2,4-D) on elongation growth (top) and ethylene synthesis (bottom) in excised segments of soybean hypocotyls. Note that maximum elongation occurred with approximately 10^{-5} M 2,4-D, but that at higher concentrations of 2,4-D the rate of ethylene production rose markedly and elongation growth decreased. It is likely that growth inhibitions in dicotyledonous species by high concentrations of auxins occur through the inhibitory effects of the auxin-induced ethylene, but this may not be so for monocotyledonous plants. (Adapted from R. E. Holm and F. B. Abeles, *Planta,* **78**, 293, 1967.)

The inhibiting effect of ethylene on stem elongation is less in the presence of light than in darkness. The reasons for this are not understood, but ethylene release can be affected, probably indirectly, by red and far-red light through a phytochrome-mediated mechanism. Thus, the results of experiments done on the effects of exogenous ethylene are likely to be influenced by the light régime, in that the response to a given concentration of exogenous ethylene will depend upon the prevailing endogenous ethylene level in the tissues.

A special case of stem elongation which involves regulation by ethylene is seen in examples of those seedlings in which the terminal part of the shoot axis is hook-shaped (Fig. 8.5). The hook is presumed to aid penetration through the soil of the delicate apical tissues of the shoot. The shape of the hook is a resultant of more rapid elongation growth on the outer convex side than on the inner concave side. Exposure of seedlings to red light causes the hook to open by an equalization of growth rates on the two sides of the stem, and far-red light reverses the effect of red light. Treating red light-grown seedlings with either auxin or ethylene results in closure of the hook by inhibition of the inner side stem tissues. It has been found that ethylene acts in an intermediary capacity in both light- and auxin-controlled hook opening and closing. That is, both far-red light and auxin induce ethylene release in the apical part of the shoot, and the ethylene inhibits extension in the inner side of the hook. Red light reduces ethylene release and causes opening of the hook. However, the mechanism of hook opening and closing appears to be more complex than this, and is not yet fully understood. For example, it is known that gibberellins can also influence hook formation and that the effects of light cannot be completely explained in terms of effects on ethylene production.

In some plants, mainly species that grow well under water (e.g. rice and *Callitriche*), rather than being inhibited, internode and root extension occurs more rapidly in the presence of ethylene. The elongation responses to ethylene shown by these aquatic plants may have evolved as an adaptation to the very much lower rates of diffusion of ethylene in water than in air. Thus, movement of emanated ethylene away from the plant surface is considerably slower in an aquatic environment, which could result in a high ethylene concentration within submerged plant tissues. Changes in sensitivity and response to ethylene may, therefore, have occurred during evolution of aquatic plants to cope with this problem.

It is clear that ethylene interacts with the other natural growth hormones in the regulation of stem elongation growth, but much research remains to

be done to unravel what is clearly a complex and sensitive control mechanism.

5.4 HORMONES AND VASCULAR TISSUE DEVELOPMENT IN STEMS

In the first section of this book we saw how the shoot apical meristem repetitively gives rise to new tissues of the longitudinal axis of the shoot, and also to lateral organs such as leaves and lateral shoots. In plants which are relatively long-lived, particularly those that attain considerable heights, it is of obvious mechanical advantage that the capacity for increasing the diameter of the longitudinal axis also be retained throughout life.

Increased stem and root girth is achieved through processes involving cell division, expansion and differentiation. In dicotyledonous plants, and in most gymnosperms, radial growth of both stem and root is an expression of the activities of the *lateral meristems*. Two types of lateral meristem are generally recognized, the *vascular cambium* and the *cork cambium* or *phellogen*. In those monocotyledonous species that undergo radial growth to any extent, cambial cells still play a part, but their activities are more restricted.

We saw earlier (p. 34) that developing buds and leaves exert a stimulating effect on the development of the vascular tissue in the internodes below. Moreover, experiments with chicory (*Cichorium intybus*) callus cultures showed that when a shoot bud was grafted on to the top of the callus, differentiation of vascular tissue, connected to the base of the bud, was induced (Fig. 5.14). The stimulus arising from the bud which caused the initiation of vascular tissue was shown to be capable of passing through a layer of cellophane placed between the bud and the underlying callus tissue. Subsequently, it was shown that auxins such as IAA can produce the same effects in the callus as an implanted bud. Cambial tissue, as well as xylem and phloem, can be induced in several types of callus tissue by the application of IAA, especially if applied in

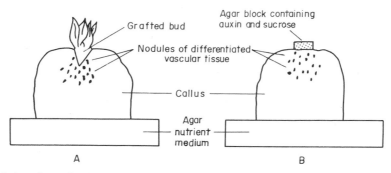

Fig. 5.14. Differentiation of vascular tissues in aseptic cultures of callus tissues. A: Vascular tissues induced to differentiate in a callus by the presence of a grafted bud of the same species. B: Induction of vascular tissue in a similar callus by an agar block containing auxin and sucrose. (A adapted from Wetmore and Sorokin, 1955; B adapted from Wetmore and Rier, 1963.)

association with sucrose. Moreover, extended application of IAA to stems and roots of pea will induce the formation of vascular tissue in the cortex (Fig. 5.15). These observations suggest that the stimulating effect of developing leaves on the initiation of vascular tissue may be mediated partly by the endogenous IAA produced by the leaves.

This conclusion is also supported by observations on the regeneration of vascular tissue following wounding. If one of the vascular bundles of *Coleus* is severed by a lateral cut, regeneration of the strand occurs through the pith (and connects the severed upper and lower ends of the bundles

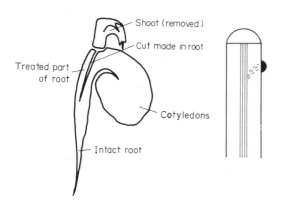

Fig. 5.15. Induction of xylem by auxin in pea roots. *Left:* 3-day-old seedling illustrating the cuts made to separate the part of the root used for experiments. *Right:* treated roots in face view. IAA applied laterally at the position indicated by black protruberance. The dotted lines indicate newly formed xylem through the cortex. (Adapted from T. Sachs, *Ann. Bot.* **32**, 781-90, 1968.)

(Fig. 5.16). The pattern of regeneration is basipetal, i.e. it starts from the upper severed end of the original vascular bundle and travels from cell to cell diagonally through the pith. The regenerated strand is formed by the differentiation of parenchymatous pith cells into xylem and phloem, but subsequently cell division occurs and a cambium is formed. It has been shown that the presence of a leaf above the cut in the stem is essential for the regeneration of the strand, but the effect of a leaf can be replaced by applying auxin to the petiole stump, suggesting that auxin stimulates the differentiation of xylem and phloem and the formation of a cambial layer in the pith cells.

The xylem elements are formed by the development of reticulate lignified thickenings on the walls of the pith cells. During the early stages of differentiation of these cells granular strands (shown to contain microtubules which lie parallel to the future thickenings in the wall) appear in the surface layers of the cytoplasm and mark out the position of the lignified bands which later develop on the walls. Moreover, the thickenings in one cell are opposite those of the next, so that the pattern can be observed to extend across the boundary wall between two adjacent cells. These observations strongly suggest that the pattern of differentiation in one cell may induce similar changes in an adjacent cell, a phenomenon known as *homeogenetic induction*. How this homeogenetic induction is achieved is not known, but it may involve other factors, in addition to IAA.

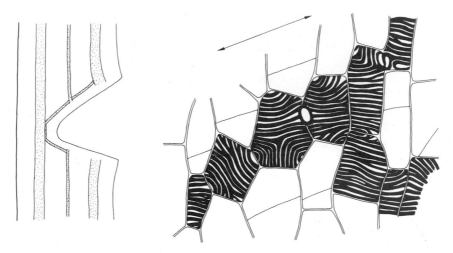

Fig. 5.16. Vascular regeneration in *Coleus*. *Left:* regeneration of connection between several vascular bundles in stem. *Right:* differentiation of parenchyma cells into reticulate xylem cells in the development of this strand. Arrow shows the direction of its development. (From E. W. Sinnott and R. Bloch, *Amer. J. Bot.* **32**, 151, 1945.)

5.4.1 The control of cambial activity in stems

As well as stimulating the initiation of vascular tissue, hormones also appear to play an important role in controlling cambial activity. The first demonstration that auxin stimulates the cambium to divide came from experiments performed by Snow in 1935 with sunflower (*Helianthus annuus*) plants which had been "decapitated" (i.e. the apical bud was excised). In such decapitated plants, it was found that the fascicular cambium of young internodes failed to divide, and also that interfascicular cambium did not form. In other words, excision of the apical bud prevented the formation of secondary vascular tissues (xylem and phloem). However, when IAA was applied to the upper cut end of the stem in decapitated plants, then normal cambial activity and secondary thickening resulted, suggesting that both the initiation of cambium in the stem, and its activity once formed, depend upon a supply of auxin from the apical bud.

On the other hand, a number of investigators have obtained results that suggest that the initiation and activity of the cambium may not always be regulated by auxin from young leaves. For example, a series of experiments by Siebers during the early 1970s showed that the formation of inter-fascicular cambium still occurred in decapitated young hypocotyls of *Ricinus communis,* or even in isolated segments of hypocotyl tissue, and he concluded that any requirement for growth hormones in cambium *initiation* must be satisfied by sources within the hypocotyl itself. However, Siebers did find that addition of IAA, GA_3, or kinetin resulted in enhanced development of the cambium once it had formed, which indicated that these types of growth hormone may normally be concerned in the regulation of cambial *activity* in the hypocotyl. Of these growth hormones, GA_3 was found to have by far the greatest stimulatory effects on cambium development and activity, and sugar, in the form of sucrose, was also important. These findings, and similar ones by other workers, suggest that one cannot regard the apical bud as the only source of auxin in the regulation of cambial initiation and activity in stems of herbaceous plants.

Rather more is known of the regulatory factors concerned in cambial activity in woody species of temperate zones. In these, there normally occurs seasonal variation in the rate of cell division activity in the vascular cambium of both shoot and root, and differences in the patterns of differentiation of cambial derivatives at different times of the year. During the winter months there is no cambial activity in such trees, but in the spring cell-division activity starts again and the newly formed cells differentiate into xylem and phloem elements.

In diffuse-porous angiospermous trees (in which all xylem vessels have a similar diameter no matter when during the growing season they develop) such as sycamore (*Acer pseudoplatanus*), cambial activity in the spring commences immediately below expanding buds and then spreads slowly *downwards,* i.e. basipetally, through the twigs to the branches, trunk and

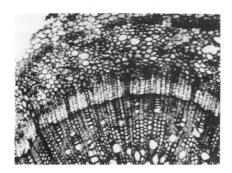

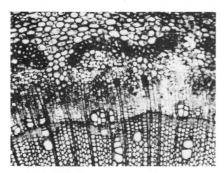

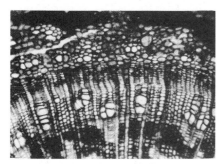

Fig. 5.17. Effect of auxin and gibberellin on cambial activity in poplar (*Populus robusta*) twigs. *Top:* twig treated with gibberellic acid (GA₃); *centre:* treated with indole-3-acetic acid (IAA); *below:* treated with IAA and GA₃ in combination.

eventually into the roots. An acropetal wave of cambial activity occurs only in the roots. Thus, the initiation of cambial activity follows the same pattern as polar movement of auxin in stems and roots. Young growing buds in spring are known to transmit relatively large amounts of auxin to the stem tissues, and it therefore appears likely that the reawakening of cambial activity in diffuse porous species is a response to auxin from the buds. This idea is supported by various pieces of evidence. Thus, disbudding prevents the onset of cambial activity, but application of auxin to the upper end of a disbudded twig results in normal basipetal spread of cambial activity.

Not only is the activation of cambium regulated by auxin, but the differentiation of its derivatives is also affected by prevailing levels of auxin. It is also known that auxin is not the only hormonal regulator of cambial activity and vascular tissue differentiation. This is most easily and convincingly demonstrated in experiments in which lengths of stem from a diffuse-porous species are taken in early spring, before bud expansion has started, removing the buds, and applying growth hormones, either dispersed in lanolin or in aqueous solution, to the upper cut ends of the stem segments. After about 2 weeks the stems are sectioned for observations on cambial activity. With no applied hormone there is no cambial division, but if IAA is applied there is limited cambial division and some differentiation of new xylem elements can be observed (Fig. 5.17). If GA₃ alone is applied, cambial division occurs, but the derivative cells on the inner (xylem) side remain undifferentiated and retain their protoplasmic contents; however, careful observation shows that some new phloem, with differentiated sieve tubes,is formed in response to GA₃. When IAA and GA₃ are applied together, there is greatly increased cambial division and normal, differentiated xylem and phloem are formed. By measuring the width of the new xylem and phloem it is possible to study the interaction of auxin and gibberellin, and other regulators, in a quantitative manner (Fig. 5.18). Experiments such as these suggest that not only can the rate of cell division in the cambium be regulated by levels of auxin and

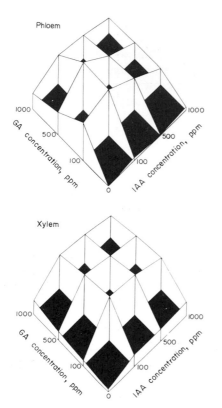

Fig. 5.18. Quantitative effects of IAA and GA$_3$ on cambial activity of poplar. The two hormones were applied in lanolin in various combinations at the concentrations shown. Vertical axis indicates width of new xylem and phloem in eye-piece units. See text for explanation. (From P. F. Wareing, C. Hanney and J. Digby, in *The Formation of Wood in Forest Trees*, Academic Press, New York, pp. 323-44, 1964.)

gibberellin, but also that the production and differentiation of xylem initials is favoured by a relatively high ratio of auxin to gibberellin, whereas phloem formation occurs when the gibberellin level is high.

This latter conclusion might be thought to indicate that the differentiation of cambial derivative cells into xylem on the inner side and into phloem on the outer side is determined by differences in the ratios of auxin to gibberellin concentrations on the two sides. However, a different conclusion is indicated by the results of experiments by Siebers, who carried out experiments with young seedlings of *Ricinus communis*. In this species, as in many other herbaceous plants, cambial zones are formed in the ground tissue ("inter-

fascicular tissue") separating the primary vascular bundles and are connected to the cambia of these bundles, so forming a continuous cambial ring.

Siebers cut out small pieces of interfascicular tissue from young hypocotyls at a stage before there was any sign of the formation of interfascicular cambium. These pieces were reversed and replaced in the hypocotyls. Later examination revealed that interfascicular cambium was initiated in the reversed pieces of tissue, but the pattern of differentiation had been reversed so that xylem formed on the *outside* and phloem the *inside*. Moreover, no direct connection was established between the interfascicular cambium and that of the primary vascular bundles. These observations suggest that although the originally complete ring of procambium at the shoot apex (p. 33) breaks down into separate strands (each of which develops into a primary vascular bundle), the regions between strands retain some propensity to develop later into cambium, even though the cells there may be morphologically indistinguishable from those of the surrounding ground tissue. Moreover, the normal pattern of differentiation of cambial derivatives (giving xylem on the inside and phloem on the outside) appears to be determined by the potentialities of the cells themselves and not by extrinsic factors, such as hormones, although the latter, especially IAA and gibberellins, are necessary for cambial division and cell differentiation.

Little is known of the role of cytokinins in normal cambial activity, but studies with isolated pea-stem sections have indicated that these hormones can also stimulate cambial division and increase the lignification of the developing xylem cells. Ethylene and abscisic acid also affect cambial activity when applied to plants, but there is not yet any evidence that these substances play a regulatory role in the natural control of cambial division and differentiation of vascular tissue.

Cambial activity is known to be affected by various non-hormonal factors, such as sugars and water availability, as well as by environmental factors, such as temperature, a fact which forms the basis of "tree-ring analysis" for dating in archaeological studies and for the study of long-term climatic changes.

5.5 HORMONES AND LEAF DEVELOPMENT

Once a leaf primordium has been initiated at the stem apex, it starts to grow and differentiate by the processes of cell division, cell enlargement and differentiation (see Chapter 2). One can reasonably assume that these processes are under the controlling influence of growth hormones, one of which would be expected to be auxin. However, it cannot be said that auxin is involved in all aspects of leaf growth. It has been found that auxins will, depending on their concentration, either stimulate or inhibit the growth of midrib and veins but have little effect on the interveinal mesophyll tissues. At the present time little is known of the hormonal control of leaf growth, other than that auxin appears to be necessary for vein growth.

It has been suggested that a growth hormone synthesized in the roots controls mesophyll growth. Thus, it is found that if young root tips are cut off as fast as they are formed in, for example, the horseradish plant (*Amoracea lapathifolia*), normal development of the interveinal tissue fails to occur (Fig. 5.19). Evidence

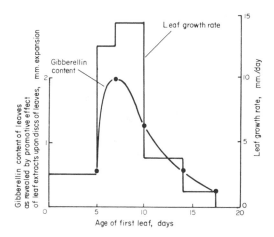

Fig. 5.20. Relationship between growth rate (histogram) and gibberellin content (graph) in first leaf of dwarf French bean (*Phaseolus vulgaris*). Maximum gibberellin content of the leaves occurred at 7 days of age, coinciding with the period of most vigorous leaf expansion. (From A. W. Wheeler, *J. Exp. Bot.* **11**, 217-26, 1960.)

that roots export cytokinins and gibberellins to the shoot may provide a partial explanation for failure of mesophyll growth in such root-pruning experiments.

It has been found that treatment of some plant species (e.g. *Triticum* and *Phascolus*) with gibberellic acid leads to a stimulation of leaf growth. Other plant species (e.g. *Pisum sativum*) do not respond in this way, and in fact mesophyll growth may be retarded following gibberellin treatment. In those species where gibberellin treatment does stimulate leaf growth, the mesophyll and vein tissues respond nearly equally. Similarly, excised leaves or leaf disks will, if floated on the surface of a solution of gibberellin or cytokinin, often expand due to growth of the mesophyll. Thus, generally speaking, gibberellins and cytokinins differ from auxins in their effects on leaf growth, in that the former two classes of hormone can promote mesophyll growth whereas auxins do not. This immediately suggests that endogenous gibberellins and cytokinins are important regulators of the growth of leaves. In fact it has been found that gibberellins are normally present in leaves, and that the concentration present is closely related to the growth rate of the leaf, so that young, rapidly growing leaves contain more gibberellin than do older leaves (Fig. 5.20). Natural

Fig. 5.19. Effect of repeated excision of root tips upon the growth of mesophyll in leaves of the horseradish plant (*Amoracea lapathifolia*). The smallest and most deeply lobed leaves are the youngest. (Photographed by Mr. J. Champion, from a plant grown by Dr. J. Dore.)

gibberellins may, therefore, be of importance in the control of mesophyll growth. We know little about endogenous cytokinins in leaves, although some experiments have shown that synthetic cytokinins can replace the need for a root system for healthy leaf growth.

In cereal and grass species, a late stage of leaf development involves unrolling of the lamina—i.e. much of longitudinal and lateral expansion of the leaf takes place with the leaf rolled into a cylindrical form with the adaxial surface innermost. The physiological basis of this leaf-unrolling process, which is under phytochrome control, is considered in more detail in Chapter 8.

We have already mentioned that certain phytochrome-controlled responses may be mediated through changes in ethylene release, and in view of the inhibitory effects which can be exerted by exogenous ethylene on leaf expansion one cannot exclude the possibility that this hormone, also, may normally participate in leaf growth and differentiation, although it should be recognized that very little evidence is available to suggest this.

The growth of leaves may, therefore, be under the controlling influence of auxins, gibberellins, cytokinins and ethylene, together with nutrients and perhaps other unknown hormonal factors, but it should be borne in mind that current knowledge is meagre and fragmentary.

5.6 HORMONES AND ROOT DEVELOPMENT

The internal mechanism of control of root development is less well defined than that which operates in stems or coleoptiles. By simple analogy with the situation as it is understood for coleoptiles and stems, one might expect that auxins and gibberellins are synthesized in the apical part of the root, and that these hormones are translocated basipetally to the region of cell elongation growth a few millimetres behind and there regulate growth and differentiation. However, no clear evidence exists that can permit one to assume that stems and roots possess similar hormonal control systems.

5.6.1 Root elongation growth

Excision of a stem apical bud, or coleoptile tip, results in a marked reduction or complete cessation of elongation growth in the lower regions. In contrast, removal of the root apex does not prevent elongation of newly formed cells in the root. It has in fact sometimes been found that removal of the root-tip results in a transitory *increase* in the rate of elongation of young root cells. At first sight this suggests that the root-tip region exerts only an inhibitory effect on extension growth of cells in the elongation zone, but this is not necessarily true.

For many years it was accepted that the root-tip is a region of auxin synthesis and that auxin moved basipetally in the root (i.e. away from the root apex) to promote elongation growth. Thorough modern analytical studies using combined gas-liquid chromatograph/mass spectrometry have demonstrated that IAA is certainly present in roots, though at extremely low concentrations compared with IAA levels in shoot tissues. Roots do appear to be very sensitive to auxins, for whereas the optimum auxin concentration for stem or coleoptile elongation is usually approximately 2×10^{-5}M (Fig. 5.8A) the corresponding value for roots is around 10^{-8}M (Fig. 5.21). In other words, roots are about 2,000 times more sensitive to exogenous IAA than coleoptiles or stems. Also, as Fig. 5.21 illustrates, promotion of elongation growth in roots by exogenous auxin is considerably less than auxin-induced elongation of stems and coleoptiles. The auxin that is present in roots is of course not necessarily synthesized in the root-tip, as it could arrive there by transport from the shoot. Studies of radioactive auxin transport in roots have so far yielded somewhat confusing results. However, most of such experiments have revealed acropetal polarity of auxin transport (e.g. Fig. 5.3) right up to the root-tip. Such acropetal polarity of auxin transport would appear to preclude the possibility of regulation of root elongation growth through the action of any auxin synthesized in the root apical region.

There is somewhat better evidence that gibberellins may be synthesized in root-tips and

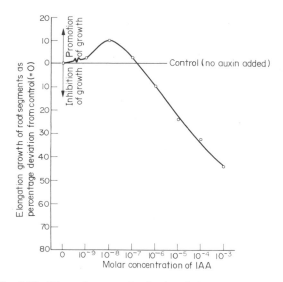

Fig. 5.21. Elongation growth of sub-apical root segments of *Lens culinaris* in response to exogenous auxin (IAA). Note that only a slight promotion of elongation occurred, and that the optimal IAA concentration was as low as 10^{-8} M (1.75×10^{-3} mg/l). Concentrations of IAA greater than 10^{-7} M inhibited growth below the control value. (From P. E. Pilet, M. Kobr, and P. A. Siegenthaler, *Rév. Gen. de Botanique,* **67**, 573-601, 1960.)

transported basipetally to the cell elongation zone, but even here the data are less than completely convincing. Applications of gibberellins to intact plants generally have little effect upon root elongation, but excised roots growing in aseptic culture sometimes grow more in length when supplied with gibberellin. Also, whereas excised roots in culture usually fail to form root hairs, they have been reported to do so in the presence of gibberellic acid. Such rather scattered observations lead one to suspect that it will eventually be found that endogenous gibberellins are concerned in the control of root elongation growth.

Exogenous ethylene inhibits root elongation as effectively as it does stem elongation (except in aquatic plants such as rice, where both stem and root elongation are promoted by applied ethylene) by what appears to be an identical mechanism. Roots also synthesize ethylene, and it is possible that supra-optimal auxin concentrations inhibit root elongation (Fig. 5.21) through enhancement of ethylene biosynthesis (Fig. 5.13) in the root tissues.

The root-tip region, particularly the root-cap, has in recent years been found to a site of production of growth-inhibitors, including ABA, that appear to play regulatory roles in root elongation growth. Almost all of this work has been concerned with the role of the root-cap in root gravitropism and is therefore discussed further in Chapter 7 (see pp. 197-198).

Cytokinin treatment of excised roots, particularly when given in combination with auxin, results in a stimulated rate of cell division, but as this does not lead to an increased rate of root elongation but to a stimulation of division of cells which are destined to differentiate into vascular tissues, the role of cytokinins in roots is considered below.

5.6.2 Vascular tissue development in roots

Almost all the increase in diameter which occurs in older regions of roots results from the activities of the root vascular cambium. As in stems, the vascular cambium in roots of dicotyledons is both fascicular and interfascicular in secondarily thickened regions. It similarly gives rise to phloem and xylem tissues. Much of the evidence which implicates growth hormones in the control of vascular cambium activity in roots has come from studies of excised roots growing in aseptic culture (see Chapter 6). In excised roots, such as those of pea (*Pisum sativum*) or radish (*Raphanus sativus*), only the primary vascular structure develops, even when auxin is supplied in the bathing nutrient medium. When, however, auxin is supplied only through the basal cut end of an excised root, then secondary thickening of the root occurs as the exogenous auxin is transported acropetally towards the root-tip. The simultaneous addition of a cytokinin greatly increases the stimulatory effect of acropetally moving auxin in excised roots.

Thus, experiments on excised roots have strongly indicated that normal control of cambial activity in roots is exerted through a supply of auxin from the shoot system. Supporting evidence for this view is provided by the fact that in temperate zone trees, cambial activity in roots does not begin in springtime until the wave of renewed cambial activity in the shoot reaches the base of the trunk. The wave of activity passes into the roots, so that seasonal cambial activity progresses acropetally in roots. This therefore again suggests that secondary thickening in roots is controlled by auxin entering from the shoot.

5.6.3 Root initiation

Root primordium initiation requires auxin. Much evidence for this has come from studies of adventitious root initiation in shoot cuttings (see Chapter 2) and from observations of regeneration in aseptically cultured callus cells (Chapter 6).

In the common horticultural practice of vegetative propagation of shoot cuttings, it is normally the case that new roots are formed at the basal end of the stem (Fig. 5.34). It is in just this region that endogenous auxin, which originates in growing leaves or buds on the cutting, would be expected to accumulate as a consequence of the polarized basipetal flux of auxin in stem tissues.

Shoot buds and root primordia may be caused to differentiate in callus cultures of plant cells by exposure to an appropriate combination of auxin and cytokinin. We may reasonably assume, therefore, that both of these growth hormones are required for the organization of a new root apical meristem. Adventitious root initiation in shoot cuttings is markedly *inhibited* by gibberellins, but the reasons for this are not yet known.

Lateral root initiation has mainly been studied in excised roots growing in aseptic nutrient culture. In general, the addition of an auxin to the medium promotes the formation of lateral roots. However, factors other than auxin also appear to be required for lateral root initiation in excised roots. These other factors are partially nutritional, but also include unidentified substances which are supplied by older root tissues. Cytokinins are also important for lateral root initiation. Thus, Torrey and others have noted that auxin-induced lateral root production in isolated pea-root segments is enhanced by a supply of a certain concentration of kinetin or adenine. It appears that lateral root initiation is at least partially determined by the relative concentrations of auxin and cytokinin present, in a similar manner to root initiation in callus cultures (p. 157). In contrast to the inhibitory effect of gibberellins on adventitious root initiation, it has been reported that, under certain conditions, gibberellic acid can promote the formation of lateral roots in excised tomato roots.

Finally, it should be mentioned that the root apical meristem exerts an inhibitory effect on lateral root initiation, a situation reminiscent of the correlative inhibition of axillary buds by the apical bud in shoots. Thus, excision of the root tip leads to an increased rate of lateral root initiation, and also to an increase in the absolute number of lateral roots formed. However, we have no good reason to believe that suppression of lateral root initiation by the root apex is a result of auxin synthesis in the latter. Indeed, there is no unequivocal evidence that auxin is synthesized in the root tip and auxin transport in roots is acropetal, the opposite situation to that pertaining in the shoot. Also, of course, auxin enhances rather than suppresses root initiation.

5.7 HORMONES AND FRUIT DEVELOPMENT

Probably because of the economic importance of fruits, the physiology of their growth, development and ripening has been very intensively studied. Most botany textbooks classify "fruits" according to a rigid and complicated system based upon morphological characters, but J. P. Nitsch has pointed out that a fruit is a physiological rather than a morphological entity. To the physiologist, and incidentally the layman, a "fruit" is simply a structure which arises by development of tissues which support the ovules. Nitsch has pointed out that such a view is valid even for seedless fruits because ovules were initially present in them.

The early growth of the ovary, which occurs during development of the flower, involves cell division but little cell vacuolation. In many species, cell division ceases at or shortly after anthesis (flower opening) and the subsequent growth of the fruit following pollination is primarily due to an increase in cell size rather than in cell number. For example, in tomato (*Lycopersicum esculentum*) and blackcurrant (*Ribes nigrum*), cell division ceases at anthesis and the whole of the subsequent growth is due to cell expansion. In such species, the final size will be a function of the number of cells already present in the ovary at anthesis. On the other hand, in some species (e.g. apple) cell division may continue for a time after pollination has occurred.

Fruit cells may enlarge by vacuolation to relatively enormous sizes, and mature fruits of water melon (*Citrullus vulgaris*) consist mainly of cells so large that they are distinguishable individually by the naked eye.

5.7.1 Fruit set

The early development of the ovule and ovary takes place along with other aspects of flower development (p. 37). In some species the ovary ceases growth at the time of, or before, anthesis (Fig. 5.22). In others, growth goes on for a time

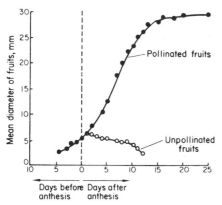

Fig. 5.22. Growth curves of ovary of *Cucumis anguria*. Growth is unpollinated ovaries ceases soon after anthesis, but pollinated ovaries show a typical sigmoid growth curve. The decrease in ovary diameter seen in unpollinated ovaries is due to "shrivelling". (From J. P. Nitsch, *Quarterly Rev. Biol.* **27**, 33-57, 1952.)

after anthesis and prior to pollination. In both cases, further growth of the ovary takes place only if pollination is effected. Should, for some reason, pollination not occur, then growth and development of the fruit ceases. Failure of the pollination mechanism usually results in the shedding from the plant of the unfertilized flower, often mediated by the formation of a separation, or abscission layer (see p. 294) in the peduncle. Successful pollination, on the other hand, is followed by rapid growth of the ovary and fruit development begins (Fig. 5.22). At the same time, the petals and stamens wither and often abscind. The start of ovule growth and withering of stamens and petals marks the start of fruit development, and this phase is often called *fruit set*. The process of pollination, whether or not it is followed by fertilization, is apparently sufficient to cause an initial stimulation of growth in the ovary and other parts of the future fruit. This is shown by the fact that in many fleshy fruits the increase in ovary growth may start before there has been sufficient time for fertilization to occur. Moreover, even "foreign" pollen, derived from an unrelated species and hence unable to effect fertilization, may cause marked stimulation of ovary growth.

The stimulatory effect of pollen on ovary growth appears to be due to the auxin which it contains. In 1909, Fitting observed that water extracts of orchid pollen were able to induce swelling of unfertilized orchid ovaries and withering of the petals, and it was shown later that pollen is a rich source of auxin. Finally, it was shown that pure preparations of IAA would, if applied to unfertilized flowers of a number of species, induce fruit set in the absence of any pollen. Among the species in which fruits can be set by auxins are tomato, pepper, tobacco, holly, figs and blackberry. In these species, fruits which have been set by treating unpollinated flowers are seedless. The production of such seedless fruits is called *parthenocarpy*. Recently it has been found that ethylene will simulate the effects of pollen on ovary swelling and petal withering in various species. Thus, it is possible that some effects of auxins on ovaries are mediated through influences on ethylene production.

Fig. 5.23. Correlation between achene development and receptacle growth in strawberry. (a) only one fertilized achene present; (b) three fertilized achenes present; (c) *left,* control fruit; *right,* fruit with three vertical rows of achenes. (d) *left,* fruit with two rows of achenes; *right,* control fruit. (e) three strawberries of the same age: *left,* control; *middle,* all achenes removed and receptacle smeared with lanolin paste; *right,* all achenes removed and receptacle smeared with lanolin paste containing 100 ppm of the auxin β-naphthoxyacetic acid. (Original prints kindly supplied by Dr. J. P. Nitsch, Laboratoire du Phytotron, Gif-sur-Yvette, France. From *Amer. J. Bot.* **37**, 211-15, 1950.)

5.7.2 Fruit growth

Although pollination may stimulate the initial swelling of the fruit, in most species further development of the fruit appears to be dependent upon the presence of developing seeds, and hence can only occur when fertilization is effected. Thus, in many fruits, such as grapes, blackcurrants, tomatoes, apples and pears, strong correlations exist between the final size of the fruit and the number of fully developed seeds it contains. In the case of the strawberry, Nitsch showed by elegant surgical experiments that receptacle growth is dependent upon achene development (Fig. 5.23).

The interaction between the growth of the ovary and that of the embryo and endosperm is shown by the changing growth rates of these different parts of the fruit at different stages of development. In some instances the growth curve for the fruit is sigmoid (e.g. in apple) and in others it is doubly sigmoid (Fig. 5.24). In peach, the changes in growth rate of the pericarp can apparently be correlated with changes in growth rate of the developing seeds. The stimulatory effect of the developing seeds upon the growth of pericarp

tissue appears to be due, at least partly, to the auxin which they produce. Developing seeds are rich sources of auxin and it has been shown that there is a declining gradient of auxin concentration within the tissues of the fruit, with the highest concentration in the seeds, less in the placenta and least in the carpel wall. This gradient is consistent with the view that auxin is produced in the developing seeds and moves outwards from there to the other parts of the fruit.

A good example of the relationships that have been found between endogenous auxins and fruit growth has come from studies of berries of the blackcurrant, which show a double sigmoid growth curve (Fig. 5.24). Two auxins were found in the berries, one acidic (possible IAA) and the other neutral (possibly IAN). These two auxins were found to be produced mainly at the times of endosperm and embryo development, which in turn coincided with the times of maximal growth rate of the fruit. It appears likely, therefore, that the first grand period of fruit growth in blackcurrant is under the control of auxins produced in the developing endosperm, and that the second grand period is induced by auxin originating in the growing embryo. Similar patterns of auxin production in relation to fruit growth have been reported for other species by several workers, but there are other conflicting reports that no direct correlation occurs in some species between auxin content and fruit growth rates.

It is important to realize that despite the evidence implicating auxins in flower and fruit growth, there is also a considerable likelihood that auxins are not the only hormones involved. It has proved impossible with many plant species to induce parthenocarpic fruit development by auxin treatment, but it has proved possible to do this by spraying the flowers with a solution of gibberellin (e.g. in members of the genus *Prunus* such as cherries, peaches and almonds). However, indicative as the effects of applied gibberellins on fruit development may be, this does not in itself provide incontrovertible evidence that internally produced gibberellins are concerned in the normal growth of fruits and seeds. Consequently, studies have been made of the gibberellin contents of various fruits and seeds at different stages of their

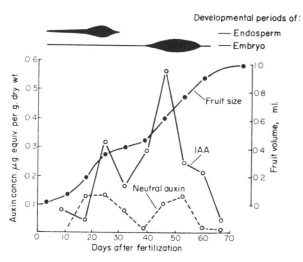

Fig. 5.24. Changes in concentration of indole-3-acetic acid (IAA) and an unidentified neutral auxin in the blackcurrant berry in relation to the double sigmoid growth curve of the berry and the main developmental periods of the endosperm and embryo. (From S. T. C. Wright, *J. Hort. Sci.* **31**, 196, 1956.)

development, and in general it has been found that young developing seeds contain relatively large amounts of gibberellins. As the seeds mature and their growth slows down there is a simultaneous fall in their gibberellin content. It appears likely that gibberellins move out from the young developing seeds in a similar manner to that suggested for auxin, and that both types of hormone are involved in the control of fruit growth.

The third type of growth hormone likely to be participating in the processes of growth in a fruit are the cytokinins. As discussed earlier (Chapter 3), cytokinins are growth hormones particularly concerned in the control of cell division, and it is, therefore, likely that the active cell division which is known to occur in young developing fruits is under the controlling influence of this type of hormone. Evidence that cytokinins are involved in fruit growth is afforded by experiments which have shown the presence of cytokinins in the young fruits of apple, banana and tomato during the stages of growth in which cell division is most rapid. Thus, we may regard fruit growth and differentiation as being under the control of several types of growth hormone, a situation which probably obtains in all phases of plant development.

5.7.3 Parthenocarpy

We have already seen that in some species parthenocarpic fruits may be produced by treating unpollinated flowers with auxins or gibberellins. In addition to the experimental production of such fruits, natural parthenocarpy may occur in certain species. Thus, horticultural varieties of bananas, pineapples, cucumbers, tomatoes and figs exist, in which seedless fruits are normally produced without the need for any exogenous hormone. In some species, fruits are formed without pollination, while in others pollination is necessary but fertilization does not occur; in others again, fertilization occurs but the embryos abort before the fruit matures. It is not known how the growth of these seedless fruits is controlled, but it seems possible that in some cases the maternal tissues, such as the placenta, may be capable of producing

auxin in the absence of normal embryos. Thus, it has been shown that the ovaries of unopened flowers of parthenocarpic varieties of orange and grape have a higher auxin content than do those of normal seeded varieties. Moreover, it has been found that young parthenocarpic fruits of cucumber contain seed-like structures, but which lack embryos and endosperm, and it is possible that these are centres of auxin production.

The production of fully developed fruits by treatment with a single application of auxin to the flowers (p. 129) also poses a number of problems, since it is not to be expected that the auxin which is applied in this way will itself be sufficient to supply the needs of the developing fruits over several weeks. However, it has been found that pollination stimulates the production of auxin by the tissues of the ovary itself in some species, such as tobacco, and it is possible that external application of auxin may similarly trigger off the production of auxin by certain tissues of the fruit, and that once this has occurred, the production of endogenous auxin will meet the requirements for the further development of the fruit. Application of auxin has been shown to lead to growth of unfertilized ovules, which develop normal looking seed coats but which contain no embryos. It is interesting to note that in some species, such as olive, hops and maize, application of auxin will stimulate initial fruit set, but further development of the fruit does not occur without pollination. Possibly, in such species the exogenous auxin does not stimulate the production of endogenous auxin necessary for further fruit development.

5.8 GROWTH CORRELATION

The various growth processes that proceed simultaneously in a plant are not independent, but are closely linked with one another. Thus, as a stem increases in length by the activities of the apical meristem, its strength is increased by activities in the cambium leading to increased girth and rigidity of the older parts, so enabling the whole shoot to stand erect. Further, as the shoot increases in size, so does the need for water and mineral nutrients increase, and this is catered for by a nicely balanced relationship between shoot growth and root growth. As the shoot increases in bulk, the size of the root system becomes proportionately larger, which allows for the additional mineral nutritional requirements of the shoot to be met. To some extent these *growth correlations,* as they are called, are explicable in terms of the availability of nutrients, or food factors, and the competition between growing regions for these substances. Thus, shoot and root growth are related to one another, and this is probably due in part to their mutual nutritional dependence; the shoot supplies the organic material which the root is unable to manufacture for itself, and in return obtains from the root the water and mineral salts to which it does not itself have direct access. In the same way, vegetative growth is very reduced when a plant is fruiting, probably mainly as a consequence of the diversion of the available food materials into the developing seeds and fruits (see Chapter 12).

This is certainly not the whole story, however. Competition for available nutrients does not adequately explain why active growth at any one time is usually restricted to only a few of the many places in a plant where it is potentially possible. An example of this is the fact that the apical meristem of a shoot is usually in a state of active growth, whereas growth of the axillary buds below it is often strongly inhibited. This characteristic of shoot growth is known as *apical dominance* or *correlative inhibition* of buds. A related phenomenon is seen in the inhibiting effect of the main root apex upon the initiation of lateral roots in the pericycle cells. Removal of the root apex causes an increase in lateral root formation in the remainder of the root, a result analogous to the effect of removing the shoot apical bud upon the growth of axillary buds. However, the mechanism by which these phenomena occur is not necessarily the same in shoot and root.

It is known that growth hormones play important roles in the correlation of growth in different parts of the plant. It is likely that all growth correlations are in one way or another affected by patterns of hormone distribution within the plant. We have already seen several examples of correlative effects in plant growth which are mediated by growth hormones, as in the stimulation of cambial activity by auxin and gibberellins arising in the buds of woody shoots, and the stimulation of fruit growth by hormones produced by the developing seeds. We shall now consider the role of hormones in apical dominance.

5.8.1 Apical dominance

The apical bud of a shoot usually grows much more vigorously than the axillary buds, despite the fact that it is apparently the least favourably situated bud in relation to the supply of organic and inorganic nutrients from the mature leaves and the root system. There is a great deal of variability between species with respect to the degree of dominance of the apical bud over the lower axillary buds. In some species, such as tall varieties of sunflower (*Helianthus annuus*), the dominance is complete and extends over almost the whole length of the stem. In others, such as tomato, the dominance of the apical bud is weaker, and the axillary buds situated only a little way below the main shoot tip grow out, resulting in a bushy shoot system.

In many species the dominance of the shoot tip becomes weaker as the plant gets older. This is seen clearly in plants such as sycamore (*Acer pseudoplantanus*) or ash (*Fraxinus excelsior*) where the early years of growth are characterized by strong growth of the leading shoot, whereas in later years a branching habit is seen. Even in herbaceous annuals there is often a weakening of apical dominance towards the end of the growing season, and in those species where the apical meristem eventually changes to produce a terminal flower, this often coincides with a release of axillary buds from correlative inhibition.

If a shoot is decapitated, i.e. the apical bud cut off, then one or more of the lower axillary buds grow out. Usually one of the outgrowing laterals becomes dominant over the others, exerting an inhibitory influence on their growth. Where this happens it is frequently the uppermost of the axillary buds which becomes the dominant shoot. Thus, some form of signal arises in an actively growing dominant shoot bud. The effects of this signal are most obviously expressed in the correlative inhibition of lateral buds and shoots, but its influence is also seen in other morphogenetic phenomenon, such as in the orientation of lateral shoots, branches, rhizomes and tubers and the pattern of their differentiation.

Some experiments conducted in the first two decades of this century yielded results which indicated that young growing leaves of the apical bud synthesized a diffusible correlative growth inhibitor that normally moved through only living cells in the plant. The discovery of auxin (indole-3-acetic acid, IAA) in the early 1930s, and the realization that the young developing leaves are the primary sources of auxin in the shoot, led to investigation of the possibility that transmission of IAA from the apical bud constitutes the correlative signal in apical dominance phenomena. It was soon found, by Thimann and Skoog in 1934, that exogenous IAA could substitute for the apical bud in maintaining inhibition of axillary buds in bean plants. This observation has been confirmed

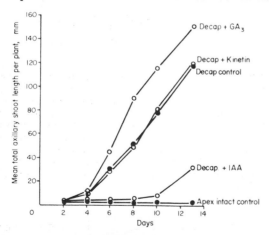

Fig. 5.25. Effect of hormones on outgrowth of axillary buds of pea plants (*Pisum sativum*) following decapitation. The hormones referred to were applied in lanolin (concentration 1000 ppm) to the decapitated stumps of the plants. The ordinates show the total growth of the three remaining axillary buds per pea plant. (I. D. J. Phillips, original data.)

innumerable times in succeeding years for many species (Fig. 5.25). A few exceptions were later reported, especially *Coleus,* in which it was found that exogenous auxin did not exert significant inhibition on axillary bud growth. Nevertheless, it has recently been demonstrated that even in *Coleus* the inhibitory effect of apically applied exogenous auxin upon axillary shoots is revealed when the plants are rather nutritionally deficient. There seems little doubt, therefore, that the synthesis of IAA in the apical region of the shoot, probably in the young expanding leaves, and its transport down the stem constitutes a basic component of the mechanism of correlative inhibition.

The way in which auxin causes the inhibition of axillary bud growth is not, however, fully understood at the present time. It appears paradoxical that auxin, which we have so far been considering as a promoter of growth, should cause an inhibition of growth in axillary buds. To resolve this contradiction it was suggested by Thimann, in 1937, that the optimal auxin concentration for bud growth is lower than for stems and that the bud is inhibited by the "supra-optimal" auxin concentration (Fig. 5.8A) normally present in the stem, as a result of its synthesis in the young expanding leaves, and basipetal transport from there. This is known as the "Direct Theory" of auxin inhibition of lateral buds. It is, of course, based on the assumption that auxin enters lateral buds from the stem.

The validity of Thimann's direct theory is considered doubtful today. One of the principal objections to the theory is that determinations of the auxin contents of inhibited lateral buds in *Lupinus, Pisum sativum* and *Syringa vulgaris* have revealed that the auxin levels, far from being supra-optimal, are, in fact, sub-optimal. Also, it has been found that applications of low concentrations of auxin to the stumps of decapitated *Lupinus* and *Phaseolus multiflorus* plants actually accelerated growth of laterals, and that only at higher auxin concentrations did inhibition occur. Consequently, it is now generally considered that auxin does not exert its inhibitory effect on lateral buds in such a direct manner as that originally proposed by Thimann.

One of the early hypotheses for apical dominance assumed that since the apical meristem is the first-formed shoot meristem in the germinating seedling, then it would continue to command a preferential supply of metabolites as these moved along their concentration gradients. This was known as the "Nutritive Theory" of apical dominance.

If the nutritive theory is correct, one might expect that the dominance of the apical bud over the lateral buds to be most clearly manifest when a plant is deficient in nutrients, for example, when growing in soil low in necessary mineral elements. This has been clearly shown to be the case for several plants, particularly flax (*Linum usitatissimum*) in which lateral growth is entirely suppressed by the terminal bud under conditions of mineral nutrient deficiency, whereas lateral growth occurs freely under conditions of high mineral (particularly nitrogen) nutrition. One can assume that when nutrients are freely available to the plant, there are sufficient "left over" after the apical bud has received its necessary quota to allow movement of nutrients into the lateral buds. The effect of the apical bud in flax plants growing under high or low nutrient conditions was exactly duplicated by IAA applications to the stump of decapitated plants.

Although it is undoubtedly true that the apical bud does obtain more available nutrients than the axillary buds, the simple "source-sink" explanation of the nutritive theory does not adequately explain why auxin can substitute for the apical bud in correlative inhibition of axillary buds. Further, many studies with isotopes such as ^{32}P-phosphate and ^{14}C-sucrose have demonstrated that nutrients do indeed move to, and accumulate in regions of high exogenous auxin concentration (Fig. 5.26). This "auxin-directed" transport of metabolites indicates that auxin production in the apical bud, and its basipetal transport, induces movement of available nutrients towards the region of highest auxin concentration, i.e. to the apical bud itself. It is not clear how this comes about. One suggestion has been that basipetally moving auxin in the stem inhibits the development of vascular connections (xylem and phloem) between axillary buds and main stele, so reducing the capacity of the axillary buds to obtain a supply of nutrients via the vascular system. This hypothesis therefore assumes that the lowering of the auxin content of the stem which occurs following decapitation in some way results in rapid development of bud-stem vascular connections. Quite a number of workers have tested this vascular connection hypothesis in recent years and, although results obtained are rather contradictory, the weight of accumulating evidence argues against the idea that lack of bud growth is attributable to deficiency in their vascular supply. Thus, in many cases it has been found that bud outgrowth commences within

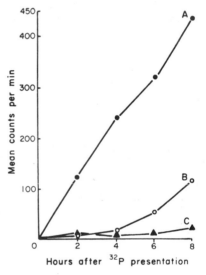

Fig. 5.26. Auxin-directed transport. The effect of IAA on the rate of accumulation of ^{32}P at the top of the stem in decapitated pea plants is shown. The radioactive ^{32}P was applied to the base of the stem just above soil level. In decapitated plants not supplied with auxin very little accumulation of ^{32}P occurred at the top of the stem (C). Application of IAA to the top of the stem immediately after excision of the apical bud (A) greatly enhanced ^{32}P accumulation. Curve B shows the effect of applying the auxin 6 hours after decapitation. (Reprinted from C. R. Davies and P. F. Wareing, *Planta*, **65**, 139-56, 1965.)

hours of decapitation of the shoot, whereas vascular connections did not develop until the buds were already growing vigorously. In other species it has been found that even completely inhibited buds appear to be served by perfectly adequate-looking xylem and phloem connections..

Other research has cast doubt upon the basic premise of the nutritive theory, in that it has been found that correlatively inhibited buds contain what appear to be perfectly normal levels of major nutrients, and that feeding inhibited buds in various ways with major inorganic and organic nutrients does not induce their growth so long as the apical bud remains intact and actively growing.

However, research over the past decade has indicated that cytokinins play an important role in apical dominance. In many plants application of a cytokinin preparation directly onto a correlatively inhibited bud can release that bud from inhibition in an intact plant (Fig. 5.27). This suggests that inhibited buds fail to develop because they lack cytokinin. We have considered elsewhere (Chapter 12) evidence that cytokinins are synthesized mainly in the root system, but that shoot tissues require adequate supplies of these hormones for their healthy functioning. It is possible then that a role of auxin from the apical bud is to direct the transport of root synthesized cytokinins so that the apical bud receives preference over the axillary buds. As considered above, correlatively inhibited buds contain low concentrations of endogenous auxin, and it is possible that it is their deficiency in auxin which prevents them from either obtaining adequate supplies of cytokinin from the roots, or synthesizing cytokinins themselves. In support of this view is the fact that buds released from dominance by direct application to them of a cytokinin normally grow for only

(a)   (b)   (c)   (d)

Fig. 5.27. Release of axillary buds from correlative inhibition in intact plants by application of a cytokinin, and subsequent treatment of the growing bud apex with auxin.
    Left to right: (a), untreated bud; (b), bud treated with kinetin only; (c), bud treated with kinetin followed 3 days later by gibberellic acid; (d), treated with kinetin followed 3 days later by indole-3-acetic acid (IAA). Only buds treated with IAA following kinetin treatment continued growing more or less normally. Gibberellic acid treatment did not effect continued bud growth following kinetin treatment. All hormones were applied directly to the buds dispersed in a mixture of 50 per cent ethanol and 8 per cent carbowax to aid their penetration of the tissues. Concentrations used were; 330 ppm kinetin, 100 ppm GA_3, and 1000 ppm IAA. The photograph was taken 7 days after the first hormone treatment. (From Tsui Sachs and K. V. Thimann, *Amer. J. Bot.* **54**, 136-44, 1967. Original print supplied by Professor K. V. Thimann.)

a short time, after which apical dominance is reimposed, and continued growth occurs only when auxin (and occasionally gibberellin as well it has been found) is subsequently applied to the bud (Fig. 5.27). A possible explanation for this feature is that the applied exogenous cytokinin is deactivated by the growing bud before its young leaves have acquired the capacity to synthesize sufficient auxin to induce cytokinin synthesis or its transport into the bud. Additional evidence that cytokinin levels limit growth in correlatively inhibited buds is supplied by cytological studies that have revealed that cell division activity is suppressed in the apical meristems of such buds. Much evidence exists that cytokinins play particularly important roles in the natural regulation of cell division activity in plant cells (Chapter 3), and application of a cytokinin to an inhibited bud does indeed result in the immediate onset there of cell division activity.

Available evidence suggests that gibberellins may not be directly involved in the control of apical dominance phenomena in the way that auxins, and cytokinins are. Thus, gibberellins do not substitute for the apical bud in the correlative inhibition of axillary buds (Figs. 5.25 and 5.28), nor do they cause the release of buds from correlative inhibition (cytokinins can do so). Application of exogenous gibberellin to an intact plant often results in an increase in apical dominance, but this is probably due to effects on internal auxin level and distribution in the gibberellin-treated plants.

5.8.2 Hormonal interaction in stolon development

A striking example of the importance of hormonal interaction in the control of shoot growth is seen in stolon development in the potato plant. The stolons are axillary shoots which normally arise from the basal nodes of the stem below the soil surface. They differ from the erect, leafy axillary shoots which arise on the aerial part of the shoot in that they have (1) only rudimentary leaves, (2) elongated internodes, and (3) poor development of chlorophyll (even in the light), and (4) they grow horizontally. Apical dominance evidently plays an important role in stolon development, since if the aerial shoot is decapitated and all axillary shoots are removed from it, the stolons will turn up and produce normal-looking, erect leafy shoots.

Stolons are not normally formed on the upper, leafy part of the aerial shoot, but they may be induced to develop by decapitating the shoot and

Fig. 5.28 Apical dominance in runner bean (*Phaseolus coccineus*) plants. Sixteen day-old plants 'decapitated' five days before photograph taken. From the left: buds subject to correlative inhibition in intact plant; buds growing out in decapitated plant to which no hormone was added; buds inhibited in plants treated with 0.1 per cent IAA in lanolin to cut end of stem; rapid outgrowth of buds of decapitated plants treated with gibberellic acid (GA$_3$) at the cut end of stem.

applying exogenous hormones. If IAA alone is applied to the stump of a decapitated shoot, the uppermost axillary shoot is partially inhibited, whereas if GA₃ alone is applied, extension of this axillary is promoted. By contrast, when IAA and GA₃ are applied together to the decapitated stem, the development of the uppermost axillary is dramatically changed and it becomes a horizontal, stolon-like structure (Fig. 5.29). If kinetin alone is applied to the decapitated stem there is no observable effect, but if kinetin or benzyladenine is applied directly to the *tip* of a natural or of an experimentally promoted stolon, the latter very rapidly changes its pattern of development and turns upwards to become a leafy shoot (Fig. 5.29). Thus, the development of an axillary shoot of potato can be controlled very precisely by manipulating the levels of auxin, gibberellin and cytokinin. It seems likely that the natural control

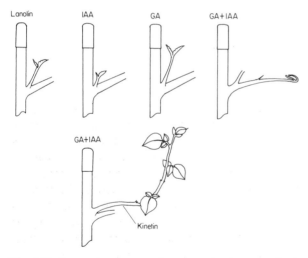

Fig. 5.29. Interaction between hormones in the control of stolon development in the wild potato species, *Solanum andigena*. Leafy aerial shoots were decapitated and hormones were applied in lanolin to the decapitated stems, as shown. The uppermost axillary shoot showed the following responses to hormone: control shoots (treated with plain lanolin) showed outgrowth of leafy axillary shoots; IAA alone caused some inhibition of growth of axillary; GA₃ alone caused some internode extension in the leafy axillary; IAA + GA₃ caused the axillary to grow as a horizontal, stolon-like shoot; when stolons were first stimulated to develop by application of IAA and GA₃, and then kinetin was applied to *the stolon tip*, the axillary shoot turned upwards and showed normal leaf expansion.(From A. Booth, *J. Linnean Soc.* **51**, 166, 1959, and D. Kumar and P. F. Wareing, *New Phytol.* **71**, 639, 1972.)

of stolon development involves a similar interaction between endogenous hormones.

From the various examples we have considered in this and the preceding chapter, it is evident that the correlated and integrated character of the plant is at least partially attributable to the presence of specific amounts of growth hormones at specific places and times. There remains, however, the problem of what it is which controls the precise production and distribution of growth hormones within the plant body.

5.9 HORMONAL CONTROL OF PLANT WATER BALANCE AND PHOTOSYNTHESIS

Various forms of stress, such as drought, flooding of the roots or salinity, have all been found to cause marked and rapid changes in endogenous hormone levels in plants. Over the past decade or so it has come to be realized that the known plant hormones are intimately involved in the regulation of cell water potential and membrane permeability, and hence control of tissue water contents and photosynthesis. The participation of growth hormones in control of internal water balance in plants was an unexpected discovery but an understanding of how they achieve this could well lead to important practical applications in agriculture and horticulture as water availability to crops is probably the major restriction on world food production.

Plants are said to experience water stress, or water deficit, when their cells and tissues are less than fully turgid. Water deficits in plants occur whenever transpiration exceeds the rate of water transport up from the roots, and this is frequently the case even if there is no visible wilting of the shoot. However, different parts of the shoot lose water to the atmosphere at different rates and therefore develop different levels of water deficit. Reasons for differential rates of water loss in the shoot clearly include factors such as surface/volume relationships of individual organs, cuticle thickness and the presence or absence of stomatal pores. Also, young growing leaves and fruits are able to compete more successfully for available water than are older leaves or fruits, and it is the mature leaves of any plant that are usually the first parts to show visible symptoms of water stress, often undergoing premature senescence in consequence. Reduction in cell turgor is now known to trigger changes in the overall hormone balance of the plant that are instrumental in

closure of stomata, reduction of root resistance to water movement, and the induction of leaf senescence.

5.9.1 Endogenous hormone levels in relation to water stress

Studies of endogenous hormones in relation to water stress have been concentrated particularly upon ABA, cytokinins and ethylene, as these substances have been found to be involved in the control of water balance of plants. The other known plant hormones, auxins and gibberellins, do not play any obvious roles in water stress phenomena.

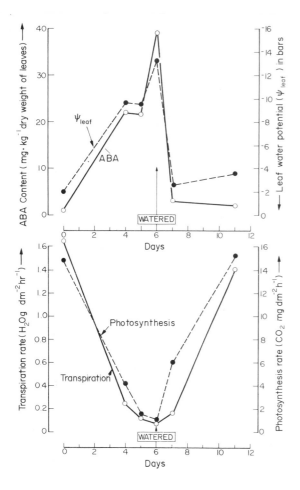

EFFECTS OF DROUGHT CONDITIONS

(a) Abscisic Acid

When plants are subjected to drying conditions and their internal water deficit rises, there is an associated marked and rapid rise in the ABA content of their shoots (Fig. 5.30). This increase in ABA results from increased *de novo* synthesis of the hormone from mevalonic acid in the stressed tissue rather than from release of free ABA from its glucose ester. There does not generally appear to be a critical threshold level of water stress beyond which ABA synthesis is initiated, but usually ABA begins to accumulate when the leaf water potential (ψ_{leaf}) decreases -2 bars below the normal daytime ψ_{leaf} value. As soon as water stress is relieved (by rewatering for example) the ABA level begins to fall, mainly by its metabolic conversion to phaseic and dihydrophaseic acids though some may be converted to its glucose ester (p. 72). The length of time it takes for the ABA level to revert to its original level is proportional to the extent of the increase which occurred during the preceding drying period.

Fig. 5.30. Changes in ABA levels in leaves of *Vitis vinifera* plants and correlated changes in leaf water potential (ψ_{leaf}), transpiration and photosynthetic rates. The plants were subjected to drought conditions for the first 6 days of the experiment but thereafter were watered again. The rather slow recovery of photosynthetic activity during the recovery period after 6 days may have been caused by phaseic acid derived from metabolism of ABA in re-watered plants. Phaseic acid has been shown to inhibit photosynthesis. (Adapted from B. R. Loveys and P. E. Kriedemann, *Physiologia Plantarum* **28**, 476-9, 1973.)

The rates of transpiration and photosynthetic activity of leaves are negatively correlated with their ABA levels (Fig. 5.30). This suggests that an effect of ABA is to induce closure of stomata. Direct measurements of leaf conductivity (i.e. the ability of the leaf surface to allow passage of gases and water vapour) and microscopic examination have confirmed that high levels of ABA causes stomata to close (Fig. 5.31). The mechanism of

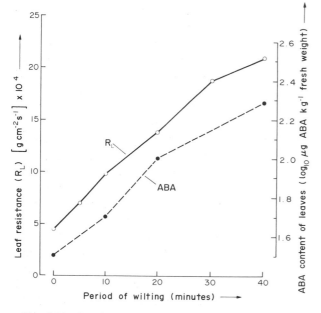

Fig. 5.31. Correlation between endogenous ABA levels and leaf resistance (R_L) to gas and water vapour movement in *Phaseolus vulgaris*. The plants were subjected a drying current of warm air for the periods of time indicated. (Adapted from R. W. P. Hiron and S. T. C. Wright, *J. Experimental Botany* **24**, 769-81, 1973.)

stomatal movements is still not fully understood, which makes it impossible at present to provide a fully satisfying explanation for the effects of ABA on stomata. However, some things have been established that shed light on the problem, particularly experimentally demonstrated effects of ABA on the permeability of guard cell membranes to potassium ions (K^+). In outline, stomatal opening takes place as a result of an influx of K^+ from surrounding cells. The extra K^+ increases the osmotic pressure of the guard cells which in turn results in water influx and the opening movement. Additionally, it is possible that the degradation of

guard cell starch which occurs at the same time that K^+ enters leads to accumulation of both sugars and organic acids. Both the sugars and organic acids would also increase guard cell osmotic pressure, and H^+ from the organic acids may well be the ion species that is exchanged for incoming K^+ to maintain electrical neutrality. Thus, ABA has been found to prevent influx of K^+ into the guard cells and the degradation of starch in the guard cell chloroplasts. It is of interest and possible significance to the control of stomatal movements that water stress-induced ABA synthesis takes place in chloroplasts and the hormone readily migrates from them to other parts of the cell and the plant. In fact, the inhibition or cessation of growth in apical parts of the plant when it is subjected to drought conditions may well be attributable to movement of ABA up to those regions from its sites of synthesis in mature leaves.

Most of the work that has been done to investigate the relationships between ABA and water stress has concentrated on the mature leaves of mesophytes, but essentially similar changes seem to occur in both xerophytes and hydrophytes. Roots normally contain much lower levels of ABA than do aerial parts of the plant, and although drought conditions do cause root ABA levels to rise they never attain the high levels seen in stems or leaves. Among the leaves of the shoot, it is the youngest ones that normally contain the highest concentrations of ABA in non-water stressed plants (Fig. 5.32). The higher levels of ABA present in youngest leaves may be the basis for the lower rates of transpiration of these compared with older leaves—i.e. stomata may be permanently closed, or partially closed, in young leaves because of their high ABA content; this suggestion has indeed received support from direct measurements of stomatal conductances of leaves of different ages in *Xanthium* plants. Droughting of plants sees increases in ABA levels of young growing leaves to an extent even greater than that which occurs in older leaves (Fig. 5.33). Although the young leaves certainly do synthesize more ABA when the plant is droughted, at least a proportion of the large increase in ABA of young

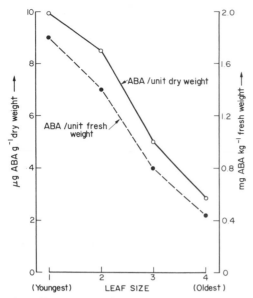

Fig. 5.32. ABA contents of *Xanthium strumarium* leaves of different ages expressed per unit fresh and dry weight. Leaf size 1 = leaves 3-4 cm long; 2 = 6.5-7.5 cm, 3 = 9-10.5 cm and 4 = 11-12 cm. The youngest leaves contain the highest concentration of ABA. (Adapted from K. Raschke and J. A. D. Zeevaart, *Plant Physiol.* **58**, 169-74, 1976.)

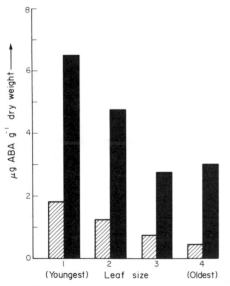

Fig. 5.33. Effects of 80 minutes transpiration (i.e. drying) on the ABA content of leaves of different ages of *Xanthium strumarium*. Leaf sizes are given in caption to Fig. 5.32. For each leaf size the cross-hatched histogram is the initial level of ABA and the solid histogram the level after 80 minutes transpiration. Note that the increase in ABA induced by drying was greater in younger than older leaves. (Adapted from K. Raschke and J. A. D. Zeevaart, *Plant Physiol.* **58**, 169-74, 1976.)

leaves may result from import of the hormone from older parts of the plant. The net effect, nevertheless, is to reduce water loss from the delicate young leaves to a minimum.

(b) Cytokinins

In contrast to ABA, cytokinin levels in plants rapidly decrease when the plant is subjected to drying conditions. There is much evidence that the root system is the main region of cytokinin synthesis in a vegetative plant and that cytokinins are exported up from the root system, *via* the xylem, into the shoot. Various experiments have revealed that dry soil conditions, or osmotic stress applied to the roots by a mannitol or polyethylene glycol solution, result in a rapid and marked fall in amounts of cytokinins being transported out of the roots to the shoot. Removal of the stress conditions induces a rise again in cytokinin export from the roots. Available evidence indicates that the decrease in cytokinin biosynthesis in roots happens when the water potential of the leaves falls, rather than being delayed until the roots themselves lose turgidity. This suggests that there is transmission of some signal from the leaves down into the root system which causes reduced cytokinin synthesis. The nature of this signal is not yet known, though it has been proposed that it is increased ABA coming from the stressed leaves.

The effects of a decreased supply to the shoot of cytokinins from the roots include leaf senescence and also stomatal closure. As is considered below, spraying plants with cytokinins promotes opening of stomata.

(c) Ethylene

Observations that increased leaf abscission can occur when plants are watered after a period of drought have led to investigations of the effects of drought on endogenous ethylene production by plants. Ethylene is known to be an extremely important hormone in the induction of abscission processes (see Chapter 12). In several species ethylene levels in the plant have been found to rise rapidly when plants are subjected to drought conditions. The mechanism by which increased ethylene synthesis occurs in response to water deficits is not known, but it is certainly highly likely that leaf senescence and the abscission of leaves and fruits that all occur in droughted plants represent responses to increased ethylene levels.

5.9.2 Effects of waterlogging

Flooding of the root system in most plants produces characteristic early responses: transitory wilting of leaves, reduction in internode extension growth rate, excessive thickening of the stem, pronounced epinastic (downward) growth curvatures of petioles, and rapid premature senescence and abscission of leaves. These phenomena reflect a drastic disturbance of the normal hormonal status of the plant.

It has been recognized for many years that waterlogging and drought induce a number of similar physiological changes in a plant and this is easily understood when one realizes that both conditions lead to impaired water uptake by plants. In the case of drought this is caused by a gradually decreasing soil moisture content, whereas under waterlogged conditions the anaerobic soil environment prevents normal root function and increases their resistance to water flow. Leaf water deficits caused by waterlogging are not, though, generally so severe as those during drought, and droughted plants do not exhibit the leaf epinasty and stem thickening phenomena associated with waterlogging. Epinasty and stem thickening in waterlogged plants result from elevated ethylene levels in the shoot, and this topic is discussed in more detail in Chapter 7 (p. 176). Leaf senescence in waterlogged plants results from both the increased ethylene level and also from a diminished supply of cytokinins from roots under anaerobic soil conditions. The effects of ethylene and cytokinins in leaf senescence are described in Chapter 12. Waterlogging of roots also reduces the supply of gibberellins from root to shoot, and this may well be an important factor in the reduction of stem extension growth under these conditions. Abscisic acid levels in shoots of waterlogged plants increase to a certain extent, but by no means as much as in droughted plants; this is probably related to the smaller leaf water deficits generated by waterlogging compared with those that occur in plants under drought.

5.9.3 Effects of external hormone applications in relation to water stress

Experiments have been conducted to examine whether the effects of water stress can simulated or alleviated by applications of growth hormone preparations. Results obtained are in line with what one would expect from studies of changes in endogenous hormones in stressed plants.

Thus, treatment of leaves with cytokinins stimulates stomatal opening and transpiration rate, so that the general effect of treatment with cytokinins is to reduce plant turgor.

Abscisic acid applications, in contrast, increase plant turgor through a stomatal closing action. This effect substantiates the inferences about the role of ABA in stomatal movements that have been derived from studies of endogenous ABA. Convincing evidence for the importance of ABA in control of plant turgor has been provided by experiments on the "wilty" tomato mutant (*Lycopersicum esculentum,* c.v. *flacca*). Wilty tomato plants contain only one tenth of the amount of ABA present in normal tomato varieties and their stomata are permanently open. However, when mutant wilty plants were sprayed daily with ABA they attained full turgor within a few days and appeared perfectly normal. Also, floating epidermal strips from leaves of various convenient species (particularly *Commelina communis*) on ABA solutions has unequivocally established that stomatal movements are very subject to control by ABA levels through inhibiting effects of the hormone on K^+ influx into guard cells and the degradation of starch in guard cell chloroplasts.

Stress phenomena induced by waterlogging can be at least partially alleviated by treatment with substances which antagonise ethylene action (e.g.

carbon dioxide), or by application of cytokinins (which reduce or prevent premature leaf senescence). These effects of exogenous substances are again what one would forecast from studies of endogenous hormone changes induced by flooding of the root system.

5.10 PLANT GROWTH SUBSTANCES IN AGRICULTURE AND HORTICULTURE

Agricultural and horticultural output per unit land area under cultivation has increased very greatly over the past half-century. This increase has been achieved through the introduction of new improved plant varieties, use of fertilizers, pesticides, fungicides and mechanization, but also, and very significantly, greatly increased use of various chemicals that affect plant development in ways other than serving as nutrients.

In general it can be said that increased agricultural or horticultural production per acre can be achieved through the adoption of two broad strategies:
(1) Changing the environment to suit the crop plants.
(2) Changing the crop plant to suit the environment in which it is grown.

Changing the environment principally involves provision of more nearly optimal amounts of water (irrigation and drainage) and of inorganic nutrients (fertilizers), and the horticultural use of glasshouses (increased temperatures and sometimes CO_2-enrichment of the air), but also, and very importantly, the use of pesticides and herbicides to reduce competition in the environment.

Changing crop plants to suit the environment has of course been historically achieved through plant breeding, but there is now a second approach based upon discoveries of various chemicals (synthetic plant Growth Regulators) that can be applied to alter the rate of plant growth or pattern of development in desired directions.

The range of chemicals in use as herbicides or growth regulators is very large and diverse. Many,

but not all, are related either structurally or functionally to the natural plant growth hormones of plants.

The fact that the mechanism of plant growth-hormone action is not understood has not prevented some of these substances becoming useful tools in agricultural and horticultural practices. The earliest commercial use of a plant hormone was the exposure of fruits to ethylene gas to accelerate ripening, a practice which is still common today. Also, we have already mentioned the usefulness of auxins and gibberellins in inducing parthenocarpic fruit development (p. 132). Another practical use for auxins is the induction of flowering, and consequently fruiting, by spraying pineapple crops with compounds such as 2,4-dichlorophenoxyacetic acid (2,4-D). The prevention of "pre-harvest drop" in apples, by synthetic auxin sprays, has also been found to have commercial value. Also, the rooting of stem cuttings is enhanced by treatment with various synthetic auxins (Fig. 5.34).

As we saw earlier (see Fig. 5.8A), a characteristic feature of auxins is that above a certain concentration, their effect is to inhibit rather than to stimulate growth. It appears that too high an auxin concentration disorganizes the delicate machinery of growth. Plants treated with an excess of auxin become distorted, with epinastically curled leaves and split stems. They may subsequently die, but not all auxins are equally toxic, and plant species also show varying sensitivities to auxins.

The reason for toxic effects of supra-optimal auxin concentrations is unknown, but an important aspect is likely to be the stimulatory effect of high auxin concentrations upon ethylene biosynthesis. Lack of complete understanding of the mechanism of their killing effects has not, however, prevented their use in the control of weeds among crop plants. The removal of weeds by mechanical cultivation is a costly and laborious business, and the discovery that spraying a synthetic auxin, such as 2,4-D, over a field can achieve the same result has proved of enormous economic value. Of course, many chemicals other

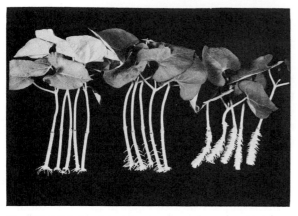

Fig. 5.34. Stimulation of root initiation in bean cuttings. *Left:* control cuttings (no auxin treatment); *middle:* cuttings treated with a solution of 5 mg/1 naphthalene acetic acid (NAA); *right:* treated with 50 mg/l NAA. (Original print kindly supplied by Dr. L. C. Luckwill, Long Ashton Research Station.)

than auxins, if applied in sufficiently high concentration, are poisonous to plants and can be used as weed-killers. The special merits of synthetic auxins such as 2,4-D are that they will kill plants when applied at relatively low concentrations, they are comparatively harmless to animals, they are non-corrosive, they are translocated within the plant to kill those parts, such as the roots, not reached by sprays and, most important, some auxins are *selective* and can, therefore, be used to kill weeds without damaging surrounding crop plants. As a general rule, cereals are fairly resistant to applied auxins, whereas a number of dicotyledonous weeds are very sensitive and are killed. Thus, the weed known as yellow charlock (*Sinapis arvensis*) can be controlled in cereal crops such as oat (*Avena*) or wheat (*Triticum*), and lawn-weeds such as daisies (*Bellis perennis*) and plaintains (*Plantago* spp.) can be killed without harming the grass.

Closely related to 2,4-D is the herbicide 2,4,5-T (2,4,5-trichlorophenoxyacetic acid) (see p. 77), which is especially valuable in the selective killing of woody perennial weeds, for which it has been widely used for many years. However, its use as a defoliant ("Agent Orange") in the Vietnam war and associated reports of harmful effects on human and animal life have led to current, and often ill-informed, controversy. A common popular misconception seems to be that 2,4,5-T is

a very toxic substance to animals, whereas there is overwhelming scientific evidence that the pure compound is of low toxicity compared with many chemicals in daily use in agriculture or other human activities. Measured LD_{50}s (LD_{50} = dose that kills 50 per cent of a population of test animals) for 2,4,5-T are of the order of several hundreds of milligrams per kg body weight. In areas of the world which have been exposed to high levels of 2,4,5-T there have not been any proven adverse effects on pregnant women or their children.

However, the synthesis of 2,4,5-T must be carefully controlled to avoid its contamination with a by-product, 2,3,7,8-tetrachlorodibenzo-*para*-dioxin (TCDD):

$$Cl \quad O \quad Cl$$

"TCDD"

(2,3,7,8-tetrachlorodibenzo-*para*-dioxin)

TCDD is an extremely toxic and heat-stable substance (LD_{50}s of 0.0006 mg/kg to 0.115 mg/kg for laboratory animals), and there is little doubt that concern over 2,4,5-T has arisen from the use of TCDD-contaminated samples in some areas. Normal application rates of 2,4,5-T approximate 1 kg/acre, and commercially available formulations must contain less than 0.1 parts per million of TCDD. Thus, normal application would disperse a *maximum* of 1 mg TCDD per acre. A 100 lb (45 kg) human being would have to absorb all the TCDD sprayed over at least 2 acres to suffer ill effects, based upon tests of TCDD on laboratory animals.

The selective action of auxins as weed-killers depends on a number of factors. Very often, susceptible plants are dicotyledonous and possess broad horizontally spreading leaves, which retain auxin solution sprayed on them, while resistant plants are often monocotyledonous with narrow, erect leaves, off which the spray droplets run. In addition, the epidermis of some plants is more easily penetrated by auxin solutions than is that of others. Another basis for selective action of a soil-applied herbicide is its water solubility. For example, a herbicide of low polarity may be held adsorbed in the surface layers of the soil, from where it can be taken up by shallow-rooted weeds which consequently die, whereas deeper-rooted crop plants are unaffected. Alternatively, with a shallow-rooted crop a more polar herbicide may be used which is readily leached downwards to be taken up by deep rooted weeds. Often more important than these factors, however, is the existence of inherent differences between the living cells of different plant species in sensitivity to synthetic auxins.

Among the most widely used synthetic auxins in weed control at the present time are 2,4-D, 2,4,5-T and MCPA (Fig. 4.1) or proprietary mixtures of these. IAA, the naturally occurring auxin of most plant species, is relatively ineffective as a weed-killer, probably due to its rapid destruction by the enzymes of the treated plants (p. 56).

Not all herbicides are selective in their killing effects, and numerous non-auxin types of herbicide are used to kill off all vegetation before seeding, or between rows of crop plants. Examples of such "total-kill" herbicides are the well-known bipyridiliums, paraquat and diquat ("Grammoxone" or "Weedol"), and the relatively recently introduced but already economically extremely valuable glyphosate [N-(phosphonomethyl) glycine] known as "Roundup" or "Tumbleweed". The mechanisms of action of non-auxin type weedkillers are not well understood, though many interfere with photosynthesis but can have numerous other deleterious effects that lead to death of plants.

Gibberellins might have at least as great a practical potential as auxins. The effectiveness of gibberellins in the induction of parthenocarpic fruits is just beginning to be realized at the commercial level, as they are often useful with species in which fruit set is not promoted by auxin. Also, gibberellins act synergistically with auxins in inducing fruit set in species such as tomato. Uses for gibberellins in the cultivation of grapes (*Vitis* spp.) include stimulation of elongation of clusters (to decrease rotting of individual berries), enlarge-

ment of berries, and increasing berry set. Strawberry plants can, by treatment with gibberellic acid, be induced to produce earlier flowers, and therefore fruits, than untreated plants. This effect can lead to commercial gain, for fresh strawberries sell for higher prices during the early part of the harvest period. Gibberellic acid has been found to delay ripening of citrus fruits on the tree, and also to improve skin colour in the fruits of certain varieties.

The effectiveness of gibberellins in speeding up germination rates of seeds of a large number of species, as well as in breaking dormancy (p. 271), has obvious practical possibilities. Gibberellins are also beginning to prove valuable in the breeding of new forest trees. Thus, economically important conifers which naturally take 10—20 years before they begin to produce "flowers" (and up to a further 20 years to set significant numbers of seed) have been induced to flower in 4—6 years by treating seedlings with gibberellins. Flowering has been induced in small seedlings of certain conifers (particularly members of the Cupressaceae and Taxidiaceae) by gibberellin treatments (Fig. 5.35) though usually far more male than female flowers are formed on such young plants.

Already economically important is the use of gibberellic acid to increase the length, and total yield, of petiole stalks in the horticultural production of rhubarb and celery (Fig. 5.36).

Probably the most important single commercial application of gibberellins yet found is in the malting industry. Malting is a process whereby barley seeds are allowed to germinate for several days. The germinated seeds are then used in the preparation of a medium for fermentation yeasts in the manufacture of beer. The purpose of the malting procedure is to allow time for stored food reserves in the endosperm of each seed to be converted to other substances more suitable as substrates for yeast growth. Thus, for example, starch reserves are converted to sugars by the action of hydrolytic enzymes such as α-amylase which are formed in the aleurone cells during germination (p. 96). Due to the stimulatory effect of gibberellins on α-amylase synthesis (p. 96), treatment of barley seed with gibberellic acid both

Fig. 5.35. Acceleration by gibberellin of reproductive maturity in trees. A: 4-month old *Thuja placata* (Western Red Cedar) in which spraying with gibberellic acid (GA₃) induced the initiation of the male cone at the top of the plant. B: 16-month old *Cupressus arizonica* (Arizona Cypress) seedling sprayed for four months with GA₃ and which bears approximately 8,000 cones, 50 of which are female. (From R. P. Pharis and J. N. Owens, *Yale Sci. Mag.* **41**, 10, 1966. Photographs by courtesy of Professor Richard Pharis.)

speeds up and provides more strict control of malting. This has the benefit of conferring considerable saving of time and production of higher yields of the malting product.

A number of plant growth-regulating chemicals in widespread use are grouped together under the general heading of *Growth Retardants*. Thus, commercial preparations known as "B-9" (= "Alar"), "AMO-1618", "Cycocel" (= "CCC" or "Chlormequat"), "Phosfon-D", and "Ancymidol" (= "EL-531") are all growth retardants. Their uses include the inhibition of shoot elongation in ornamental plants (e.g. chrysanthemums, poinsettias, *Coleus*, petunias, etc.) resulting in more compact and desirable plants. Similarly, they are used to shorten and strengthen the stems of some crops, particularly wheat, and thereby reduce "lodging" (the bending over of shoots by heavy rain and wind which can cause severe harvesting difficulties and premature grain germination). There are many other applications of these growth retardants, such as to increase

flowering, fruit-set, rooting of cuttings, tuberization, bulbing, resistance to drought, cold and salt. The physiological and biochemical mechanisms by which these useful effects are achieved are generally ill-understood, but at least some of them result from the capacity of growth retardants to inhibit gibberellin biosynthesis in plants (see p. 61). There are a few other non-phytotoxic growth inhibitory chemicals used in agriculture and horticulture, such as maleic hydrazide and the "morphactins" (derivatives of fluorene-9-carboxylic acid, such as "IT 3465"), but these do not appear to act by affecting gibberellin biosynthesis:

Maleic Hydrazide

"IT 3465"—a "morphactin"

(methyl-2-chloro-9-hydroxyfluorene-9-carboxylate)

Fig. 5.36. Effect of gibberellic acid (GA_3) sprays on celery yield. Left, GA_3 at 100 mg.dm^{-3}. Right, untreated control. (Photograph by courtesy of Dr. T. H. Thomas, National Vegetable Research Station, Wellesbourne, U.K., and Dr. A. Whitlock, Arthur Rickwood Experimental Husbandry Farm, Ely, U.K.)

It is highly likely that some synthetic cytokinins will be widely used in future to delay the process of senescence in flowers and vegetables, but their potential in this area has not yet been fully explored.

As already stated, ethylene has been used for some years for accelerating ripening processes in a number of fruits (an aspect of senescence), particularly citrus fruits such as oranges and lemons. Moreover, demonstrations that this gas is a most potent stimulator of premature leaf fall have raised the possibility of its use as a defoliant in crops such as cotton, pea, beans, etc., where mechanical harvesting of fruits is expedited if leaves are not present. However, because of its gaseous nature, ethylene does not readily lend itself to applications to outdoor crops. Because of this, a search has been made for other chemicals which promote leaf abscission. A number of such substances are now known, and the most effective of these are those which enhance ethylene synthesis in the plants to which they are applied. A recent development is the search for chemicals which can be sprayed on to plants, but which are themselves broken down in plant tissue to yield ethylene. An example of such a substance is 2-chloroethanephosphonic acid, known commercially as "Ethrel" or "Ethephon":

$$Cl\,CH_2\,CH_2\,\overset{\displaystyle O}{\overset{\displaystyle \|}{P}}\!\!\begin{array}{l}OH\\[-2pt]OH\end{array}$$

2-chloroethanephosphonic acid

This compound undergoes chemical decomposition at pHs higher than 4·1. Because plant cells normally have a pH above 4.1, an aqueous solution of Ethrel will on entering the tissues be degraded, liberating ethylene:

$$ClCH_2CH_2-\overset{\displaystyle O}{\overset{\displaystyle \|}{\underset{\displaystyle \underset{\displaystyle O^-}{|}}{P}}}-OH \;+\; OH^- \xrightarrow{\text{pH} > 4.1} CH_2=CH_2 \;+\; \underset{\text{(ethylene)}}{}\; \overset{\displaystyle O}{\overset{\displaystyle \|}{\underset{\displaystyle \underset{\displaystyle O^-}{|}}{P}}}-(OH)_2 \;+\; Cl^-$$

(2-chloroethanephosphonic acid in ionized form)

Ethrel therefore induces physiological responses in plants identical to those elicited by ethylene, and is gradually being introduced to agriculture and horticulture. At the present time it is already in widespread use to increase the rate of latex flow from rubber trees, and to hasten fruit ripening and leaf abscission. It is also coming into use to control branching and suckering of both crop and ornamental plants.

There is little doubt that practical uses of synthetic plant growth hormones and related compounds will in the future be more extensive and varied than at present. Full realization of their potential will, however, not be achieved until our understanding of their functions and modes of action in plants is greatly increased.

FURTHER READING

General

Ashton, F. M. and A. S. Crafts. *Mode of Action of Herbicides,* Wiley-Interscience, New York, 1973.

Abeles, F. B. *Ethylene in Plant Biology,* Academic Press, New York and London, 1973.

Albersheim, P. The walls of growing plant cells, *Scientific American,* **232**(4), 80-95, 1975.

Audus, L. J. *Plant Growth Substances,* 3rd ed., L. Hill Ltd., London, 1972.

Hill, T. A. *Endogenous Plant Growth Substances,* Edward Arnold, London, 1973.

Morey, P. R. *How Trees Grow,* Edward Arnold, London, 1973.

Nitsch, J. P. Perennation through seeds and other structures: Fruit development. In: F. C. Steward (ed.), *Plant Physiology—a Treatise,* Academic Press, New York, pp. 413-501, 1971.

Pilet, P. E. (Ed.). *Proc. 9th Int. Conf. Plant Growth Regulating Substances, Plant Growth Regulation,* Springer-Verlag, Berlin-Heidelberg, 1977.

Pimental, D. and M. Pimental. Food, Energy and Society, Edw. Arnold, London, 1979.

Weaver, R. J. *Plant Growth Substances in Agriculture.* W. H. Freeman, San Francisco, 1972.

Wilkins, M. B. (Ed.). *Physiology of Plant Growth and Development,* McGraw-Hill, London, 1969.

More Advanced Reading

Articles by: P. B. Dickenson, J. W. Dicks, J. P. Hudson, J. K. Kapoor and J. N. Turner, L. C. Luckwill, L. G. Nickell, M. R. Parham, R. P. Pharis and S. D. Ross, T. H. Thomas and P. F. Wareing. In: *Outlook on Agriculture* 9, No. 2 (Growth Regulation Issue), 1976.

Audus, L. J. Correlations. *J. Linn. Soc. (Bot.)* **56**, 177, 1959.

Brian, P. W. The gibberellins as hormones, *Int. Rev. Cytology,* **19**, 229, 1966.

British Plant Growth Regulator Group Monographs. No. 1, Opportunities for Chemical Plant Growth Regulation, 1978. No. 2, The Effect of Interactions between Growth Regulators on Plant Growth and Yield, 1978. No. 3, Differentiation and the Control of Development in Plants—Potential for Chemical Modification, 1979. No. 4, Recent Developments in the Use of Plant Growth Retardants, 1980. Published by British Plant Growth Regulator Group, A.R.C. Letcombe, Wantage, U.K.

Catesson, A. M. Cambial cells, Chapter 10. In: A. W. Robards (ed.), *Dynamic Aspects of Plant Ultrastructure,* McGraw-Hill, London, 1974.

Champagnat, P. Physiologie de la croissance et l'inhibition des bourgeons: Dominance apicale et phenomenes analogues. In: *Encyc. Plant Physiol.* **15**, (1).

Coombe, B. G. The development of fleshy fruits. *Ann. Rev. Plant Physiol.* **27**, 207-28, 1976.

Crane, J. C. Growth substances in fruit setting and development, *Ann. Rev. Plant Physiol.* **15**, 303, 1964.

Goldsmith, M. H. M. The polar transport of auxin, *Ann. Rev. Plant Physiol.* **28**, 439-78, 1977.

Goodwin, P. B. Phytohormones and growth and development of organs of the vegetative plant. In: *Phytohormones and Related Compounds—A Comprehensive Treatise,* Vol. II, (eds. D. S. Letham, P. B. Goodwin and T. J. V. Higgins), Elsevier-North Holland Biomedical Press, Amsterdam, pp. 31-173, 1978.

Goodwin, P. B. Phytohormones and fruit growth. In: *Phytohormones and Related Compounds—A Comprehensive Treatise,* Vol. II, (eds. D. S. Letham, P. Goodwin and T. J. V. Higgins), Elsevier-North Holland Biomedical Press, Amsterdam, pp. 195-214, 1978.

Goodwin, P. B., B. I. Gollnow, and Letham, D. S. Phytohormones and growth correlations. In: *Phytohormones and Related Compounds—A Comprehensive Treatise,* Vol. II, (eds. D. S. Letham, P. B. Goodwin and T. J. V. Higgins), Elsevier-North Holland Biomedical Press, Amsterdam, pp. 215-249, 1978.

Hall, M. A. The cell wall, Chapter 2. In: M. A. Hall (ed.), *Plant Structure, Function and Adaptation,* MacMillan Press, London, 1976.

Jones, R. L. Gibberellins: Their physiological role. *Ann. Rev. Plant Physiol.* **24**, 571-98, 1973.

Luckwill, L. C. Hormonal aspects of fruit development in higher plants. *Symp. Soc. Exp. Biol.* **11**, 63, 1967.

Moreland, D. E. Mechanisms of action of herbicides, *Annual Review of Plant Physiology* **31**, 597-638, 1980.

Nitsch, J. P. Plant hormones in the development of fruits, *Quart. Rev. Biol.* **27**, 33, 1952.

Pallos, F. M. and J. E. Casida (Eds.). Chemistry and Action of Herbicide Antidotes, Academic Press, New York, 1978.

Phillips, I. D. J. The cambium, Chapter 10. In: M. M. Yeoman (ed.), *Cell Division in Higher Plants,* Academic Press, London and New York, pp. 347-90, 1976.

Phillips, I. D. J. Apical dominance. In: *The Physiology of Plant Growth and Development* (M. B. Wilkins, ed.), McGraw-Hill, London, pp. 163-202, 1969.

Phillips, I. D. J. Apical dominance, *Ann. Rev. Plant Physiol.* **26,** 341-67, 1975.

Preston, R. D. Plant cell walls, Chapter 7. In: A. W. Robards (ed.), *Dynamic Aspects of Plant Ultrastructure,* McGraw-Hill, London, 1974.

Sachs, R. Stem elongation, *Ann. Rev. Plant Physiol.* **16,** 73, 1965.

Scott, T. K. Auxins and roots, *Ann. Rev. Plant Physiol.* **23,** 235-58, 1972.

Street, H. E. The physiology of root growth. *Ann. Rev. Plant Physiol.* **17,** 315, 1966.

Torrey, J. G. Root hormones and plant growth, *Ann. Rev. Plant Physiol.* **27,** 435-59, 1966.

Wareing, P. F. Growth substances and integration in the whole plant. *Symp. Soc. Exp. Biol.* **31,** 337-365, 1977.

Wareing, P. F., C. E. A. Hanney and J. Digby. The role of endogenous hormones in cambial activity and xylem differentiation. In: *The Formation of Wood in Forest Trees* (M. M. Zimmerman, ed.), Academic Press, New York, 1964.

Wightman, F. and G. Setterfield (Eds.). *Biochemistry and Physiology of Plant Growth Substances,* Runge Press, Ottawa, 1969.

Wittwer, C. H. Growth regulants in agriculture, *Outlook on Agriculture* **6,** 206-217, 1971.

Wright, S. T. C. Phytohormones and stress phenomena. In: *Phytohormones and Related Compounds—A Comprehensive Treatise,* Vol. II, (eds. D. S. Letham, P. B. Goodwin and T. J. V. Higgins), Elsevier-North Holland Biomedical Press, Amsterdam, pp. 495-536, 1978.

6

Aseptic Culture
Methods in Studies
of Differentiation

6.1 HISTORICAL ASPECTS OF PLANT TISSUE CULTURE

In the preceding and later chapters of this book various aspects of morphogenesis are considered and a number of experimental approaches to the subject are described. No one aspect of botanical research is able to provide a unified picture of dif-ferentiation, and studies of this subject must aim at the unification of information derived from the fields of morphology, physiology, biochemistry and biophysics. This is a difficult task, for in the intact plant there are complex interactions between the various processes underlying growth and differentiation. Consequently, it is desirable to reduce the complexity of the system so that the controlling processes may be more easily identi-fied and studied. This can be done to varying degrees, by isolating embryos and other parts of a plant in order to eliminate certain complicating correlative influences during studies of their behaviour. In general, isolated embryos and plant parts must be carefully nurtured to keep them alive and free from infection by micro-organisms. This normally entails the use of *aseptic culture techniques*. Such methods often facilitate the maintenance of isolated embryos, organs and tissues for considerable periods of time. However, the culture of isolated plant tissues, organs and embryos presented many difficulties to earlier workers which were only gradually surmounted.

The smallest viable unit of a plant one can at present envisage as reproducing, growing and developing in culture is a single cell. As long ago as 1902, Haberlandt attempted to grow single plant cells in aseptic culture, but, for various reasons which we now understand, his attempts were unsuccessful.

Following Haberlandt, other workers established methods which would allow the growth of isolated plant organs and tissues in culture. Excised roots were the first plant organs to be successfully brought into aseptic culture, and work by White in the 1930s demonstrated that given appropriate nutrients, such roots would grow and differentiate normal root tissues. Work by Gautheret (1939) and others established that isolated portions of storage tissue, e.g., from carrot roots, could be kept alive and grown in aseptic culture. *Callus cultures* derived from such isolated tissues lend themselves to studies of the effects of nutrients, vitamins and hormones upon cell division, differentiation of vascular tissues, and the inception of organized meristematic regions within the predominantly parenchymatous tissue. By definition, a callus is a mass of proliferating tissue consisting predominantly of parenchymatous cells, but in which differentiation may occur under suitable conditions.

When a callus is grown in agitated liquid culture, cells at the surface are often broken away and float free in the medium to give a *liquid suspension culture.* Usually, such free cells do not divide when retained in a medium suitable for callus growth, for propagation of free plant cells demands a more elaborate medium that does a callus. Media have, however, been devised which are capable of supporting proliferation of free plant cells.

6.2 ORGAN CULTURE

6.2.1 Root culture

It has proved possible to grow several types of plant organ in aseptic culture, including roots, shoot apices, leaves, flower parts and fruits. The nutrient requirements for such organ culture vary considerably from species to species and according to the type of organ in question, but certain general requirements can be recognized. Intact higher plants are autotrophic; that is, they are able to synthesize all the organic substances required for their own life from carbon dioxide, oxygen and mineral nutrients. However, since most aseptic cultures are unable to carry on photosynthesis, it is clear that they will require at least a carbon source, usually supplied in the form of a sugar such as sucrose or glucose. In addition, aseptic cultures require the same mineral nutrients as the intact plant, including both macronutrients (nitrogen, phosphorus, potassium and calcium) and micronutrients (Mg, Fe, Mn, Zn, etc.).

In addition to those requirements for a carbon source and mineral nutrients, it is found that most isolated organs have also a requirement for certain special organic substances. Thus, most isolated roots grown in aseptic culture require to be supplied with certain vitamins, particularly thiamin (vitamin B_1) and sometimes pyridoxin (vitamin B_6), nicotinic acid and others. It appears that in the intact plant, certain vitamins are synthesized in the leaves and that roots are dependent upon the shoots for the supply of these substances, which they are unable to make for themselves. Tomato roots require only sucrose, mineral nutrients and thiamin, and given these they will grow successfully in culture for many years.

Excised roots of some monocotyledonous plants fail to grow even when supplied with a full complement of B-vitamins and other vitamins. In some of these cases (e.g., rye), an exogenous auxin supplement to the nutrient medium allows growth to proceed. In order to maintain the culture, the excised roots must be regularly subcultured on to fresh medium, by excising a piece of root bearing a lateral, which then proceeds to grow rapidly and maintain the culture.

Excised roots of most species produce only root tissues in culture, but there are some species the cultured roots of which regenerate shoot buds as well as further roots, e.g., *Convolvulus,* dandelion (*Taraxacum officinalis*) and dock (*Rumex crispus*).

6.2.2 Culture of shoot apices and leaves

Isolated shoot apical meristems and leaf primordia can also be grown in aseptic culture. These frequently produce adventitious roots and can eventually develop into complete plants. The shoot apices and leaves of vascular cryptograms, such as ferns, are relatively more autotrophic than those of angiosperms. Thus, even a small fern apex can be grown on a medium containing only a carbohydrate source and mineral nutrients. Small angiosperm apices (less than 0·5 mm in diameter) require a general source of organic nitrogen and certain specific amino acids and vitamins, in addition to the basic medium, but larger apices will grow on a simple medium. The simpler requirements of large apices may be due to the fact that they carry larger leaf primordia, which apparently can supply some of the requirements of the apex for vitamins and other organic nutrients.

An economically valuable application of the aseptic culture of shoot tips is in the production of virus-free stocks of some crop plants. Many normally vegetatively—propagated plants, such as potatoes, raspberries, carnations, geraniums, rhubarb, apple and other fruit trees, gradually accumulate viruses that reduce their vigour. However, the shoot tips often remain free of infection, can be excised, grown in aseptic culture and propagated into virus-free plants.

Isolated young leaves of the fern, *Osmunda cinnamomea,* and of sunflower (*Helianthus annuus*) and tobacco (*Nicotiana tobacum*), have been successfully grown on a simple medium containing only sucrose and inorganic salts (Fig. 2.12). Such isolated leaf primordia continue to grow and develop into normally differentiated leaves, although they are usually very much smaller than normal leaves developed on the plant (p. 31).

6.3 EMBRYO CULTURE

We have earlier (p. 22) given an account of the development of embryos as it occurs naturally within the embryo sac. Experimental approaches to the study of plant embryogenesis have in recent years made extensive use of aseptic culture of isolated young embryos.

The easiest method of culturing an embryo is to allow it to develop *in situ* within the ovary or dissected out ovule. The ovule, if placed on a suitable nutrient medium, is able to support the development of the zygote to maturity. The presence of placental tissue aids the development of the ovule and embryo, and consequently it is easier to maintain ovules in aseptic culture when they are left within the ovary. The physiological requirements of ovules do not appear to be species specific, since young fertilized ovules of widely different species have grown to mature seeds following transplantation on the placenta of *Capsicum* fruits.

Although embryos can be grown from the zygote stage to maturity when they are left inside a cultured ovule, they show complex nutrient requirements when isolated from the ovule. The mature embryo is, of course, autotrophic and will grow if provided with the normal conditions necessary for germination, viz. adequate water and oxygen and a favourable temperature. However, if embryos are excised at younger stages it is found that they will not grow, even if supplied with sucrose and mineral salts.

A useful technique for the culture of young embryos was introduced in 1941 by van Overbeek, who found that they could be grown from a quite immature stage by supplementing sucrose and mineral salts with coconut milk, which is a liquid endosperm. This endosperm allows the development of the coconut palm embryo, as it contains a complex range of substrates necessary for the growth of the embryo, and it is also very effective in supporting the growth of embryos of other species. Attempts have been made to identify the active components of coconut milk and they are now known to include sugar-alcohols such as myoinositol, leucoanthocyanins, cytokinins, and probably also auxins and gibberellins. By using this technique it has been possible to develop to maturity embryos of certain species isolated at an early stage of development.

As embryos develop they appear to become less heterotrophic, as has been demonstrated by observations on isolated embryos of *Capsella* and *Datura,* which showed that globular embryos did not survive at all in culture, but that early heart-shaped embryos (see Fig. 2.2) would develop further if supplied with a nutrient medium containing sugar, inorganic salts, vitamins and coconut milk.

With slightly older heart-shaped embryos the coconut milk may be replaced by a source of reduced nitrogen such as L-glutamine, and with more highly developed embryos even the glutamine can be omitted. Thus, there appears to be a progressive increase in the synthetic abilities of the embryo during its development.

On the other hand, there is more recent evidence that the concept of a decrease in heterotrophic nutrition with increase in embryo age may not be strictly true. Even young globular embryos of *Capsella* have been cultured to maturity in a relatively simple medium containing no organic nitrogen source such as L-glutamine, nor high sucrose concentration. But, in addition to the usual inorganic salts, vitamins and 2 per cent sucrose, the medium did have to contain a balanced mixture of very low concentrations (about 10^{-7} M) of an auxin (e.g., IAA), a cytokinin (e.g., kinetin) and adenine sulphate. It appears that, in the case of *Capsella* at least, it is the balance of growth hormones in the immediate environment rather than substrate level metabolites, which plays a determining role in embryogenesis.

6.4 TISSUE CULTURE

In contrast to the techniques of organ culture, tissue culture involves the aseptic culture of an isolated homogeneous mass of cells. All plant organs consist at the time of their excision of a number of different tissue types, and therefore represent more complex systems than do isolated individual tissues. In studies of the physiology and biochemistry of morphogenesis it is desirable to work with as simple a system as possible, and clearly culture of isolated tissues would appear to represent a simplification of experimental material in comparison with organs in culture. Fortunately, tissues from many sources can be maintained in culture for an indefinite period, and afford enormous but as yet largely untapped, possibilities for research in physiology and biochemistry. Such cultures have already proved very valuable for cer-

tain biochemical studies. For example, cultures of sycamore (*Acer pseudoplatanus*) cambial tissue have been extensively used for studies on cell-wall metabolism, and we shall see below several examples of the value of tissue culture for studies on differentiation.

When small pieces of root phloem parenchyma of wild carrot (*Daucus carota*), or pith parenchyma of tobacco (*Nicotiana tabacum*) stem, or even chlorophyll-containing palisade cells from leaves of *Arachis hypogea* and *Crepis capillaris* are place on a suitable medium, they can not only be kept alive but can be induced to grow. That is, mature parenchymatous or mesophyll cells, which, if left undisturbed in the plant body, would undergo no further cell division, can be made to divide mitotically, giving rise to an undifferentiated *callus*. An extreme example of retention of the capacity for cell division in mature plant cells was provided by cultivation of a callus from medullary ray tissue excised from a region adjacent to the pith in 50-year-old lime (*Tilia*) trees. These cells had matured a full half-century earlier, and yet their continued potential for active cell division was revealed under suitable conditions in culture.

It seems likely, therefore, that any living, nucleated, plant tissue can give rise to a proliferating undifferentiated callus when excised and placed on a suitable culture medium. However, great difficulty is often experienced in establishing a callus from a previously untried source of tissue, for there is apparently considerable diversity in nutritional requirements of tissues from different species, or even from different locations within one plant. In general, it has proved easier to culture tissues consisting originally of non-green parenchyma, such as phloem or pith parenchyma. The establishing of green, photosynthesizing, callus growths from chloroplast-containing leaf cells has come much later.

In addition to the usual macro- and micro-inorganic nutrients, and an organic carbon source, isolated tissues are frequently found to require (1) an organic source of reduced nitrogen, which may be supplied as amino acids or, in some species, as the amide of glutamic acid, L-glutamine; (2)

vitamins, including thiamin, nicotinic acid and pyridoxine, and (3) the sugar alcohol, myo-inositol. In addition, an auxin, such as 2,4-D, and sometimes a cytokinin, are required. The fact that it is necessary to supply hormones to callus cultures, whereas they are not normally required by organ cultures, may indicate that the organized meristems of organ cultures may be centres of hormone biosynthesis, whereas the parenchymatous tissues from which callus cultures are derived do not have the capacity for hormone synthesis.

It is of interest to note that photosynthetic callus growth from palisade mesophyll cells of *Arachis hypogea* does not need an external supply of any vitamins. This can be related to what was said earlier concerning the normal production of vitamins in the leaves of plants, and their supply to other regions such as the roots.

Repeated sub-culturing of some tissue cultures, such as those of carrot, grape, *Scorzonera*, tobacco, and other plants, leads to a spontaneous and irreversible change in that they acquire the capacity to synthesize excess quantities of auxin. Thus, a tissue which when first brought into culture requires an exogenous supply of auxin and cytokinin, later on, after subculture, becomes autotrophic for these hormones. Such long-established callus cultures are then said to be *habituated,* or *anergized,* and closely resemble tumorous as opposed to normal plant tissues. For example, plants infected with the crown-gall bacterium (*Agrobacterium tumefaciens,* synonym *Phytomonas tumefaciens*) exhibit tumourous (callus-like) growths at the points of infection. By appropriate heat treatment, the bacteria can be killed and removal of a portion of one of these treated tumours into aseptic culture leads to the production of a massive undifferentiated callus, which is completely self-sufficient in auxin and cytokinin. Thus, infected or habituated cells have undergone a permanent change in that they are able to synthesize substances which they were unable to produce before. This capacity is transferred from one cell generation to the next and is perhaps brought about by transfer of bacterial DNA to the plant cell (see p. 322). *Agrobacterium* plasmid is under investigation as a potential tool for the introduction of "useful genes" into crop plants.

6.5 SUSPENSION CULTURE OF PLANT CELLS

A suspension culture consists of cells and small aggregates of cells dispersed and growing in a moving liquid medium. The principal problem encountered initially in attempts to make isolated single cells and small aggregates of cells divide in culture is that such isolated cells are "leaky". For a number of reasons, particularly their large surface areas exposed to the liquid medium, in contrast to callus cells surrounded by like cells, isolated plant cells in an agitated liquid medium tend to lose substances required for cell division to the medium. Consequently, the nutritional requirements of free plant cells and small cell aggregates may be more complex than those for callus cultures of the same species, since it is necessary to supply the substances which tend to be lost by the cells into the medium. For some types of cell culture it has been possible to determine the precise nutrient requirements so that they can be grown on a defined medium, which usually included the various constituents already listed for callus cultures (p. 154). In other cases, it has not yet been possible to grow cell cultures in a defined medium and it is necessary to add coconut milk, which must, therefore, include certain, as yet unknown, special nutrient factors.

Although all the nutritional problems involved in growing colonies of cells from different sources have not yet been solved, we are able now to see the general picture. What is striking is the very close similarity between nutritional requirements for suspensions of plant cells and for development in isolated young embryos (p. 153). In the intact plant these requirements are met by surrounding tissues—particularly, in the case of embryos, by the endosperm.

Plant cells growing in suspension culture look very much alike from whatever species they originated. The principal characteristics of cells in

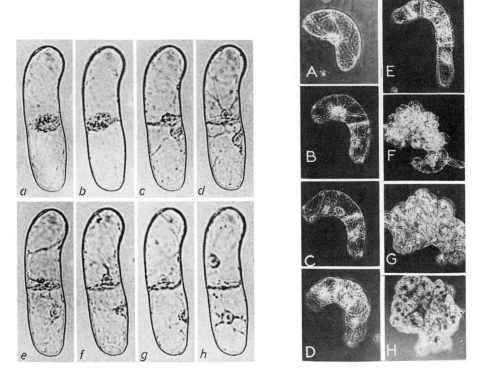

Fig. 6.1 *Left:* Cell division in a single isolated cell from *Phaseolus vulgaris.* Note the large vacuole, prominent cytoplasmic strands and large nucleus with nucleolus. Time lapse between pictures a, b and c, 30 minutes; between c, d, e, f, g and h, 60 minutes. (Reprinted by permission of the Rockefeller University Press from the *Journal of General Physiology,* **43,** 843, 1959—60. Print supplied by Dr. Ludwig Bergmann.)

Right: Development of a cluster of cells from a single cell of tobacco stem-pith isolated in aseptic culture. A. Single cell 1 day after placing in culture medium. B—H. Stages in the formation of a mass of cells, from the single cell in A. (From W. Vasil and A. C. Hildebrandt, *Science,* **150,** 889—92, 1965. Print supplied by Dr. A. C. Hildebrandt.)

suspension culture are: (i) numerous and large vacuoles, even in cells capable of division; (ii) prominent cytoplasmic strands which show active streaming movements; (iii) a large nucleus with nucleolus (Fig. 6.1 *Left*). In addition, a variety of cell types coexist in a given cell suspension, only some of which are free single cells. Some free cells divide and give rise to clusters of smaller, more dense cells. Others increase in size and divide with the formation of internal cross-walls to produce either a filament of cells or, in some cases, a new free cell by a process analogous to "budding" of yeast cells in culture (Fig. 6.1 *Right*). Thus, cells of higher plants in suspension culture do not have the morphology of cells in the tissue from which they were derived. Furthermore, they evidently have a different pattern of metabolism, for they usually do not contain typical storage products.

Ideally, the initial inoculum of isolated cells in a culture vessel will consist exclusively of individual free cells. Only rarely is this ideal attained, by, for example, filtering the suspension through a sterile gauze possessing a pore diameter too small to allow the passage of cell clusters. By such means a very dilute suspension of free cells (5 or less cells per cm^3 of nutrient medium) can be set up. Under suitable conditions this suspension of cells will multiply so that in 2 or 3 weeks' time the cell density will have risen to approximately 100,000 per cm^3. Microscopic examination of the cell population at this time reveals that not all the cells are now free—i.e. cell clusters of various sizes and shapes are usually present in addition to single cells. Various studies have shown that formation of a multicellular cluster in a suspension culture of plant cells takes place by repeated division of one cell, the daughter cells of which do not separate. Separate free plant cells do not aggregate into clusters in the way that some cultured animal cells do. In most research with cell suspension cultures, the original inoculum contains not only free cells but also small aggregates of cells and even dead cells and cell debris.

6.6 REGENERATION STUDIES WITH ASEPTIC CULTURE

Plants show remarkable capacities for regenerating whole organisms from isolated pieces of shoot, root, or leaf, and even from relatively unorganized tissue such as callus. Some more general aspects of regeneration will be discussed later, but we shall first describe experiments on regeneration which have been carried out with callus and suspension cultures.

In callus cultures cell division occurs randomly in all directions and gives rise to an unorganized mass of tissue; thus, there are no clearly defined axes of polarity in a callus. By contrast, in a shoot or root meristem we have a highly organized tissue structure, in which, as we have seen, quite well-defined patterns of division can be recognized. It has been found that under certain cultural conditions shoot and root meristems can be formed within a callus, so that whole new plants may be regenerated.

6.6.1 Root and bud regeneration in callus cultures

We have already made mention in Chapter 3 (p. 64) of Skoog's studies on the interacting influences of auxins and cytokinins on the growth of tobacco pith-derived tissue cultures. Skoog observed that cytokinin and auxin interacted to initiate cell division, but he also found that these same two types of growth hormone could interact to initiate organized meristems. Thus, it was

discovered that if the *proportions* of auxin and cytokinin were varied, then the pattern of meristem formation was altered. When the proportion of auxin to cytokinin was relatively high, there was differentiation of some callus cells into root primordia. A higher concentration of cytokinin relative to auxin caused cells to differentiate into shoot apical meristems. Subsequent growth of the root and shoot primordia led to the callus cultures having the appearance shown in Fig. 6.2. Thus, small changes in the auxin-cytokinin ratio could (a) initiate meristems and (b) channel differentiation of these into either shoot or root apical meristems.

Control of root or shoot-bud formation in callus cultures by variations in auxin-cytokinin balance has now been demonstrated for tissues of several origins, though there have also been many failures in attempts to regenerate buds or roots from cultured tissues of a range of plant species. The interacting effects of the hormones in this phenomenon can be modified by other factors, such as sugar and phosphate levels, sources of nitrogen, and other constituents of the medium such as purines. There is, however, no doubt that auxin and cytokinin may regulate not only the initiation of organized meristematic centres in the callus, but also the type of meristem formed. Nevertheless, stimuli other than auxin and cytokinin are involved in apical meristem initiation in callus cultures. For example, initiation of lateral roots in pea-stem segments is inhibited by red light. Thus a phytochrome-based mechanism (Chapter 8) may be involved in the initiation of root apical meristems.

6.6.2 Embryo formation in aseptic cultures of plant cells and tissues

Totipotency of plant cells has perhaps been most spectacularly demonstrated by the regeneration of whole plants from embryos formed in cultures of both somatic cells and those of male generative cells (i.e. pollen grains). The formation in tissue cultures of plant embryos of at least superficially normal appearance and behaviour

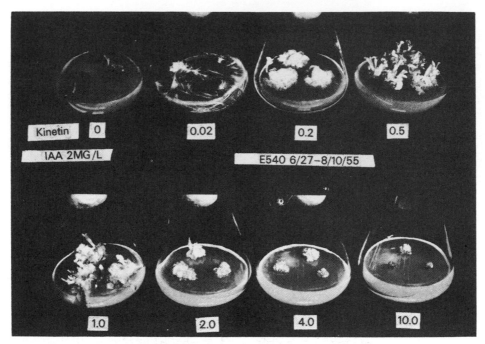

Fig. 6.2. Effect of a range of concentrations (0—10 mg/l) of kinetin on growth and organ formation in tobacco-pith derived callus cultured on nutrient agar containing, in all flasks, 2 mg/l indole-3-acetic acid. All cultures 44 days old. (Reprinted from *Symp. Soc. Exp. biol.* **II**, 1957. Print kindly supplied by Professor F. Skoog.)

was first observed by Steward in 1958 and Reinert in 1959, who by imposing a sequence of changes in the composition of the nutrient media caused callus cultures of carrot-root phloem parenchyma to give rise to embryos which were very similar to normal embryos and which on transfer to a suitable medium developed into whole carrot plants. Embryos formed from cells of origins other than a fertilized egg are often referred to as either *adventive embryos* or *embryoids*. The changes in the nutrient media required to bring about adventive embryo formation principally involved alterations in the balance of auxin and cytokinin.

Since that first report, other workers have obtained adventive embryos in callus cultures, in suspension cultures, and in cultures of isolated anthers and pollen grains (Fig. 6.3). Regeneration in cultivated and wild varieties of carrot (*Daucus carota*) has been extensively investigated, particularly with respect to embryogenesis in aseptic culture (Fig. 6.5), but it has been found that the capacity for adventive embryo formation is widely distributed in the plant kingdom. Nevertheless, one cannot assume that all cells of a plant, or all species, retain their totipotency, for numerous unsuccessful attempts have been made to obtain embryogenesis, and/or organogenesis, in aseptic cultures of tissues from many species. It is possible that only true diploid cells (or haploid in the case of pollen grains) can undergo embryogenesis, and that failures to induce adventive embryo formation may have been associated with polyploidy in cultured cells and tissues.

The formation of an adventive embryo is marked by the appearance of an organized group of cells possessing longitudinal polarity and also with, at a very early developmental stage, a shoot and a radicular pole at opposite ends. The adventive embryo is formed without any connection with vascular tissues of the mother plant or callus, and this contrasts with the pattern of formation of monopolar buds and roots, as these are always connected to such vascular tissue.

Adventive embryo formation can, under appropriate experimental conditions, be a con-

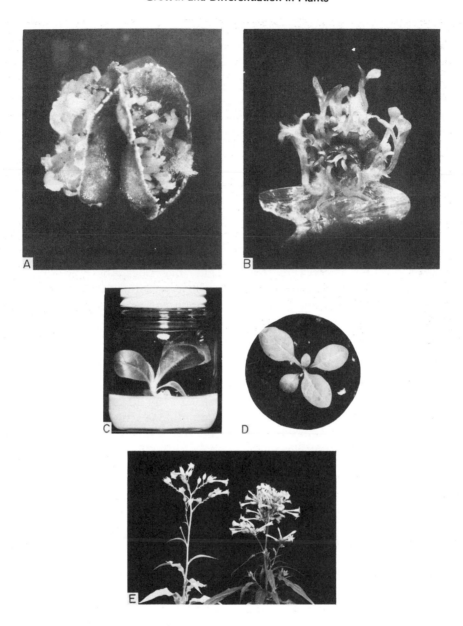

Fig. 6.3. Development of embryoids (adventive embryos) from pollen grains during anther culture of *Nicotiana tabacum* c.v. "White Burley". A. Anther cultured for 28 days at 25°C in continuous light, showing a large number of germinating embryoids, each of which has developed from one pollen grain. B. The same anther 7 days later. At this stage the plantlets can be readily teased apart and transplanted individually as shown in C. The transplanted plantlets can later be transferred to a mixture of peat and sand in a normal plant pot (D). E. Inflorescences of, on the *left*, a normal diploid tobacco plant, and on the *right*, of a haploid plant grown from a pollen grain as shown in A—D. It is interesting to note that haploid plants usually develop larger inflorescences than do diploid plants. (From N. Sunderland, "Pollen and Anther Culture", Chapter 9 in *Plant Tissue and Cell Culture*, ed. H. E. Street, Blackwell Scientific, 1973. Original prints supplied by Dr. N. Sunderland.)

tinuous process, and a culture may therefore contain, side by side, embryos of all stages of development. Adventive embryos induced in tissue cultures appear to be formed from single cells, usually at the upper surface of the callus. The surrounding callus cells may act as a "nurse-tissue" to the embryo in its very early stages of development. There is some evidence that the formation of adventive embryos in suspension cultures of plant cells similarly involves a nurse-tissue, in that there is first the production of a multicellular aggregate (i.e. a small callus) by repeated division of an originally single cell, following which embryos may be initiated from single cells at the surface of the aggregate. On the other hand, it has been claimed by Steward and his co-workers that isolated free single cells can develop directly by segmentation without callus formation to pro-embryos and that these then go on to form embryos and normal plantlets. Steward has, in fact, argued that one of the requirements for the initiation of adventive embryos in suspension cultures is that cells must be set free from associations with their neighbours and thus be able to grow independently. However, various studies in recent years have shown that adventive embryos can form from individual cells at the surface of a callus. More microscopic studies of adventive embryos will undoubtedly resolve the question as to whether or not a single plant cell, not in contact with other cells, can develop directly into an embryo. Recent electron-microscope studies have, however, indicated that embryogenic cells in callus from *Ranunculus scleratus* possess protoplasmic strands, linking them with adjacent callus cells during embryogenesis. This does indicate that cells do not have to be physiologically isolated from other cells for their development into an adventive embryo, and in addition strengthens the suggestion that nurse-cells are in some way necessary in early embryogenesis. Similarly, the concept of nurse-cells is supported by the reports that adventive embryos form in suspension culture from one cell of a cell aggregate, with the remaining cells of the aggregate in some way supporting the initial development of the embryoid. Nevertheless, it is true that only "exposed" cells of a callus, or a cell

in suspension culture, actually become embryogenic, even though they may require the support of "nurse-cells", and that the mass of cells in a callus do not form embryoids but may do so if they are separated into a suspension culture.

Many of the earlier experiments on adventive embryo formation in aseptic culture involved the use of media containing the liquid endosperm of coconut. At one time it was considered that coconut milk contained special substances which were essential for the initiation of embryos in suspension cultures. More recent research has, however, shown that cells in suspension culture can form embryos on synthetic, chemically defined, media in the absence of liquid endosperm. Thus, one does not need to invoke unknown nutritive or hormonal "embryo-inducing" factors in liquid endosperms. Adventive embryo formation is achieved experimentally by successive changes in the nutrient media, an important aspect of which includes the nitrogen to auxin ratio. Because of this a number of workers utilize a sequence of media changes which culminates in transfer to an auxin-free from an auxin-containing medium, and this can be sufficient to induce embryogenesis in the culture.

6.6.3 Pollen- and anther-culture

Over the past few years, the techniques previously developed for adventive embryo formation from somatic cells have been used to produce embryos and even entire whole plants from pollen grains. Pollen grains are produced naturally in relatively large numbers and in an easily accessible form for the research worker. Each grain consists of just a few cells (five in gymnosperms; three in angiosperms) and possesses a unique genome derived from the process of meiosis in microsporogenesis. In true diploid species each gene in the pollen grain is present as a single copy only, and therefore formation of an adventive embryo from the pollen grain results in a haploid plant in which every gene can be expressed in the phenotype. This fact has extremely important connotations in the practice of plant breeding, for experimentally produced mutant genes can be evaluated very much more quickly and easily in haploids. In addition, the exposed adventive embryos from pollen or anther cultures and derived plantlets also lend themselves to convenient and uniform mutation-inducing treatments such as irradiation with X-rays.

The techniques required to induce embryogenesis in pollen and anther cultures are still in the process of development, and to date only a limited number of species have been successfully propagated in this way. The first experiments on pollen culture took place in the 1950s, when Tulecke was able to cause pollen of certain gymnosperms to proliferate into callus. Not until the mid-1960s was angiosperm pollen brought successfully into aseptic culture by Guha and Maheshwari, and then only by culturing the whole anther of *Datura innoxia*. However, Guha and Maheshwari also found in 1967 that development in cultured anthers of *D. innoxia* led to the formation of haploid plants, with one pollen grain giving rise to one plant. Since that time aseptic cultural conditions have been developed which permit growth and embryogenesis in pollen (sometimes within the excised anther) in various other angiosperms (Fig. 6.3).

Methods of anther culture usually involve aseptic excision of anthers, and their transfer to the surface of sterilized agar medium, or flotation on the surface of liquid medium, or onto filter-paper bridges over liquid medium. An important aspect of the procedure is the selection of anthers containing pollen at an appropriate stage in microsporogenesis. Maximum yield of adventive embryos in *Nicotiana tabacum* anther culture is obtained when the excised anthers contain pollen at the stage when the vegetative cell is in the process of rapid cytoplasmic synthesis, but results obtained with other species have indicated that the critical stage may vary from one species or cultivar to another. Furthermore, it may also be influenced by aseptic culture conditions and the environment under which the donor plants were grown. Culture media used in pollen and anther culture are similar to those used in the aseptic culture of somatic tissues, and as with these it is the hormonal component of the medium which is critical. For anthers of most species, both auxin and cytokinin are included, although in a few instances either one or the other is sufficient. Cytokinin can be replaced, if desired, by coconut milk.

Embryogenesis in angiosperm anther culture occurs by development of the vegetative cell of each pollen grain. The initial stages of adventive embryo formation varies from species to species. In some cases the vegetative cell divides repeatedly, and there is complete suppression of pollen-tube development. Gradually the derivatives of the vegetative cell can be seen to be organized as an embryo which further develops into a plantlet. In other cases, embryogenesis appears to occur where the normal unequal division resulting in a vegetative and generative cell is replaced by an equal division (p. 22).

Under certain cultural conditions, and depending on species, pollen in cultured anthers may also divide in manners which give rise to a callus. Shoot buds and root initials often form from such a callus and further plantlets can be obtained in this way. Although much research remains to be done, it is already clear that pollen and anther culture techniques will be very important in future breeding of new crop and ornamental plants.

6.6.4 Isolation and culture of plant protoplasts

Since the mid-1950s methods have been devised which allow the isolation of protoplasts from somatic and reproductive cells. A protoplast is, of course, a plant cell from which the cell wall has been removed, and in aseptic culture isolated naked protoplasts behave in many ways rather like animal cells in culture. Under suitable cultural conditions they grow and divide, but can also be induced to regenerate cell walls.

Plant protoplasts can be isolated from their parent tissues by either mechanical or enzymatic means. Mechanical isolation of protoplasts has been achieved by first plasmolysing the cells in a hypertonic plasmolyticum and then carefully cutting away the walls by micro-surgical procedures.

Such a technique therefore cannot be used on non-vacuolated cells such as those of meristems, and for this and other reasons enzymatic methods of protoplast isolation have been adopted by many workers in this field. The enzymes used are ones which degrade components of plant cell walls, and most commonly a pectinase (to separate cells) and a cellulase (to degrade the cellulosic walls) are used either sequentially or as a mixture. Once isolated, plant protoplasts have to be kept in a liquid culture medium, the osmotic potential of which closely matches that of protoplasts; otherwise irreversible damage can be caused by the protoplasts bursting or shrinking excessively.

Isolated protoplasts in aseptic culture have very great potential uses to research workers. For example, uptake of viruses and various macromolecules by plant cells is very much more readily studied in the absence of cell walls. However, a most exciting prospect lies in the successful fusion of protoplasts of differing origins to produce new plant species by somatic hybridization. The breeding of new plants has always been restricted by the necessity of effecting sexual fertilization. Failure to obtain viable hybrid embryos from interspecific and intergeneric crosses can be due to a number of causes, arising from disruption of the normal sexual processes, e.g., by failure of the pollen tube to penetrate the embryo sac, breakdown of the endosperm, etc. Such difficulties could be avoided if sexual reproduction can be by-passed by direct fusion of vegetative cells. Successful fusion of isolated plant protoplasts has already been achieved (Fig. 6.4), and appropriate methods have already been devised to allow some fused protoplasts to regenerate a whole new plant, either by adventive embryo formation or by root and shoot-bud initiation from a derived callus.

In addition to providing a means of circumventing natural incompatibility mechanisms in genetic hybridization, the mixing of the cytoplasms and organelles of different plant species is also of potential, but as yet unrealized, practical importance. For example, one possible development would be the fusion of protoplasts between different divisions of the plant kingdom, which may, for example, provide one means for the creation of cereal plants containing a characteristic of blue-green algae in being able to utilize atmospheric nitrogen rather than being dependent on nitrogenous fertilizers added to the soil. The fusion of plant and animal cells has been already achieved, and there is the possibility that such procedures could lead to the creation of plant-animal hybrids.

6.7 REGENERATION IN SHOOT AND ROOT CUTTINGS

Although not normally performed in aseptic culture, one of the most obvious examples of regeneration is seen in the common horticultural practice of vegetative propagation of plants by taking shoot or root cuttings, and allowing them to develop adventitious roots and/or buds. In shoot cuttings, a callus is frequently formed at the base of the cutting, as a result of divisions originating in the cambium, and from such a callus root primordia arise. However, adventitious roots may also be formed in normal tissues of the stem—usually in the pericycle, but in some species in the cambial zone. In root cuttings, both roots and buds commonly arise from callus formed from parenchyma in the younger phloem.

The ease with which roots can be formed on shoots varies enormously; cuttings from plants such as bean will produce roots if simply left with their lower ends immersed in water, whereas those from other species will do so only rarely, even under what appear to be the most favourable conditions. In those species which do produce roots, it is generally true that a piece of stem which possesses a bud or leaves will form roots at its base, but that a disbudded and defoliated stem piece produces roots much less readily or not at all. This suggests that a substance is formed in the buds and leaves which moves downward and stimulates root formation at the base of the stem. The existence of such a root-initiating substance in young leaves has been proved by the demonstration that extracts of young leaves do stimulate the

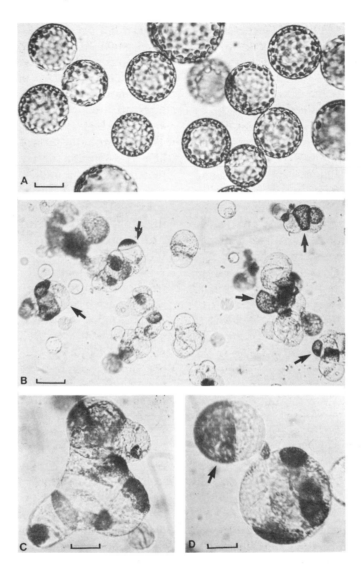

Fig. 6.4. Isolated plant cell protoplasts and inter-species fusion. A. Protoplasts isolated from leaf mesophyll of *Petunia hybrida* following treatment of the leaf with cell-wall-degrading enzyme (bar represents 25 μm). B. Protoplasts isolated from leaf mesophyll of *Petunia hybrida* and colourless leaf epidermis of *Nicotiana tabacum* which have been treated with polyethylene glycol to induce aggregation and fusion. Several fusing inter-species aggregates are visible (arrows) (bar represents 50 μm). C. Large aggregate containing fusing protoplasts of *Petunia hybrida* and *Nicotiana tabacum* (colourless epidermal protoplasts) (bar represents 25 μm). D. Inter-species heterokaryon which has resulted from the fusion of several protoplasts of *Petunia hybrida* and of *Nicotiana tabacum*. The heterokaryon has rounded off and extensive mixing together of the cytoplasms is beginning to take place. Two very closely adhering protoplasts of the two species are also visble (arrow) (bar represents 20 μm). (Original prints supplied by Professor E. C. Cocking.)

rooting response of stem cuttings. Further, it has been found that the application of auxins to cuttings has a similar effect to that of extracts of young leaves (Fig. 5.34), suggesting that the stimulatory effect of buds and leaves upon rooting is probably due to the production of auxin by these organs. The effect of auxin in rooting of cuttings is to increase the rate of formation and absolute number of adventitious root initials. This is, therefore, another example of cell division and differentiation being activated by auxin.

The formation of roots in stem cuttings normally occurs at the *basal end* of the stem. This is true even if the cutting is inverted, so that the morphological lower end is uppermost (p. 17). The stem therefore shows polarity in the initiation of roots. The fact that auxins are known to stimulate the formation of root initials and also that auxins move in a basipetal manner (p. 107), lead one to believe that the polarity shown in root formation is a consequence of the movement of auxin to the morphologically lower tissues, where its arrival triggers off the processes of root initiation. In fact, if an auxin is applied to the apical end of a stem cutting, then callus formation and subsequent root formation is stimulated at the base of the cutting. If applied basally, then roots are again stimulated there.

If a stem section is dipped in a solution of cytokinin it may react by producing many buds at the morphologically upper end of the stem, but few or no roots, the opposite effect to that elicited by dipping in an auxin solution. Nevertheless, as we have seen (p. 157), the stimulatory effect of auxins upon root formation may not be revealed unless the responding tissues also contained an appropriate concentration of cytokinins, since root formation involves active cell division. Root cuttings behave similarly to stem cuttings with regard to polarity of root and shoot bud initiation, and the effects of auxins and cytokinins (p. 157). The fact that buds and roots are formed at opposite ends of isolated segments of stem or root appears to be the result of movement of auxin and cytokinin in opposite directions, a preponderance of one or the other accumulating at either end, causing either buds or roots to be initiated. In-

deed, it has been found that if cuttings of chicory (*Cichorium intybus*) roots are placed under moist conditions, which favour regeneration, then certain changes occur in the distribution of the endogenous hormones within the cuttings, so that high auxin concentrations are found at the basal ends, and high cytokinin levels at the apical ends. These changes occur *before* there is any observable regeneration of buds and roots and hence they may play an important role in the pattern of regeneration. Certainly these observations are consistent with the findings that bud and root regeneration in callus cultures are associated with high cytokinin and high auxin levels, respectively (p. 157).

At the present time we know little of the translocation patterns of cytokinins in plants. Kinetin itself is apparently not translocated readily, for it remains at, or very close to, the place to which it is applied on a plant. Naturally occurring cytokinins may well behave differently though, as there is some evidence that cytokinins are synthesized in roots and translocated up into the shoot system (p. 112). Certainly it is known that some synthetic cytokinins are translocated quite readily in plant tissues.

Auxins are not the only factors concerned in root formation. A supply of sugar is necessary, as well as other nutrients. The stimulating effect of leaves on initiation in stem cuttings may be due in part to their production of nutrients, and perhaps also to other hormonal substances more specific in promoting root formation in conjunction with auxin.

In contrast to the fairly ready regeneration of roots in shoot cuttings, excised roots of most species growing in sterile culture normally form only further root tissues, including the initiation of lateral roots, and only relatively rarely are shoot buds initiated. Auxin is in some way involved in the formation of lateral roots as well as of adventitious roots. Immersion of the main root of a dicotyledonous seedling in a solution of an auxin results in a reduction of main root extension but a stimulation of lateral root initiation. The subsequent growth of the newly produced lateral roots is also inhibited by the auxin solution. Thus auxins, at other than very low concentrations, stimulate the formation of roots but inhibit their subsequent elongation. The result is that a root immersed in a solution of auxin becomes stunted, and possesses rows of newly emerged but suppressed lateral roots.

6.8 GENERAL ASPECTS OF REGENERATION

Sinnott has defined regeneration as "the tendency shown by a developing organism to restore any part of it which has been removed or physiologically isolated and thus to produce a complete whole". This broad definition includes a wide variety of phenomena, but we can distinguish a number of general aspects of regeneration. Firstly, we have seen that we can apply the term to the initiation of shoot and root meristems in a disorganized mass of callus, which may be growing in aseptic culture or may form at the surface of a cutting in response to wounding. This is a remarkable phenomenon, even though it may be so familiar that we come to accept it as commonplace. We have no conception as to the nature of the factors operating whereby in a mass of disorganized callus a high degree of organization emerges, but we have already suggested (p. 34), that the apical meristem of the shoot or root is a stable configuration which, as it were, "crystallizes" out under certain conditions.

It is important to realize that in regeneration we see in operation the processes which determine normal development. When the normal course of development is disturbed by wounding or in other ways, compensating events occur which tend to restore the normal situation. Thus, it would seem that the normal form of the plant represents an equilibrium state, and that when this equilibrium is disturbed, built-in control mechanisms operate to restore the equilibrium. This phenomenon is well illustrated in the regenerative properties of shoot and root meristems. We have already seen that if a shoot apex of *Lupinus* is divided by two vertical cuts at right angles, then each segment of

the original apex is able to regenerate into a normal apex (p. 34). Similar experiments have been successfully carried out with root apices. Further examples are seen in the regeneration of vascular tissue (p. 121) and in the formation of a phellogen when the surface of a stem is cut or damaged.

So far we have discussed the problems presented by the spontaneous development of organized meristems within unorganized meristematic tissue. A further problem concerns the resumption of cell division in previously differentiated, non-dividing cells, which follows wounding. In some cases regeneration of root primordia takes place in callus tissue which has developed at the cut basal surface of a shoot cutting, while in other cases the root primordia may be formed by the resumption of cell division in stem tissues, such as the pericycle. In either case, however, it is clear that the isolation of a piece of stem or other organ results in renewed cell division, and the question arises as to what causes this cell division. There is some evidence that the wounding of plant tissues results in the release of "wound hormones", which stimulate cell division. Whether such substances are involved in all cases of cell division following wounding is not clear. In any case, however, it is clear that certain differentiated cells of the stem or root become "dedifferentiated" when they resume meristematic activity.

The phenomenon of regeneration provides strong evidence that the process of differentiation in many types of plant cell does not involve any loss in their genetic potentialities, so that they remain "totipotent".

Although the totipotent behaviour of individual cells of a number of plant species has been demonstrated experimentally (Fig. 6.5), it is nevertheless wise to be cautious in assuming that all living, nucleated, plant cells are totipotent. Until regeneration of whole plants has been seen to occur from isolated cells of all types known to occur in the plant body (e.g. parenchyma, palisade and spongy mesophyll cells, companion cells), the case cannot be regarded as proven. Even so, as we have mentioned earlier (p. 157), plants have been regenerated from adventive embryos formed from cells derived from the root, hypocotyl, stem,

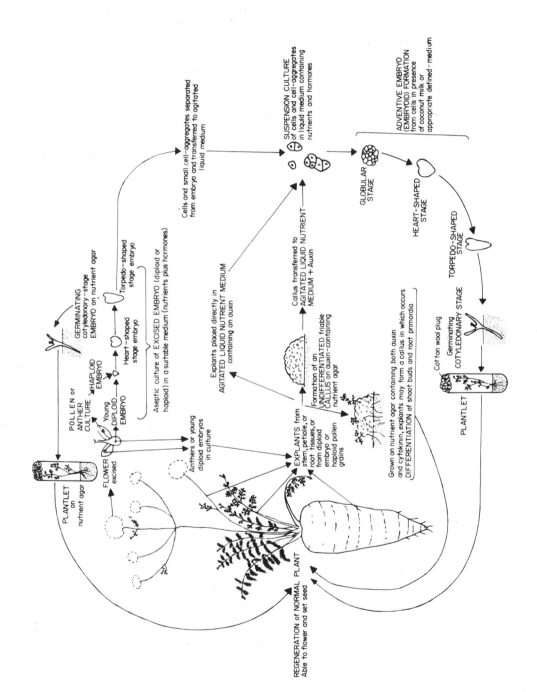

Fig. 6.5. Diagram to illustrate alternative pathways of regeneration of a whole new plant from tissue explants derived from various parts of a carrot plant. The regeneration cycles can be repeated indefinitely. (Adapted, with additions, from F. C. Steward *et al.*, *Brookhaven Symp. Biol.* **16**, 73, 1963.)

petiole, embryo, and even pollen. Cells from the lamina of green leaves do appear to be more recalcitrant in demonstrating their presumed totipotentiality. A number of workers in the field of plant cell and tissue culture, particularly Steward, nevertheless consider that any free cell from a higher plant will, if provided with the right stimuli (nutritional and hormonal), regenerate a whole new plant taking one of the alternative routes toward organization described above.

It seems clear that the developmental activities of most callus cells are restricted in some way, and that further restraints are imposed should differentiation of vascular tissues, shoot-buds and root initials occur. Thus, the cells of undifferentiated callus are usually capable of unlimited division, but if a bud is regenerated, then the cells which become part of the leaf primordia are subject to considerable restraints with respect to the planes of their divisions, and they are no longer capable of unlimited division so long as they remain part of the leaf. We do not know how a cell becomes restricted when it is a part of a tissue system, but possibly some control over each cell is exerted by its neighbours through the plasmodesmata, which connect the protoplasts of adjacent cells.

The discovery that the regeneration of buds and roots by callus tissues can be regulated by the relative concentrations of auxin and cytokinin in the culture medium has led to the suggestion that these hormones play an important role in "organ formation". However, it should be noted that the primary effect of the hormones is upon the initiation of shoot and root apical meristems, and we do not find callus cultures giving rise directly to organs of determinate growth, such as stems and leaves, although these may be formed as a result of the subsequent growth of the meristems. Unless we are prepared to call a shoot meristem an "organ", which seems inappropriate, it is not strictly accurate to say that cytokinins promote organ formation, although it is perhaps more justifiable to say this of the initiation of root primordia in response to auxin.

Nevertheless, the initiation of two kinds of apical meristem in response to different hormone levels is a highly interesting effect and raises the question as to whether differences in the levels of endogenous auxins and cytokinins in different parts of the plant may play a role in the normal control of patterns of development. The observation that there is an increase in the levels of endogenous cytokinins in the apical ends, and of auxins at the basal ends of cuttings of chicory root (p. 164) and that these changes precede the initiation of adventitious buds and roots, suggests that these hormones are important in natural regeneration. We shall return to a discussion of the possible role of growth hormones in morphogenesis in Chapter 13.

FURTHER READING

General

Butcher, D. N. and D. S. Ingram. *Plant Tissue Culture,* Edward Arnold, London, 1976.
Laetsch, W. M. and R. Cleland (Eds.). *Papers on Plant Growth and Development,* Little, Brown & Co., Boston, 1967.
Steward, F. C. *Growth and Organization in Plants,* Addison-Wesley, Reading, Mass., 1968.
Steward, F. C. The control of growth in plant cells, *Sci. Amer.* **209,** 104, 1963.
Sunderland, N. Pollen plants and their significance, *New Scientist,* **47,** 142—4, 1970.

More Advanced Reading

Cocking, E. C. Plant cell protoplasts—isolation and development, *Ann. Rev. Plant Physiol.* **23,** 29—50, 1972.
Cocking, E. C. Growth substances and protoplasts. In: *Proc. 9th Int. Conf. Plant Growth Regulating Substances, Plant Growth Regulation* (ed. P. E. Pilet), pp. 281—285, Springer-Verlag, Berlin—Heidelberg, 1977.
Gresshoff, P. M. Phytohormones and growth and differentiation of cells and tissues cultured *in vitro.* In: *Phytohormones and Related Compounds—A Comprehensive Treatise,* Vol. II (eds. D. S. Letham, P. B. Goodwin and T. J. V. Higgins), pp. 1—29, Elsevier—North Holland Biomedical Press, Amsterdam, 1978.

Murashige, T. Plant propagation through tissue cultures, *Ann. Rev. Plant Physiol.* **25,** 135—68, 1974.

Skoog, F. and W. Miller. Chemical regulation of growth and organ formation in plant tissues cultured *in vitro, Symp. Soc. Exp. Biol.* **11,** 118, 1957.

Steward, F. C. and H. Y. M. Ram. Determining factors in cell growth, *Adv. in Morphogenesis,* **1,** 189, 1961.

Steward, F. C., A. E. Kent and M. O. Mapes. The culture of free plant cells and its significance for embryology and morphogenesis. In *Current Topics in Developmental Biology* (ed. A. Monroy and A. A. Moscona), Vol. I, Academic Press, New York, 1966.

Sunderland, N. Anther culture as a means of haploid induction. In: K. J. Kasha (ed.), *Haploids in Higher Plants: Advances and Potential,* University of Guelph Press, pp. 91—121, 1974.

Torrey, J. G. The initiation of organized development in plants, *Adv. in Morphogenesis,* **5,** 39, 1966, Various articles by: H. E. Street, E. C. Cocking and P. K. Evans, M. M. Yeoman and H. E. Street, N. Sunderland, M. M. Yeoman and P. A. Aitchison, P. J. King and H. E. Street, J. Reinert, D. N. Butcher, and D. S. Ingram, In: H. E. Street (ed.), *Plant Tissue and Cell Culture,* Botanical Monographs, Vol. 11 Blackwell Scientific, Oxford, 1973.

SECTION III

Environmental Control of Development

Introduction

The preceding sections of this book considered mainly the internal processes which are involved in and may control growth and differentiation in plants. However, land plants are subject to a wide variety of external influences, including light, temperature, water stress, gravity and so on, which may modify the growth and differentiation of the plant to a greater or a lesser degree. Some of these external influences may have deleterious effects upon the plant and may constitute a hazard, as with freezing temperature or drought. Animals also have to survive under these conditions but they are frequently able to *avoid* the stresses by migration or hibernation, whereas land plants, being sedentary, have little scope to take avoiding action and must *tolerate* the environmental stresses if they are to survive.

However, apart from the direct effects of environmental stress on plant tissues, variations in certain other external factors may be advantageous to the plant by providing it with information about the environment which facilitates adaptation of the plant to its local situation. Thus, gravity never constitutes an environmental stress, but the capacity of the plant to detect the direction of gravity provides information whereby it can regulate its orientation in relation to the soil surface. In such cases the environmental factor acts as a *signal* or *stimulus,* and the plant responds in an active, "programmed" manner, which enables it to become better adapted to its environment. Thus, it is useful to distinguish between:

(1) effects of environmental factors which are direct and non-adaptive (e.g., freezing injury); and

(2) adaptive, programmed responses, in which the environmental factor acts as a signal or stimulus as in growth movements, photoperiodism, photomorphogenesis and dormancy.

Adaptive responses to the environment are numerous and varied and we shall be considering a number of these in the remaining parts of the book. In the first chapter of this section of the book we shall largely confine ourselves to a consideration of growth curvatures, particularly those that occur in response to directional stimuli, such as gravity and uneven illumination.

7

Growth Movements

7.1 CHARACTERISTICS OF PLANT MOVEMENTS

A characteristic property of living organisms is that they have the capacity to perceive and respond to changes in external environmental conditions. The environmental factor constitutes a *stimulus* and the resulting observable change in the plant is called the *response*. Many types of response to the environment are shown by plants, but certain external stimuli elicit movements of various kinds.

All plants have the capability of movement. In lower plants, such as many algae, fungi and bacteria, movement of the whole organism occurs. In higher plants the capability of movement is restricted to individual organs or parts of the whole organism. The process of straight extension growth itself can perhaps be regarded as such a movement, in that, for example, the root tip moves through the soil as a result of growth. Other types of movement result from *differential* growth rates; that is, the organ shows different rates of growth on opposite sides and this results in the bending of the organ in one direction. Such movements are termed *growth movements*. Not all plant movements are growth movements. Some are brought about by reversible turgor changes in tissues at the base of each leaf and there is often a specialized structure, the pulvinus, at which bending occurs due to reversible changes in turgor. (A pulvinus is the swollen base of a leaf or leaflet which contains a high proportion of thin-walled parenchyma.)

The occurrence of a response, such as a growth movement, following exposure to an environmental stimulus is the only way we have of knowing that a plant has perceived the particular environmental stimulus. One must presume that the stimulus has acted upon some *receptor, or sensor,* within the plant and that the sensor is changed in a way such that a sequence of processes is initiated that ultimately results in the observable response. The whole sequence of events may be expressed in outline as:

Stimulus→Perception→Transduction→Response

Perception involves, as already indicated, alteration of a receptor sensitive to the particular environmental stimulus. *Transduction* in the above scheme is the term used to cover all the events that connect stimulus reception to the observed response. Thus, transduction includes: (a) the control of specific biochemical and/or biophysical processes in the receptor cell(s) by the activated sensor, (b) signal transmission from the receptor site to a spatially separated region of response (transmission may involve movement of the signal from cells containing receptor sites to cells that show the response, or signal transmission may be purely intracellular—i.e. the same cell may both perceive and respond to the stimulus), and (c) dose-response relationships.

We have already mentioned (Chapter 3) one growth movement, that of phototropism, where movement of an organ occurs in response to unilateral illumination. A tropic movement of this type is thus a response to an external stimulus. In the case of phototropism the stimulus is light, but other external stimuli such as gravity, water, chemicals, heat or mechanical contact can also induce growth movements.

Where the direction of the response is related to the direction of the stimulus, we speak of a *tropic* response, or *tropism*; but in many cases the direction of movement does not bear a direct relation to that of the stimulus, and we then speak of a *nastic* response. Thus, a curvature of a shoot towards the more illuminated side is a phototropic curvature, while the opening or closing of flowers with a change of light intensity all round is a *photonastic* one. All tropic responses are induced by directional or unilateral stimuli, whereas in nastic responses the stimulus may be diffuse. Another example of a nastic growth response is seen in the opening and closing of certain flowers, such as those of crocus, in response to changes

in temperature (thermonasty). When the temperature rises growth is faster on the inner side of the petals, so that the flower opens, whereas the reverse is true when the temperature falls. Not all nastic movements are growth movements, some being brought about by reversible changes in the turgor of specialized cells in pulvini. Vivid examples of nastic movements not involving growth are provided by *nyctinasty* in leaves ("sleep movements") of many species, such as French bean (*Phaseolus vulgaris*), clover (*Trifolium repens*), wood sorrel (*Oxalis acetosella*) and the sensitive mimosa (*Mimosa pudica*). Leaflets of nyctinastic plants assume a vertical (i.e. closed) position in darkness and revert to horizontal orientation (open) upon illumination. The closing movement to the "sleep" position can be either upwards (e.g. clover) or downwards (e.g. wood sorrel).

Some growth movements in plants do not obviously result from any external stimulus, even though environmental factors may nevertheless have some influence on them. Examples of spontaneous (autonomic) movements that arise from causes within the plant itself include *nutations* and *epinastic movements*.

In the case of growth movements that occur in response to an external stimulus, it is possible to measure the speed of response. It is usually the case that after exposure to the stimulus there follows a *latent period* (or lag period) before differential growth begins. What it is that happens during the latent period is of course what sets in motion differential growth in the two sides of the organ, and plant physiologists therefore have a keen interest in elucidating the physiological and biochemical events during the latent period.

Growth curvature of a plant organ results from differential elongation of its convex and concave sides. As emphasized by Firn and Digby (1980) although measurement of the angle through which the organ is displaced by the curvature is an expression of the net differential in growth along the two sides, it does not distinguish between the following possible ways in which that differential was achieved:

(a) An increase in growth rate of one side, with no

change in the other side; (b) A decrease in growth rate of one side, with no change in the other side; (c) A differential acceleration of growth of the two sides; (d) A differential retardation of growth of the two sides; (e) Acceleration of one side and deceleration of the other side.

From the studies that have been made of the changes in growth rates of opposite sides of bending plant organs, it seems clear that different plants and organs execute growth curvatures through the operation of more than one of the alternative processes outlined in (a)—(e) above.

A final general question that presents itself in connection with environmentally-induced growth movements, is whether or not there is always separation of the region of perception of the external stimulus and the region of response (i.e. during transduction is there always intercellular transmission of a signal from cells containing the stimulus perceptor apparatus to cells that respond to the signal?).

All these general considerations and questions must be borne in mind when we consider the following examples of plant growth movements.

7.2 CIRCUMNUTATION

An example of a nutational movement is seen in the growth of a plant stem, for it does not grow straight upwards, but performs a series of rhythmic movements which result in the shoot tip oscillating about the longitudinal axis. Roots execute similar oscillating movements as they grow downwards. This particular type of nutation, called circumnutation, is particularly pronounced in both shoots and tendrils of climbing plants, and may confer a biological advantage in the finding of a support.

It is known that circumnutations occur as a result of different rates of extension growth on different sides of the organ. For circumnutations to occur, the rates of elongation around the organ must vary rhythmically, and hence it is an example of oscillatory processes in plants. Several different mechanisms have been proposed to explain circumnutation, involving processes such as (a)

autonomous oscillatory changes in endogenous growth hormone concentrations in tissues around the organ, (b) gravitropic "overcompensation", and (c) combinations of autonomous processes and gravitropic overcompensation. It is not possible on available evidence to decide which, if any, of these models for circumnutation is correct.

7.3 EPINASTIC MOVEMENTS

The angles in the vertical plane at which leaves, and also lateral branches, are held along the stem axis in dicotyledons and gymnosperms are determined by the relative (differential) rates of elongation growth along the inner (adaxial) and outer (abaxial) sides of the organs. *Epinasty* of leaves, for example, is seen when the petioles show a growth curvature; the upper (adaxial) side elongating more rapidly than the lower (abaxial) side, resulting in a downward movement of the lamina as the leaf grows (Fig. 7.1). Conversely, *hyponasty* (upward movement) results from a relatively faster growth rate in the abaxial side of the organ. Epinastic movement of a leaf is achieved through a growth curvature that moves basipetally from the lamina base along the petiole as the leaf develops towards maturity. At maturity the leaf reaches a final orientation position in the vertical plane and no further epinastic or hyponastic bending occurs under normal conditions (though phototropic *lateral* movements of the leaf may still occur—see p. 188).

Epinasty is known to result primarily from an internal stimulus (i.e. it is an autonomic growth movement) in that the *direction* of bending is not influenced by external factors. The autonomic nature of epinasty distinguishes it from all tropisms, which are orientated with respect to the direction of the external stimulus, and also from other nastic movements, such as nyctinasty and thigmonasty which are controlled primarily by external stimuli. Nevertheless, environmental factors can markedly influence the extent of epinastic bending, for even matured leaves can be induced to undergo further epinastic movements by

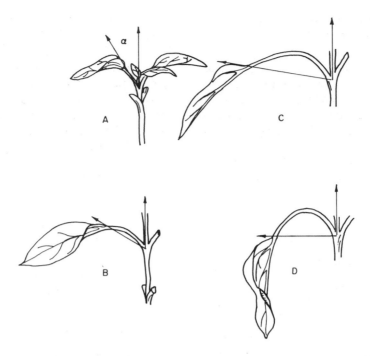

Fig. 7.1. Principal changes in orientation of the lamina and petiole during the growth of the leaf in *Helianthus annuus*. A. Young leaf emerging from the terminal bud. B. Leaf during the phase of rapid growth. C. Mature leaf. D. Leaf approaching senescence. α= orientation angle. (Reprinted from J. H. Palmer and I. D. J. Phillips, *Physiol. Plant,* **16**, 572-84, 1963.)

various external factors, such as gravity, soil aeration, or gases in the atmosphere (Fig. 7.2).

The physiology of epinastic and hyponastic leaf movements has been a subject of research for many years, and there is clear evidence that the known plant growth hormones are involved. Ethylene and auxin are particularly concerned in leaf orientation movements in the petioles of dicotyledonous species, but there is still some uncertainty about how these two hormones exert their influences. Exposure of plants to exogenous ethylene or auxins induces epinasty of leaves (and of other organs such as lateral shoots). Epinasty always seems to be associated with high ethylene levels. Auxins are known to induce ethylene biosynthesis in plants (p. 119 and Fig. 5.13) and induction of epinasty by auxins appears to be mediated through effects on ethylene production.

The importance of ethylene and auxin in the control of epinastic growth movements has also been convincingly demonstrated by studies of the effects of anaerobic waterlogged soils on plant development. Various symptoms of "flooding-injury" usually appear in the shoot quite quickly after the roots are placed under anaerobic conditions, with epinastic movements beginning within a day or so of flooding (Fig. 7.2). Enzymatic conversion of the immediate precursor of ethylene in plants, 1-aminocyclopropane-1-carboxylic acid (ACC) (p. 68), is an oxygen-requiring reaction. In a plant subjected to flooding ACC generated in the anaerobic roots is unable there to be converted to ethylene but is translocated up into the shoot, where the availability of atmospheric oxygen permits its transformation to ethylene. The role of auxin in these events is not quite clear although, as we have seen, auxin has a stimulating effect on ACC and ethylene formation (p. 119 and Fig. 5.13). It is possible, therefore, that auxin, derived originally from the apical region of shoot, enters the roots where it stimulates the formation of ACC.

7.4 THIGMOTROPISM AND THIGMONASTY

The familiar sight of tendrils of a pea plant, or many climbing plants, twining around a support is a good example of a response to contact or mechanical stimulus. Tendrils of some plants are *thigmotropic* (the direction of bending is determined by which side of the tendril surface is stimulated by touch) while those of other species coil in the same direction regardless of which side is stimulated and are therefore designated as *thigmonastic* tendrils. In morphological terms tendrils are either modified leaves (e.g. pea) or modified stems (e.g. *Vitis vinifera,* the grape). Their function is to hold the plant to supporting structures through their coiling movements. The observable response of a tendril begins quite quickly after tactile stimulation, the lag phase (latent period) varying from approximately 0.5 to 35 minutes depending on species. In some species the tendrils form adhesive pads after making contact

Fig. 7.2. Epinastic responses of tomato plants to waterlogged soil conditions ("Waterlogged"), to high concentrations of atmospheric ethylene ("100 ppm Ethylene"), or to low concentrations of oxygen around the roots in plants growing in nutrient solution ("1% Oxygen"). (Photographs by courtesy of Dr. M. B. Jackson, A. R. C. Letcombe Laboratory, Wantage, U.K.)

with a surface (e.g. *Parthenocissus tricuspidata*) rather than coiling tightly around the support as in species such as pea.

In coiling tendrils the earliest bending response near the point of stimulus after the initial lag phase occurs quite rapidly, but this can then be followed by a slower but continuous process of bending along the length of the tendril (both acropetally and basipetally) so that it is gradually thrown into a series of coils. The initial rapid reaction appears to involve a turgor-based mechanism based upon solute movements into and out of cells near the region of tactile stimulus. In the case of pea tendrils this early rapid reaction is enhanced by light or ATP, and some workers have suggested that an ATPase-sensitive contractile-protein may be involved in early pea tendril coiling. There is not, however, convincing experimental evidence for this suggestion.

Auxin and ethylene do appear to be involved in tendril movements as they can induce coiling in the absence of any contact stimulus but their precise roles are uncertain. Auxin-induced coiling occurs if the whole tendril is floated on a solution of IAA or if only the tip is dipped in an IAA solution (Fig. 7.3), and contact-stimulated tendrils of pea plants evolve ethylene at a rate several times greater than do other parts of the pea shoot.

Tendrils must possess a receptor that is sensitive to touch (*a mechanoreceptor*) which when stimulated causes a sequence of processes, including those involving hormones, that culminate in the morphological response. Although the details of mechanoreceptors in plants are by no means understood, touch-sensitive organs such as tendrils do show certain morphological features that appear to be associated with stimulus perception. Scanning electromicroscopy has revealed the

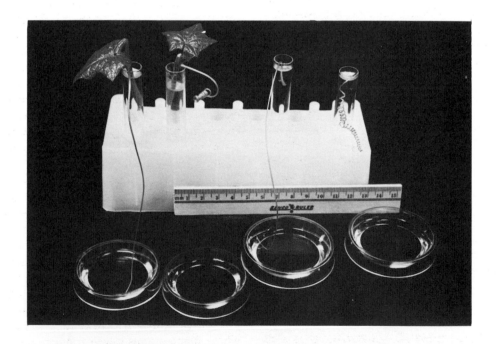

Fig. 7.3. Ability of auxin (IAA) to substitute for contact stimulus in induction of tendril coiling. Tendrils of *Marah fabaceus* 15 hr after their tips were placed in either water (first and third from the left) or in 150 mg/dm³ IAA solution (second and fourth from left). Coiling of the IAA-treated tendrils has caused them to lift out of the solution. (Photograph by courtesy of Dr. Leonora Reinhold, Hebrew University, Israel.)

presence of various dome-shaped cells called tactile papillae (e.g. in *Eccremocarpus scaber*), warty protrusions of epidermal cells called tactile blebs (e.g. in *Luffa cylindrica*). Many species, though lacking these specialized cell types, do have a wrinkled or corrugated thigmosensitive epidermal surface.

Mechanoreception may be assumed to occur at the cell surface, particularly the cell membrane, which is deformed by mechanical forces, and electrostatic interaction between a solid stimulus and the cell surface may also be involved. Plasmodesmatal connections are particularly well developed between the mechanoreceptor cells and surrounding cells, and this feature may be of significance in the transmission of the local effects of the stimulus to other responding parts of the tendril.

Mechanical stimulation, particularly rubbing, of most plants (i.e. those not previously thought to be especially touch-sensitive) also results in growth responses, with the stems of many vascular plants elongating at a slower rate and increasing in diameter more than normal, resulting in a much shorter more stocky shoot. The name *thigmomorphogenesis* has been given to this and related developmental responses to mechanical stimulation.

Thigmomorphogenetic responses are associated with enhanced ethylene synthesis in the mechanically stimulated tissues, suggesting that they represent ethylene-induced developmental processes. Supporting this view are findings that inhibitors of ethylene synthesis prevent thigmomorphogenesis, whereas ethylene or ethylene-generators such as 2-chloroethane phosphonic acid (p. 147) can induce thigmomorphogenesis in non-mechanically stimulated plants.

7.5 CHEMOTROPISM AND HYDROTROPISM

The tropic movement of a plant organ in response to an external chemical stimulus is named *chemotropism*. An example is seen in pollen-tube growth down the style towards the ovules, for the directional stimulus for the elongating pollen-tube is provided by certain chemicals present in the ovule and ovary wall, though their precise nature is not known.

It is sometimes said that roots respond to differences in soil moisture content by growing towards regions of greater water potential, and the term *hydrotropism* has been applied to this. However, there is no unequivocal evidence for true hydrotropism in plants. In other words, it appears unlikely that roots are able to detect and respond tropically to a gradient in water potential in their environment.

7.6 PHOTOTROPISM

Phototropism is the term applied to the phenomenon whereby a plant organ responds to a directional ("unilateral") light stimulus by undergoing a directional, or differential, growth response. As we saw earlier (p. 52), it was studies of phototropism which led to the discovery of auxins. Phototropism differs from photomorphogenesis (Chapter 8), in that photomorphogenic responses are neither dependent on a *directional* light stimulus nor do they show characteristics that are related to the direction from which the photomorphogenic light stimulus may be received.

In general, stems and other aerial portions of plants are positively phototropic (i.e. they bend towards the light source), while roots and other underground organs are negatively phototropic (they bend away from the light source). There are, however, many exceptions to these rules; for example, some tendrils and stems are negatively phototropic, and many roots non-phototropic or even positively phototropic when young, becoming negatively phototropic only later on. There is no doubt, nevertheless, that phototropism is of great importance in determining the direction in which plant organs develop under natural conditions. Phototropic responses, by definition, can occur only in those parts of a plant that retain the capacity for growth, particularly elongation growth. Thus, one sees phototropism in the young growing stems, leaves and roots of higher plants, and also in the sporangiophores of some fungi, in the sporophores of mosses, and chloronemata of ferns. The majority of studies of phototropism have, however, been made on etiolated coleoptiles of grass seedlings (particularly of oat, wheat, maize, and barley) and on fungal sporangiophores. The reasons for this are the convenience which such organs offer in experimentation, and their great sensitivity to directional illumination. Although such work has yielded much valuable information on phototropism, one cannot be confident that the derived concepts are directly applicable to other organs, such as green leafy shoots, growing under natural daylight conditions.

As in all environmentally induced adaptive responses, phototropism involves first of all *perception* of the stimulus (directional light) which is followed by development of the *response* (directional growth). One can discuss separately the mechanisms of perception and response, but keeping in mind that the two processes are linked together in the plant.

The earliest major advance in our understanding of phototropism was taken by Charles Darwin (1880) in his studies of the phototropic responses of coleoptiles (Chapter 3). In particular, he demonstrated that the tip of the coleoptile was the region of perception of a directional light stimulus, and that the differential growth response occurred lower down. Although it has subsequently been found that more basal parts of coleoptiles may also show some sensitivity to directional light, Darwin's basic results and conclusions have been amply confirmed so far as coleoptiles are concerned. Thus, the coleoptile tip is the region of maximum photosensitivity, and following receipt of a directional light stimulus it transmits basipetally some influence which elicits differential growth in the elongating parts of the coleoptile.

7.6.1 The nature of the photoreceptor in phototropism

(a) *Dose-response relationships.* Early in the twentieth century, before the discovery of auxin,

experimental work on phototropism concentrated upon biophysical aspects. As early as 1909 it was established by Blaauw that in both grass coleoptiles and *Phycomyces* sporangiophores, the Bunsen-Roscoe Reciprocity Law holds over a rather wide range of light intensity and time. The reciprocity law states that when only a single photoreceptor is operating, then the photochemical effect of light remains the same if the *quantity,* or dose, of light (i.e. irradiance level × duration of irradiation) remains the same. Blaauw's observations therefore indicated that a single photoreceptor is operative in phototropism, and led logically to investigations of *action-spectra* for phototropism with a view to identifying the photoreceptive pigment concerned in the perception of directional light.

The dose-response relationships for phototropism are, nevertheless, much more complex than first appeared the case. Thus, in the case of etiolated coleoptiles it has been found that with increase in the quantity of stimulus there is an increase in the bending response towards the light source until a maximum is reached (with approximately $0 \cdot 1$ J m^{-2} light energy), above which, with increasing quantity of stimulus, the response falls off until at a certain value the initial "positive curvature may even be reversed and a "negative" curvature (i.e. away from the illuminated side) occur. With still greater quantities of light stimulus the curvature may again become positive. (Fig. 7.4). The Bunsen-Roscoe Reciprocity Law has been found to be valid for only the first positive and first negative curvature responses.

(b) *Photropic action-spectrum studies.* A major step in elucidating any photobiological process is the identification of the primary photoreceptor pigment in the responsive organism. Usually, the first thing to be done in the pursuit of the photoreceptor is to determine an action spectrum for the photobiological process. Research to identify the phototropic photoreceptor has therefore naturally centred around attempts to match the action spectrum for phototropism with the absorption spectra of likely photoreceptor pigments.

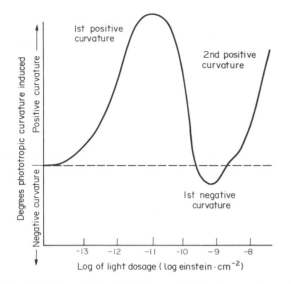

Fig. 7.4. The relationship between phototropic curvature and dosage of unilateral blue light in the *Avena* coleoptile. Coleoptiles of other cereals show a similarly shaped curve, except that there may not be a clear 1st-negative curvature. (Adapted from W. R. Briggs, in *Photophysiology*, Vol. 1 (ed. A. C. Giese), pp. 223-71, Academic Press, New York and London, 1964.)

Results obtained do not yet permit a definite decision to be taken on the identity of the pigment. The action spectrum for the first-positive curvature in coleoptiles (Fig. 7.5) shows that maximum curvature occurs in response to light of the blue wavelengths (maxima at about 445 and 474 nm, and a shoulder in the 425 nm region). Another, much lower, maximum occurs near 370 nm (the near ultra-violet). There is a sharp cut off in activity at about 500 nm with no activity at wavelengths longer than this.

Accurate measurements have shown that the action spectrum for positive phototropism in sporangiophores of the fungus *Phycomyces* is almost identical to that for *Avena* and other cereal coleoptiles. This indicates that the same phototropism photoreceptor is probably involved in these very diverse organisms. Such action spectra suggest that the phototropism photoreceptor is a yellow pigment. The two strongest possibilities are a carotenoid or a flavin pigment. The absorption spectra of ß-carotene and riboflavin are shown in Fig. 7.6, and it can be seen that although both have absorption maxima in the blue wavelengths,

neither of the absorption spectra are good matches with the action spectrum for phototropism (Fig. 7.5).

Thus, action spectrum studies in relation to naturally occurring pigments such as carotenoids and flavins have not produced a definite answer on the question of the identity of the phototropism receptor. Recent research is, nevertheless, leading to the view that the phototropism photoreceptor may indeed be a flavoprotein complex of a flavin (perhaps riboflavin) and a b-type cytochrome, and that this flavin/b-cytochrome complex is membrane-bound. The work which initiated this currently attractive hypothesis was first conducted in 1974 and 1975 on the cellular slime mould *Dictyostelium discoideum* and on the fungi *Phycomyces blakesleeanus* and *Neurospora crassa*.

It was found that blue light (maximum approximately at 465 mm) brings about a photoreduction of b-type cytochrome in these lower organisms, which is a region of radiation *not* absorbed to any significant extent by cytochrome itself. Further work indicated that a flavin absorbs the blue light, and close association between the flavin and cytochrome results in the latter being photoreduced. Preliminary cell-fractionation and other

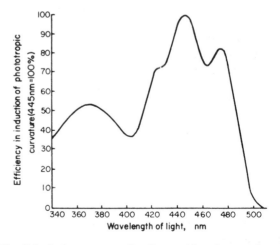

Fig. 7.5. Action spectrum for first-positive phototropic curvature in the *Avena* coleoptile. Maximum curvature occurs with unilateral light of the blue wavelengths. (From K. V. Thimann and G. M. Curry, *Comparative Biochem.* 1, 243-306, Academic Press, 1960.)

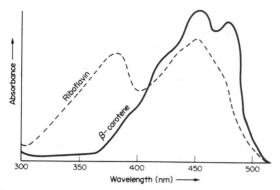

Fig. 7.6. Absorption spectra for riboflavin and ß-carotene. Neither spectrum matches the action-spectrum for phototropism (see Fig. 7.2), but when the determination on riboflavin is made in a lipoidal solvent then its absorption spectrum more nearly resembles the action-spectrum.

studies on *Neurospora* and on *Zea mays* coleoptiles have suggested that a similar flavin/b-cytochrome photoreceptor is present in the cells of both organisms and is associated with the plasma membrane. Further detailed research can tell us whether this flavin/cytochrome complex is the photosensor in phototropism.

Phytochrome, the red and far-red absorbing photoreceptor concerned in the numerous photomorphogenic phenomena in plants (Chapter 8), does not appear to be directly concerned in phototropism. However, it is known that exposure to red-light (non-directional) does affect the sensitivity of plant organs to directional blue light, and the effect of red can be reversed by exposure to far-red light. Such results indicate that phytochrome is the pigment which mediates red light effects in phototropism, but it is not known how phytochrome is linked to the phototropism system.

7.6.2 Transduction in phototropism of etiolated coleoptiles

A. Hypotheses Based On Lateral Asymmetries In Auxin Levels

Three main hypotheses have been put forward which account for transduction in phototropism on the basis of unequal levels of auxin in the illuminated and shaded sides of the organ:

1. The Choldny-Went Hypothesis

This was put forward independently by Cholodny and Went in the 1920s. The hypothesis proposed that a unilateral light stimulus induces *lateral transport of auxin* molecules across the most photosensitive region of a coleoptile (the tip region) towards the darker side, so that a higher concentration of auxin occurs in the shaded than the illuminated half of the tip. The hypothesis therefore envisages that greater amounts of auxin move down from the shaded half of tip into the sub-apical tissues of that side. These sub-apical tissues would consequently elongate more rapidly than the "auxin-starved" tissues below the illuminated half of the tip—i.e. a positive phototropic curvature ensues.

2. The Photodegradation of Auxin Hypothesis

Exposure of solutions of IAA to light results in photo-oxidation of IAA to hormonally-inactive products (p. 56), particularly in the presence of blue-light absorbing pigments such as riboflavin. Thus, it has in the past been suggested that photodegradation of auxin may occur in the illuminated side of a coleoptile tip, causing a differential auxin concentration between the two sides which in turn results in differential growth.

3. The Photoinhibition of Auxin Synthesis Hypothesis

This proposes that the rate of auxin synthesis is depressed in the illuminated half of a coleoptile tip.

Note that if either hypothesis 2 or hypothesis 3 above is valid, then unilateral light stimulation of a coleoptile tip would result in *a decrease in the total amount of auxin present,* whereas hypothesis 1 forecasts that exposure of the tip to unilateral light should have *no effect on total auxin* content.

B. Direct Light-Growth Reaction Hypothesis

The essential elements of this hypothesis are:
(a) light can have a direct effect on the growth rate of cells;
(b) the direct effect of light on growth is proportional to level of irradiance;

(c) auxin is not necessarily directly involved in the light-growth reaction.

Let us now first consider the three hypothesis that are based on lateral asymmetries in auxin levels. We consider the Direct Light-Growth Reaction Hypothesis of phototropism separately (7.6.3 below).

EVIDENCE RELATING TO HYPOTHESES OF PHOTOTROPISM INVOLVING LATERAL ASYMMETRIES IN AUXIN CONCENTRATION

The original evidence in favour of the classical Cholodny-Went theory was based upon work by both Boysen-Jensen and Went in 1928. Boysen-Jensen performed a simple experiment with oat coleoptiles, in which a small thin mica sheet was inserted across half the cross-sectional area just below the tip to act as a barrier to auxin flow. When the barrier was below the illuminated side

of the tip a normal positive phototropic curvature developed, but a barrier positioned below the shaded side of the tip prevented curvature (Fig. 7.7a). These results suggested that phototropism involves transmission of auxin from the shaded side of the tip, but did not provide any evidence for lateral displacement of auxin as suggested by Cholodny and Went. However, further mica barrier experiments were carried out by Boysen-Jensen, in which the barrier was inserted vertically into the tips of coleoptiles which were exposed to unilateral light. A positive phototropic response occurred only when the vertical barrier was parallel to the light rays (Fig. 7.7b), which indicated that a phototropic curvature of a coleoptile could develop only when lateral transport of auxin in the tip region was not impeded by a barrier.

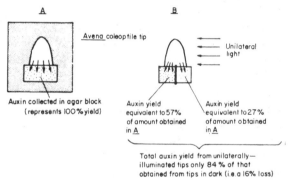

Fig. 7.8. Went's original experiment demonstrating that an oat coleoptile tip, when exposed to unilateral light, transmits more auxin to an agar block below the shaded side than to the block below the illuminated side (B). The total auxin yield from illuminated tips was 16 per cent less than that from tips maintained in darkness (A). This apparent loss was probably not significant. (From F. W. Went, *Rec. Trav. Bot. Neerl.* **25**, 1-116, 1928.)

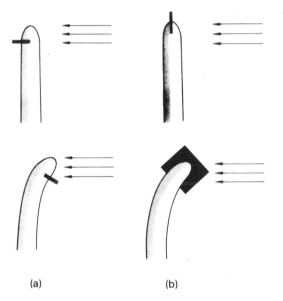

(a)   (b)

Fig. 7.7. Two experiments performed by Boysen-Jensen in 1928, which indicated that auxin moves preferentially down the shaded half of a unilaterally illuminated coleoptile (a), and that this involves lateral transport of auxin from the illuminated to the shaded side (b). Only when the route of basipetal auxin movement in the darkened side was unimpeded did a growth curvature appear (a). Similarly, a mica sheet inserted vertically into the tip region at right angles to incident light prevented a phototropic response, presumaby because auxin could not migrate laterally. A similar mica sheet inserted parallel to the light rays did not prevent curvature (b).

More direct evidence for lateral auxin transport in phototropically stimulated coleoptile tips was provided by Went in 1928, who made measurements of the quantities of auxin which could be collected in agar blocks placed below the illuminated and shaded sides of cut coleoptile tips (Fig. 7.8). It was found that unilateral light increased the proportion of total auxin in the agar blocks positioned below the shaded side. Later, Briggs and his co-workers carried out a series of experiments in 1957 using corn coleoptiles (*Zea mays*), which essentially confirmed earlier work by Boysen-Jensen and Went in that the results showed that the lateral differential in auxin concentration occurring in phototropically stimulated tips could be a consequence only of light-induced lateral transport of auxin from the light to the dark side. Briggs collected in agar blocks the auxin which diffused from coleoptile tips which had been subjected to various treatments, and then determined the quantity of auxin present in the blocks by means of the Went *Avena* curvature test. The amount of auxin produced by the tips kept in total darkness was similar to that produced by tips exposed to light at levels of irradiance capable of evoking a first-positive curvature (Fig. 7.4), whether or not the tips had been completely separated into two vertical halves by a thin piece of glass (Fig. 7.9a-d). This result argues against both the hypothesis of photodestruction of auxin

and the hypothesis of photoinhibition of auxin synthesis. Further, it was found that when a tip was stood on two separate agar blocks and only partially bisected at its base by a glass sheet, and was exposed to unilateral light incident at right angles to the glass plate, then significantly more auxin diffused into the agar block below the half of the tip remote from the light source (Fig. 7.9c). When, however, the whole of a coleoptile tip was bisected and separated by glass and exposed to the same conditions as a partially separated tip, then it was found that there was no difference in the amounts of auxin diffusing into the separate agar blocks below the "light" and "dark" halves (Fig. 7.9f). If either auxin destruction or inhibition of synthesis was responsible for the observed differential in the partially split tips one would have expected that a total glass barrier, as in Fig. 7.9f, would make no difference to the differential distribution of auxin observed in the partially split tips. The fact that a total glass barrier completely prevented the establishment of an unequal distribution of auxin led Briggs to conclude that the observed differential distribution of auxin that occurs in phototropically responding coleoptile tips is a consequence of lateral movement of auxin towards the dark side, and is not a result of photodestruction of auxin or of photoinhibition of synthesis.

In summary, there is a considerable body of evidence which indicates that when an etiolated coleoptile tip is illuminated at low light dosages from one side, there is first of all *perception* of the light-stimulus, followed by *transverse migration* of endogenous auxin molecules within the tip. In the case of positive phototropic responses auxin moves towards the darker side, which in turn means that more auxin is transmitted to the region of response in the coleoptile below the darker half of the tip resulting in greater elongation growth of that region and consequent bending of the whole coleoptile towards the light source.

There is, therefore, a substantial body of evidence, derived from several decades of research, that appears to have demonstrated that phototropism in etiolated coleoptiles is mediated through light-induced lateral transport of en-

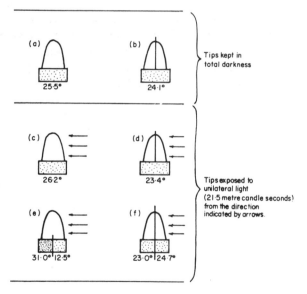

Fig. 7.9. Auxin diffusion into agar blocks from variously treated coleoptic tips of *Zea mays*. The figure under each agar block indicates the degrees curvature produced by that block in the Went *Avena* curvature test for auxin. The vertical line running through some of the tips and agar blocks represent an impervious glass barrier. Twice as much auxin was obtained in (e) and (f) than in (a)—(d), because each agar block had been in contact with six half-tips, the equivalent of the three whole tips placed on the agar blocks in (a) to (d) inclusive. (From W. R. Briggs, R. D. Tocher and J. F. Wilson, *Science*, **126**, 210-12, 1957.)

dogenous auxin (i.e. the Cholodny-Went Hypothesis for phototropism). Despite this it is only realistic to have reservations about the Cholodny-Went Hypothesis of phototropism in etiolated coleoptiles. The grounds for caution can be summarized under five headings as follows (much the same criticisms also apply to phototropism in green-non-etiolated, plant organs):

(1) All quantitative measurements of endogenous auxin from coleoptile tips in relation to phototropism have been obtained by use of bioassays, which are subject to large experimental error.

(2) The difference in concentration of auxin in the two halves of a phototropically stimulated tip has never been found to be greater than a factor of five and far more commonly a two to threefold difference has been found. Bearing in mind that the dose-response curve for auxin promotion of growth is log-linear (see Fig. 7.10) it is reasonable to entertain doubts as to whether a 2—3 fold difference in lateral auxin concentration could account for cessation of growth in one side of a coleoptile and its maintenance, or even acceleration, on the other side (see Fig. 7.10).

(3) Experiments with radioactive auxins have indicated that under certain conditions phototropic curvatures of etiolated coleoptiles can take place without lateral displacement of auxin, which of course is contrary to the prediction of the Cholodny-Went Hypothesis.

(4) The latent period for phototropic responses varies from 6 to 30 minutes depending on species and experimental conditions.
There is a significant question mark over the possibility of a sufficient lateral differential in auxin concentration developing within the latent period to cause the onset of differential growth in such times.

(5) There does not appear to be a simple quantitative relationship between amount of phototropic bending and the relative endogenous auxin contents of the two sides of a coleoptile.

Recognition of these weaknesses of the classical Cholodny-Went Hypothesis for phototropism have led in recent years to new approaches to the whole problem of phototropism (and also of geotropism as is discussed later), with a revival and intensification of interest in hypotheses of phototropism involving direct effects of light on cell growth (see p. 186 and 188).

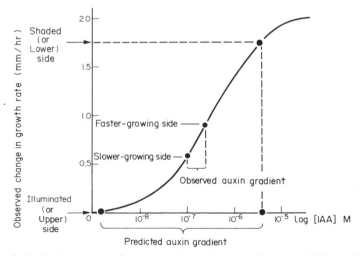

Fig. 7.10. The log-linear relationship between auxin concentration and growth. The lateral differentials in auxin concentrations which have been reported in tropically bending shoots are indicated ("Observed Auxin Gradient"), as are the observed changes in growth rate of the opposite sides of a tropically bending shoot. Also indicated is the lateral differential in auxin concentration which would be predicted to be necessary if such a concentration were solely responsible for the asymmetry in growth rates. (Adapted from A. B. Hall, R. D. Firn and J. Digby, *J. Biol. Education* **14**, 195-199, 1980.)

7.6.3 Evidence relating to the direct light-growth reaction hypothesis

In the early years of this century it was noticed by Blaauw that in both grass coleoptiles and *Phycomyces* sporangiophores, blue light administered symmetrically results in transient changes in growth rates. This has been called the "light-growth reaction". The action spectrum for these light-growth reactions matches the action spectrum for phototropism for the particular organism. Some workers, including Blaauw, concluded that phototropism can be explained completely in terms of the light-growth reaction.

Thus, it has been sometimes envisaged that in the oat coleoptile, where blue light initially suppresses growth, a unilateral irradiation with blue light would set up the type of growth asymmetry required for commencement of a positive curvature (i.e. towards the light source). However, it is difficult to understand what would happen after the first 20 minutes, when the transitory growth inhibition is converted to a transitory increase in growth rate, but it is possible that the time relationships are different under conditions of unilateral irradiation. Thus, the involvement of a light-growth reaction in phototropic responses of coleoptiles or other higher plant organs has not been convincingly demonstrated, but we shall return later to evidence that suggests direct effects of light on cell elongation when discussing the physiology of phototropism in green plants, as opposed to etiolated coleoptiles which have been the subjects of so much of the research into phototropism for the past century.

7.6.4 Phototropism in green (de-etiolated) plants

The great majority of experiments so far conducted on phototropism have been concerned with the behaviour of etiolated organs, particularly coleoptiles, and there is uncertainty as to how relevant the findings and derived concepts are to the

behaviour of green leafy plants. It is of course a commonplace observation that green dicotyledons and monocotyledons, as well as etioled coleoptiles, show phototropic bending responses to uneven illumination (Fig. 7.11).

Morning

Noon

Evening

Fig. 7.11. Phototropic growth movements in a single sunflower (*Helianthus annuus*) plant photographed from the north side at different times of the day. Such "sun-following" movements are sometimes termed "heliotropic" movements, but the phenomenon is that of phototropism. (From H. Shibaoka and T. Yamaki, *Sci. Papers Coll. Gen. Education, Univ. Tokyo,* **9**, 105-26, 1959. Photographs by courtesy of Dr. Hiroh Shibaoka.)

Individual leaves of green plants also show phototropic movements, and these involve differential growth rates of the shaded and illuminated sides of the leaf. In phototropic movements of green leaves a supply of auxin is essential for a bending response to occur in the petiole. Thus, excision of the lamina (the region of auxin synthesis) prevents a phototropic response in the petiole, and application of exogenous auxin to the cut distal end of a debladed petiole restores its capacity to bend.

There seems little doubt that phototropism is of importance in plant ecology, as it involves responses to the light environment that regulate the orientation of plants and their individual leaves for the most efficient trapping of available light energy for photosynthesis. However, very few experiments indeed have been done on light-grown plants—either monocotyledonous or dicotyledonous, but available evidence indicates that light-grown plants are much less sensitive to directional blue light than are etiolated coleoptiles, responding only to energies corresponding to the second-positive response of etiolated coleoptiles (Fig. 7.12). Nothing is known of the

photoreceptor pigment(s) in phototropism of green plants. An added complication in leafy shoots, particularly in dicotyledonous plants, is that it has been found that most of the auxin and gibberellin required for elongation of the stem comes from the young expanding leaves near the apex. In other words, the lamina of a young leaf exports auxin and gibberellin via the petiole into the stem. Consequently, any phototropic curvature of such a stem may be a result of some alteration in the distribution, or quantity, of hormones coming from the leaves. In the case of sunflower plants (*Helianthus annuus*), the leaves are arranged in pairs on opposite sides of the stem (decussate arrangement) and it has been found that each leaf of a pair will, if both are illuminated to the same degree, supply equal quantities of auxin to the stem. If, however, one leaf of a pair is more brightly illuminated than the other because of its orientation in relation to the incident light (Fig. 7.13), then the leaf receiving a higher intensity of light produces a greater quantity of auxin than its partner. This perhaps results in the side of the stem beneath the brightly illuminated leaf receiving more auxin and consequently growing at a more rapid rate than the other side, causing the stem to execute a positive phototropic curvature, until the position is reached whereby both leaves receive light at equal angles of incidence (Fig. 7.13). Similarly, it has been found that if one side of the lamina of a leaf is covered with opaque material such as aluminium foil, then the side of the petiole on that side elongates more rapidly and consequently bends to displace the lamina in the opposite direction. One interpretation of such results is that the shaded part of the lamina transmits more auxin (and perhaps gibberellin) to the corresponding side of the petiole than does the other, illuminated, half of the lamina to its side of the petiole.

Thus, there is reasonable circumstantial, and some direct, evidence that differential growth in the opposite sides of phototropically responding green stems and leaves involves differences in the lateral distribution of auxin levels in the tissues. Such lateral differentials in auxin concentration do not, however, appear to arise by transverse

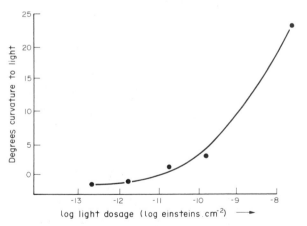

Fig. 7.12. Light-grown (green) plants show phototropic responses only to relatively high energies of irradiance. The graph is the dose-response curve for phototropism in green (i.e. non-etiolated) radish seedlings. Unilateral monochromatic blue light (460 nm) was applied at the rate of 6 erg.cm^{-2}. sec^{-1} for varying times to give the indicated light dosages. Note that a significant positive phototropic response occurred only with a stimulus at energy levels as high as those that induce the second-positive response in etiolated coleoptiles (see Fig. 7.4). (Adapted from M. Everett, *Plant Physiol.* **54**, 222-5, 1974.)

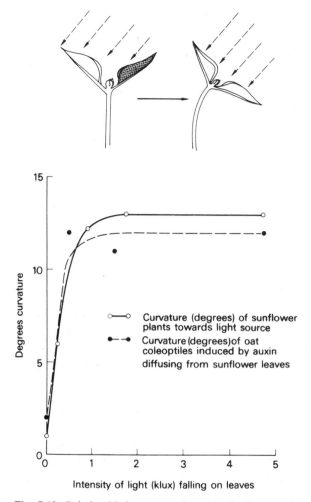

Fig. 7.13. Relationship between auxin production by opposite leaves and phototropic growth curvature in the stem of *Helianthus annuus* plants. *Above.* When light is not falling evenly on each member of a pair of leaves, a phototropic curvature takes place until both leaves are equally illuminated. *Below.* The effect of light intensity on auxin production in leaves, and on induction of phototropic growth curvatures. (From H. Shibaoka and T. Yamaki, *Sci. Papers Coll. Gen. Education, University of Tokyo,* **9,** 105-26, 1959.)

amounts of endogenous gibberellins are transported along the darker (more rapidly elongating) side of a sunflower stem than through the tissues of the illuminated side. Also, the more slowly elongating illuminated sides of phototropically responding hypocotyls of sunflower seedlings have been found to contain a greater proportion of the growth-inhibitor xanthoxin (p. 71) present than the more rapidly elongating shaded sides (Table 7.1). The physiological significance of lateral differentials in xanthoxin levels remains to be established.

Thus, whilst shoots of green plants exhibit phototropic responses to unilateral blue light, available evidence is inconclusive as to whether differentials in lateral distribution of auxin are always involved. Where lateral auxin distribution does appear to be important (such as in petioles) there is no evidence for differential lateral transport of auxin being concerned.

Relatively high irradiance levels of diffuse blue light have rapid (less than 5 minutes latent period) inhibitory effects on straight elongation growth of green stems. The magnitude of the inhibitory effect of blue light is irradiance dependent, but the duration of the latent period is not affected by irradiance level. These features of the effect of blue light on straight elongation growth in de-etiolated plants have logically led to the recent suggestion that phototropism is probably a result of a direct blue light-inhibition of elongation growth in cells on the side of an organ exposed to unilateral light. The pigment system (perhaps a flavoprotein) involved in blue light effects is probably the same in photomorphogenetic phenomena induced by high light irradiances (the High Irradiation Reaction, see Chapter 8) and in phototropism. The name *cryptochrome* is sometimes given to this blue-light photoreceptor. The rapidity of the blue light action suggests that differences in hormone transport are not involved in phototropism. If, as seems likely, the direct blue light-reaction proves to be the basis of phototropism in plants, the similarity to the much better defined mechanism of phototropism in sporangiophores of the fungus *Phycomyces* (p. 186) is obvious and of considerable interest.

lateral transport of auxin within the responding organ (as would be forecast by the Cholodny-Went Hypothesis (p. 182).

Measurements of endogenous auxin concentrations in green stems exposed to unilateral light have indicated no convincing difference between illuminated and shaded sides (Table 7.1). On the other hand, there is some evidence that greater

Table 7.1 *Lateral distribution of endogenous IAA and endogenous xanthoxin (as percentage of total in each side) in straight (non-phototropically stimulated), bending and fully curved hypocotyls of* **Helianthus annuus.** *(Data from J. M. Franssen and J. Bruinsma, in* **Tropic Responses of Plants** *(Eds.), J. Digby and R. D. Firn, Abstracts of Papers on Tropic Responses in Plants, p. 16, S.E.B. Conference, York, 1979, and from J. Bruinsma et al., J. exper. Bot.* **92,** *411-18, 1975.)*

	Straight Hypocotyls		Bending Hypocotyls			Fully responded Hypocotyls		
	Left side (%)	Right side (%)	Light side (%)	Shaded side (%)	Curvature (in°)	Light side (%)	Shaded side (%)	Curvature (in°)
IAA*	49.5	50.5	52.0	48.0	17	48.0	52.0	21
Xanthoxin**	48.5	51.5	66.5	33.5	15.5	59.0	41.0	34.5

*One experiment involving 55 plants.
**Mean of two separate experiments, each involving approximately 50 plants.*

7.7 GRAVITROPISM (GEOTROPISM)

The term gravitropism is applied to plant growth movements induced by either a gravitational stimulus or by mass acceleration (e.g. by centrifugation). It is because these types of growth movement are induced not only by the earth's gravitational pull but also by mass acceleration that the traditional prefix "geo" has now been replaced by "gravi". Growth of an organ towards the centre of the earth (or in the direction of the vector of mass acceleration) is called positive gravitropism, and growth away from the centre of the earth (or against the mass acceleration vector) is termed negative gravitropism. Positive and negatively gravitropic organs, such as the stem and root of the main plant axis, which align themselves parallel to the direction of the gravitational pull, are said to be orthogravitropic. When the axis of an organ come to lie at right angles to the direction of the gravitational field it is said to be diagravitropic (e.g., rhizomes of Solomon's seal, couch grass, etc., or stolons of potato and strawberry). Where an organ becomes oriented at intermediate angles (i.e. between 0° and 90°, or between 90° and 180° from the vertical), it is said to be plagiogravitropic (e.g. lateral branches are very often so). Most main roots are positively

gravitropic. The rhizomes of many mosses are sometimes positively gravitropic, but, in general, positive gravitropism is not well developed in lower plants. Negative gravitropism is shown by the stems of higher plants, by the sporangiophores and sporophores of many fungi, and by the foliage shoots of mosses. While many rhizomes and stolons are diagravitropic, lateral stems and lateral roots of the first order and foliage leaves are commonly plagiogravitropic. Lateral shoots and lateral roots of a higher order generally possess little gravitropic sensitivity and are consequently said to be *agravitropic.*

Moving a plant from its usual vertical position to a horizontal one causes gravity to act across the width of stem and root. This results in growth curvature responses, whereby the stem bends to grow upwards and the root bends to grow downwards. This can easily be demonstrated with young seedlings.

Gravitropism of shoots resembles phototropism in many respects: both, (a) are directional growth curvature responses to directional stimuli, (b) have similar latent periods, (c) involve bending along all or most of the length of the responding organ as a consequence of cessation or reduction of elongation one on side and an acceleration on the other, and (d) sensitivity to the stimulus is not confined to the apical region. In roots, however, sensitivity to gravity is completely or largely confined to the root apical region.

7.7.1 Gravity perception

Some sort of *graviperception* mechanism must exist in plants, which senses the direction of gravity in relation to the orientation of the organ. It has been known for many years that the gravitropic response is a "threshold" phenomenon, in that a gravitational stimulus has to reach a certain minimum level, specific to the particular organ, in order to evoke a gravitropic bending response. The quantity of stimulus is equal to the gravitational force multiplied by the time for which it acts. For a given force, the period of exposure that is required to elicit a just detectable response is called the *presentation time.* Such threshold phenomena suggest that geoperception in plants involves the movement of free-falling bodies, or *statoliths,* which must move a certain distance to trigger the geotropic response mechanism. At a given temperature the presentation time is proportional to the inverse of the quantity of gravitational stimulus applied; that is, reciprocity holds for the gravitropic stimulus as well as for the phototropic stimulus.

Mathematical analyses have been made to evaluate putative statoliths, and these have indicated that cellular inclusions as small as mitochondria could move rapidly enough within the cytoplasm in response to gravity to account for known presentation times. However, the kinetics of graviperception match most nearly the kinetics of gravity-induced displacement of starch grains in plant cells (Fig. 7.14). In fact, light- and electron-microscope studies of *statocytes* (gravity-sensing cells that contain statoliths) have revealed sedimentation of starch grains in response to gravity (Fig. 7.15), and also that the sedimentable starch grains are membrane-bound as *amyloplasts* (an amyloplast is a modified plastid, containing two to several starch grains). The term *statenchyma* is applied to a tissue composed of statocyte cells. Not only are the positions of amyloplasts changed following reorientation of statocyte cells,

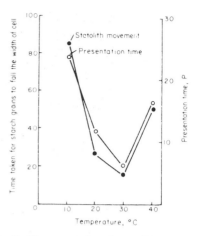

Fig. 7.14. Positive correlation observed between presentation time in gravitropism of *Lathyrus odoratus* seedling stem, and time taken for starch grain sedimentation at different temperatures. (From L. Hawker, *Ann. Bot.* **47**, 505-15, 1933.)

but other cytological events have been observed, particularly alterations in the distribution of the endoplasmic reticulum. Although other organelles (in the *Chara* rhizoid barium sulphate crystals appear to serve as statoliths; Fig. 7.16) may function as statoliths in some tissues, particularly those that do not contain amyloplasts, there is considerable circumstantial evidence that in most higher plant organs gravitropic responses are initiated by the sedimentation of amyloplasts under the influence of gravity. Thus, plant organs deficient in starch (either naturally so or experimentally "destarched" by treatment with solutions of cytokinin and gibberellin, or low temperature) have been found to require very much longer presentation times for a geotropic response to occur.

Where in plants are statocyte (statolith-containing) cells located? Until very recent years it has been generally assumed that graviperception can occur only in the apical regions of coleoptiles, shoots and roots.

In the case of gravitropism in seedling roots, it has been established for several species that the important graviresponsive amyloplasts are located in the central cylinder of root-cap cells. By micromanipulation it is possible to remove the root cap from otherwise intact roots, and this abolishes gravitropic responsiveness in most

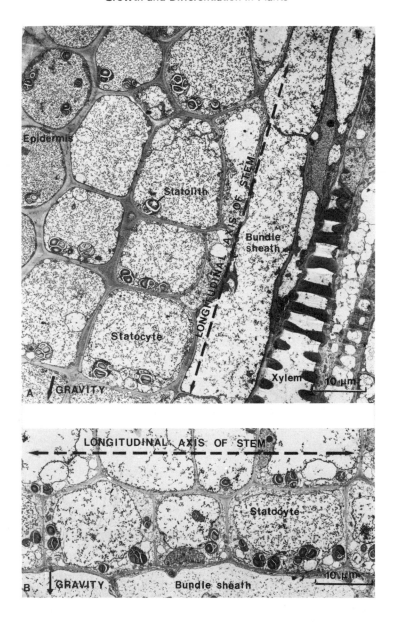

Fig. 7.15. Effect of gravity on distribution of amyloplasts (statoliths) in statocyte cells of the leaf sheath base of the shoot of *Echinochloa colonum*. A: Longitudinal section of a vertical shoot, in which the statoliths are sedimented at the basal ends of the statocytes. B: L.S. of shoot held horizontally for 2 hours (the apical ends of the cells lie to the left) in which it can be seen that the statoliths have sedimented to the now lowermost lateral wall of each statocyte. (Photographs by courtesy of Dr. Mary Parker, Plant Breeding Institute, Cambridge.)

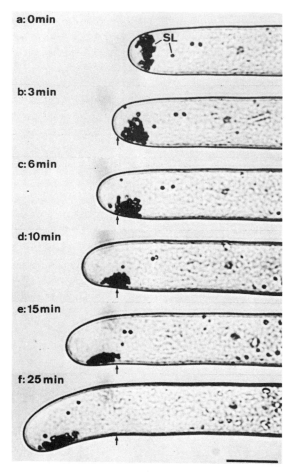

Fig. 7.16. Time-lapse photographs of a *Chara* rhizoid at various times after displacement from vertical to horizontal orientation. Note sedimentation of the barium-sulphate crystallite statoliths followed by downward bending of the rhizoid tip. The arrow indicates the same point on the cell wall in each photomicrograph. (Photographs by courtesy of Professor Dr. A. Sievers and Dr. D. Volkmann. From, *Encyclopedia of Plant Physiology N.S.*, Vol. 7, eds. W. Haupt and M. E. Feinleb, Springer-Verlag, Berlin/Heidelberg/New York, pp. 567-72, 1979.)

species. Only roots that contain additional sedimentable amyloplasts in the root apical zone, or which rapidly form such amyloplasts after root-cap removal, retain some ability to respond to a gravitational stimulus after depriving them of the root cap.

Although graviperception does appear to be restricted to the apical zone of *roots,* it is now quite clear that the capacity for graviperception

exists along the *whole* length of the stem of a dicotyledonous plant or of a coleoptile. In the mature shoots of grasses and cereals, gravitropic bending occurs only at the nodes. Experiments with isolated grass and cereal nodes have shown that the capacities to both perceive the direction of action of gravity and to show a differential growth response are contained within the node itself. Thus, in all types of shoot structures studied there is no longitudinal separation of regions of graviperception and graviresponse, and therefore, gravitropism in shoots does not involve the longitudinal signal transduction mechanism which for many years has been frequently assumed to occur (the situation in roots is different—see below).

However, the apical region of a dicotyledonous plant stem or coleoptile is the source of hormones (particularly auxin) which are required for the elongation growth of cells in the regions of the growth axis that bend during gravitropism. Thus, although it is true that a freshly decapitated dicotyledonous plant stem will execute a gravitropic bending response, it is able to do so because it already contains sufficient auxin derived from the excised apical parts. Consequently, the apical region of a dicotyledonous plant shoot or of a coleoptile transmits auxin that is required for growth of cells in gravitropically responding regions. There is, though, no evidence for any type of shoot structure that *graviperception* is restricted to apical regions. The question as to whether a *lateral asymmetry* in auxin concentration is involved in gravitropism is discussed later (p. 194).

Although it is quite well established that graviperception in gravitropism involves sedimentation of statoliths, usually amyloplasts, little is known of the nature of the biochemical or physiological changes induced within statocyte cells by organelle displacement. However, several strands of evidence suggest that sensitivity to gravity in plants involves sensitivity of cell membranes to pressure. Most evidence indicates that membranes of the endoplasmic reticulum (ER) are of importance in root-cap statenchyma, but that in shoot statenchyma the plasmalemma may well be the sensitive membrane. The pressure is presumed to be exerted by sedimenting statoliths (usually amyloplasts in higher plants). A diagrammatic representation of the effects in statocyte cells of re-orientation of *Lepidium* roots from vertical to horizontal are shown in Fig. 7.17. One can see from this model that since the ER complex in these root statocytes shows polar distribution (the ER lies mainly at the apical end of each statocyte cell), then turning the root to a horizontal position results in a reduction of pressure by the amyloplasts on the ER in cells on the upper side of the root and an increase in the lower side cells. These changes in pressure would actually occur even without movement of the statoliths, though in fact the amyloplasts do sediment within a few minutes. In statocyte cells of other species of plant, and particularly in shoot statenchyma, there is not always such polarity of ER distribution, but in these similar pressure changes occur with respect to the plasmalemma and statolith movement can also displace the ER membranes in various ways.

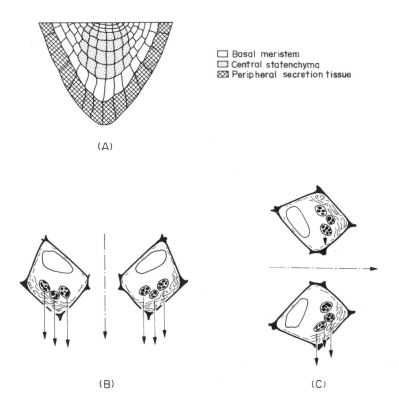

☐ Basal meristem
▨ Central statenchyma
⊠ Peripheral secretion tissue

(A)

(B)   (C)

Fig. 7.17. Graviperception in *Lepidium* roots. A: diagram of a longitudinal section through the root-cap illustrating the central cylinder that contains sedimentable amyloplasts and appears to serve as statenchyma. Note the geometry of the cell layers. B and C: cells of symmetrically located lateral statocytes in vertical (B) and horizontally positioned (C) roots. The dashed arrow in each case points towards the root tip. In vertical orientation of the root the pressure exerted by the amyloplasts on the endoplasmic reticulum (which lies mainly at the apical end of each cell) is equal in statocytes on either side of the root axis (equal arrow lengths), but in horizontal orientation pressure by amyloplasts on the ER is different either side of the longitudinal axis (unequal arrow lengths). (Adapted from D. Volkmann and S. Sievers, *Encyclopedia of Plant Physiology*, N. S., Vol. 7, ed. W. Haupt and M. E. Feinleb, Springer-Verlag, Berlin/Heidelberg/New York, pp. 573-600, 1979.)

Electron microscopy has demonstrated that ER membranes are certainly sensitive to the pressures exerted by amyloplasts, for imprints of sedimented amyloplasts can be clearly seen in the surface of ER membranes. Also, the ER cisternae are compressed and distances between successive elements of the stack are reduced to a minimum. There is, therefore, rather convincing evidence that the ER complex may function as a gravi-sensor.

The connection between effects of sedimenting statoliths on pressure-sensitive membranes such as the plasmalemma or those of the ER in statocyte cells, and the changed direction of growth which occurs in eventual consequence, is an intriguing problem that we will now consider. Although the mechanism of graviperception appears to be fundamentally similar in shoots and roots, the mechanism by which the activation of perceptor membranes by statoliths is translated (i.e. transduced) into growth curvatures seems to differ in shoots and roots. Thus, transduction in gravitropism of roots and shoots is best considered separately.

7.7.2 Transduction in shoot gravitropism

In the principal types of shoot organ studied (coleoptiles, hypocotyls and internodes) it is generally the case that negative gravitropic curvatures occur through a slowing down or complete cessation of elongation growth along the whole upper side and an acceleration along the lower side. The mature grass and cereal node differs slightly in that in the vertical position it does not elongate at all, but when it is placed horizontally the lower side elongates to cause upward bending and cells of the upper side becomes compressed. The overall growth patterns in graviresponding roots are more varied and complex and are described later (p. 196).

The latent period for negative gravitropism can be as short as 10 minutes (i.e. upward curvature can start 10 minutes after the organ is moved from a vertical to horizontal orientation).

As discussed previously, graviperception in a stem or coleoptile occurs along the whole length of the growth axis (i.e. sensitivity to gravity is not confined to the apical region). Thus, one cannot be sure on currently available evidence that there are separate statenchyma cells and cells that show differential growth responses. Thus, it is possible that the same cell can both perceive the gravitational stimulus *and* respond by an alteration in its growth rate. Nevertheless, one cannot ignore the existence of a considerable body of information that has for many years been interpreted as supporting the classical Cholodny-Went Hypothesis (see p. 182). This hypothesis, as applied to gravitropism, states that in a vertical organ auxin which is synthesized in the apical region is transported basipetally in a *symmetrical* manner so that the subapical regions grow straight, whereas if the organ is displaced to a horizontal position there takes place a downward lateral displacement of auxin so that auxin is *asymmetrically* distributed between upper and lower halves. In shoots, the greater concentration of auxin in the lower side is, on the basis of the Cholodny-Went Hypothesis, regarded as being the cause of faster elongation growth of that side and resultant upward curvature of the whole organ.

Although never clearly stated in the Cholodny-Went Hypothesis, it has been generally assumed that the apical region of the coleoptile would be where greatest lateral (downward) transport of auxin would occur. This idea led early workers such as Dolk, in 1930, to concentrate attention on the lateral distribution of endogenous auxin in coleoptile tips. Dolk placed excised coleoptile tips of *Avena sativa* and *Zea mays* horizontally, and collected the auxin which diffused from upper and lower halves into agar blocks pressed against the cut surfaces. The amounts of auxin in the agar receiver blocks were measured by means of the *Went* Avena curvature test (Fig. 3.1). A number of other workers have repeated Dolk's experiments and obtained similar results (Fig. 7.18), which therefore are in agreement with the Cholodny-Went Hypothesis. However, since in all these experiments auxin measurements were made by bioassays of the upper and lower agar receiver blocks, it is possible that what was measured was a *net* growth promoting effect of auxin and a growth inhibitor. That is, the results of Dolk and those illustrated in Fig. 7.18 could equally well be interpreted in terms of equal amounts of auxin moving into upper and lower agar blocks, but with an *upward* lateral displacement of a growth inhibitory substance coming from the coleoptile tip.

Therefore, there does not exist unquivocal evidence for downward lateral displacement of endogenous auxin in gravistimulated plant organs. More positive evidence for the Cholodny-Went Hypothesis has however been obtained in experiments on the movement of radioactive auxins in coleoptiles. When [^{14}C]-IAA or [^{3}H]-IAA was applied to either the apical cut ends of isolated coleoptile segments, or *via* a micropipette to the apical 0.2 mm of intact coleoptiles, it was found that horizontal orientation caused a downward displacement of radioactivity and that the lateral transport of auxin depended on respiration. Other experiments with radioactive auxins on dicotyledonous plants have yielded much less convincing evidence for downward lateral transport of auxin playing a role in the transduction processes of gravitropism. In the case of hypocotyls of *Helianthus annuus* slight downward movement of [^{14}C]-IAA has been reported, but no lateral displacement of [^{14}C]-IAA was found in horizontally positioned internodes of the same species. Endogenous gibberellin-like activity has been found to be asymmetrically distributed in

gravitropically responding internodes of *Helianthus annuus,* with higher levels in the lower, more rapidly elongating, side. Just as for similar observations of endogenous auxin, the results could be equally explained by the presence of an interfering growth inhibitor that is displaced upwards. Studies with radioactive gibberellins on stem segments indicated that the asymmetry found for endogenous gibberellin-like substances did not result from lateral displacement of the hormone.

The validity of the Cholodny-Went Hypothesis can be questioned on two further grounds:

(a) The kinetics of gravitropism in relation to the known velocity of auxin transport (approx. 1 cm/hr^{-1}) and the lag period for auxin promotion of cell growth (usually approx. 10 minutes). The latent period for gravitropism can, as previously stated, be as short as 10 minutes. It is reasonable to question whether during this period there is sufficient time for both a lateral differential in auxin concentration to develop and for auxin to initiate increased growth in the cells for the lower side.

(b) Whether the magnitude of experimentally measured asymmetries in auxin concentration between upper and lower tissues are sufficiently large to account for observed differentials in growth rates of upper and lower tissues. The lateral differentials in auxin concentration found in horizontal shoot organs have most commonly been of the order of 60-70 per cent

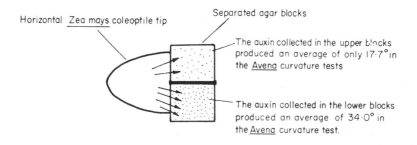

Fig. 7.18. Greater amounts of auxin activity are obtained from the lower side of a horizontally positioned coleoptile tip than from the upper side. The amount of auxin present in each agar block was measured by means of the Went *Avena* curvature bioassay (see Chapter 3). (From B. Gillespie and W. R. Briggs, *Plant Physiol.* **36**, 364-8, 1961.)

of total auxin on the lower side and 40-30 per cent on the upper side (i.e. ratios lower/upper of 1.5 to 2.3). It seems doubtful that a two-fold difference in auxin concentration could be the sole basis for the measurable differences in growth rates of upper and lower tissues (see Fig. 7.10).

In summary, there are substantial arguments against the long-established Cholodny-Went Hypothesis of gravitropism in shoots. What can be offered to replace it and which will provide a basis for future research into the mechanism of transduction in gravitropism? One hypothesis postulated by Firn and Digby over the past few years is based on the idea that the outer cell layers (perhaps the epidermis alone) of a stem or coleoptile both perceive gravity and are also the site of the gravitropic growth response. In essence, these workers propose that the rate of elongation growth of a stem or coleoptile is determined by the growth rate of the outer cells, and that the latter grow at rates dependent on their orientation with respect to the gravitational vector. There is experimental evidence that the outer cell layers of dicotyledonous shoots do play a very important part in controlling straight shoot extension growth. Thus, "peeling" (i.e. stripping of the epidermal cell layers) isolated stem segments is known to result in enhanced straight extension growth. Such peeled segments do, though, fail to undergo a gravitropic bending response when placed horizontally. An inference from these observations is that the outer cell layers normally limit the rate of shoot extension growth and that they must be present for graviperception. Firn and Digby suggest that peripheral cells of a stem grow more rapidly when their outer face is directed downwards than when they face upwards. Immediate future research to test these ideas will be aimed at maintaining such isolated superficial cell layers *in vitro,* and measuring their growth rates with different orientations. Also, of course, it is necessary to explore whether cells of this type contain sedimentable organelles that possess the required characteristics and behaviour to qualify as statoliths. If they do not, then it must be demonstrated what other type of graviperception mechanism they may possess.

7.7.3 Transduction in root gravitropism

The overall patterns of elongation growth in gravitropically responding roots appear to vary from species to species and with experimental conditions. Also, in contrast to shoot gravitropism where a particular pattern of differential growth is established after graviperception and that same pattern is maintained throughout the period of response, during root gravitropism the types of differential growth may change with time. There can, for example, be first of all a slowing of growth on the lower side and an acceleration on the upper side, but later on the rate of elongation of the upper side can fall to a level lower than before the root was placed horizontally. In a number of species it has been found that moving a root from a vertical to horizontal orientation reduces the *overall* rate of elongation growth.

Until recently, concepts of the transduction mechanism in root gravitropism were similar in principle to the mechanism suggested by Cholodny and Went in the 1920s for phototropism and gravitropism in coleoptiles. That is, it has been considered that auxin is synthesized or released from the root apex and undergoes lateral displacement towards the lower side of the organ. Because root elongation is inhibited by much lower concentrations of auxin than are supra-optimal for coleoptile or stem elongation (Chapter 5), it was thought that the auxin concentration on the lower side of a horizontally positioned root increases to become supra-optimal and consequently inhibitory to the growth of the auxin-sensitive root tissues. On the basis of this theory, the upper side of a horizontal orthogravitropic root would contain auxin at a more nearly optimal level and consequently grow at a more rapid rate than the inhibited lower side, resulting in a downward curvature of the root. However, current confusion as to the possible role of auxin in root elongation growth, and the problem of whether or not auxin is synthesized in root tips (Chapter 5), means that it is not possible to ascribe a role to auxin in root gravitropism. Nevertheless, an open mind should be kept on the problem, for there exists experimental evidence that downward lateral displacement of auxin can occur in gravitropically responding roots.

Over the past few years, it has been found that the root cap is not only the site of graviperception in root geotropic behaviour, but that in addition it appears to be a source of growth inhibitors, including abscisic acid, that play regulatory roles in root elongation. Both indirect and direct experimental evidence has been obtained to establish this. Thus, for example, numerous experiments by several groups of workers on *Zea mays* roots involving root-cap removal, removing half root caps, and the insertion of glass barriers, which have demonstrated that the root cap exerts an inhibitory effect on root elongation (Fig. 7.19) and that under the influence of gravity this influence becomes asymmetrically distributed to cause greater inhibition of the lower half of a horizontal root (Fig. 7.19). Other direct analytical work has revealed the presence of several growth inhibitors in root caps, including abscisic acid (ABA). Careful studies are currently going by several research groups attempting to relate the transport and concentrations of these inhibitors, and endogenous IAA, to positive gravitropism in roots. It is too early to reach definite conclusions, but there is certainly accumulating evidence that endogenous growth inhibitors are produced in the root cap from where they are transported basipetally and may become asymmetrically distributed across the region of extension growth in a horizontal root. Some recent studies have revealed that the concentration of ABA increases more in the upper side of gravitropically responding *Zea mays* roots (which would not, of course, be expected since the upper side grows faster than the lower side), but that another as yet unidentified inhibitor from the root cap is preferentially distributed towards the slower growing lower side of the root (Table 7.2). In these same experiments, and in others by different workers, a greater concentration of IAA was also found on the lower

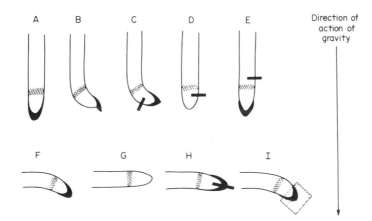

Fig. 7.19. Diagrammatic representation of some of the experiments which have indicated that the root cap is the source of growth inhibitor which is involved in the gravitropic response mechanism in roots. The root cap is shown in black, and the elongation zone of the root is shaded. A. Vertical intact root grows downwards. B. Removal of half the root cap results in bending towards the remaining half cap regardless of the direction of gravity. C. Insertion of a glass barrier between half the root cap and the elongation zone has the same effect as removing half the cap. D. A similar barrier in the absence of the cap has no effect. E. A barrier positioned behind the growing zone is without effect. F. Intact horizontal root executing normal downward gravitropic curvature. G. Removal of the root cap abolishes gravitropism (because it appears to be both the region of graviperception and the source of growth regulating substances). H. A horizontal glass barrier through the root cap and apex abolishes, or largely removes, gravitropism in a horizontal root. I. A glass barrier similar to that in H, but orientated vertically, does not prevent the development of a gravitropic curvature. (Adapted from M. B. Wilkins, *Current Adv. Plant Sci.* 6(3), 317-28, 1975.)

side of the horizontal roots (Table 7.2). However, it is not possible to conclude that this lateral asymmetry in auxin concentration is responsible for inhibition of elongation growth on the lower side of horizontal roots. The reason for this is that the data shown in Table 7.2 were obtained from experiments on one of several varieties of *Z. mays* whose roots show positive gravitropism only after they have been exposed to light, and as can be seen in Table 7.2 there was no difference in the lateral asymmetry in IAA concentration in roots in darkness or that had received light, whereas the unidentified inhibitor was preferentially distributed towards the lower side only in light-exposed roots. One cannot, though, dismiss the possibility that ABA may play a role in root gravitropism, as in another *Z. mays* variety that requires light for root gravitropism it was found that addition of ABA to intact completely dark-grown roots resulted in them bending downwards in response to a gravitational stimulus.

There is, therefore, increasing evidence that lends support to the concept that the root cap possesses both the capacity to detect the direction in which gravity is acting (i.e. it is the site of graviperception—see also p. 192 and Fig. 7.17) and to produce and transport growth regulators in such a way as to control the direction in which the root grows. Further work will establish the chemical natures of the important growth regulating substances from the root cap.

7.7.4 Non-orthogravitropic behaviour of plant organs

We have considered above the physiology of orthogravitropism in shoots and roots. As was however described in the introduction to the subject of gravitropism (p. 188) many plant organs show other types of gravitropic behaviour. The mechanisms by which rhizomes and stolons grow horizontally (diagravitropism) at a constant depth below the soil surface have received rather little experimental investigation and are therefore not considered further here. Similarly, virtually nothing is yet known of the basis for a given organ changing during its development from being negatively to positively gravitropic, or vice versa. Examples of such reversals in gravitropic behaviour are provided by flower and fruit stalks between the flower-bud and mature fruit stages (e.g. in *Papaver*, *Fritillaria* and *Tussilago*).

Correlative influences from other parts of the plant can also modify an organ's gravitropic behaviour. A good example of this is seen in the

Table 7.2. *Lateral distribution of IAA, ABA and an unidentified growth inhibitor in horizontal roots of* Zea mays *(c.v. Golden Cross Bantam) grown either in total darkness or after exposure to 5 minutes red-light. Only the light-treated roots showed a positive gravitropic bending response. (Data from T. Suzuki, N. Kondo and T. Fujii,* Planta 145, 323-9, 1979.)

	Side of root	IAA present (ng.g^{-1} fresh wt.)	Ratio lower/ upper	ABA present (ng.g^{-1} fresh wt.)	Ratio lower/ upper	Unidentified inhibitor present (%inhibition of *Zea* root growth)	Ratio lower/ upper
Dark	Upper	13.9		35.7		27	
			2.9		1.0		0.7
	Lower	40.0		35.6		19	
5 min Red light	Upper	14.5		77.8		20	
			3.4		0.6		2.1
	Lower	49.0		48.4		42	

influence that the apical bud of the main, or-thogravitropic, shoot has upon the orientation of plagiogravitropic lateral organs such as leaves and shoots. Removal of the apical bud of the main axis results in an upward (hyponastic) movement of both leaves and branches (see also pp. 175-176). One or more of the lateral branches usually become orthogravitropic and grow vertically up-wards. Thus, it is clear that plagiogravitropic behaviour of laterals is at least partially determined by a correlative influence from the main apical region of the shoot. Exogenous auxin can substitute for an excised apical bud in maintenance of plagiogravitropism in lateral bran-ches (see pp. 175-176). A further example of the effects of the apical region of the shoot, and growth hormones, is provided by the effects of auxin, gibberellin and cytokinin on stolon orienta-tion and development in plants such as *Solanum andigena* (p. 137 and Fig. 5.29).

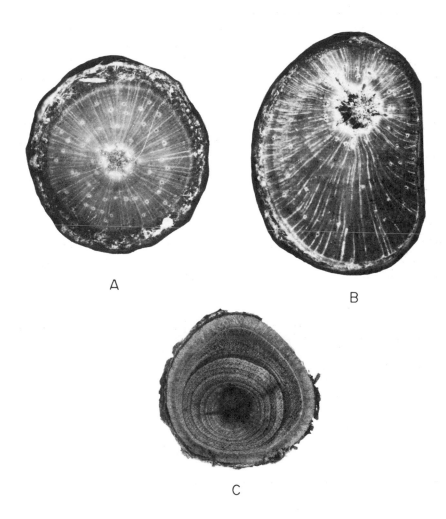

A

B

C

Fig. 7.20. Reaction wood formation in response to a gravitational stimulus. A and B: transverse sections through stems of a conifer (*Pinus radiata*); A, vertical control and B, held horizontal for one month. Reaction wood ("compression" wood) formed on lower side. C: T.S. of an angiosperm tree stem (*Eucalyptus geniocalyx*) held horizontally for 2 years, in which reac-tion wood ("tension" wood) formed only on the upper side. (Photography by courtesy of Dr. G. Scurfield. From, *Science* **179** (4074), 647-755, 1973.)

7.8 GRAVIMORPHISM

A number of important aspects of the influence that gravity has on plant development fall under the general heading of gravimorphic effects. The term gravimorphism is used to categorize the morphogenetic, or developmental, effects that gravity can have in addition to gravitropisms. Reaction-wood formation in plagiogravitropic branches of trees and shrubs is an obvious example of a gravimorphic effect (Fig. 7.20). Others include the marked tendency for lateral buds to grow only from the upper sides of horizontal or plagiogravitropic shoots, the buds on the lower side remaining inhibited (Fig. 7.21), and the promotion of flower-bud initiation in horizontally trained branches of fruit and other trees. The physiological basis of gravimorphism has received some study, but it is not yet possible to present a clear picture of mechanisms concerned.

Fig. 7.21. Gravimorphism in bud outgrowth from a horizontal stem of a woody plant. Buds along the upper side grow whilst those on the lower side are inhibited. (Photograph by courtesy of Dr. G. Scurfield.)

FURTHER READING

General

Audus, L. J. *Plant Growth Substances,* Vol. 1. Hill Ltd., London, 1973.

Darwin, C. *The Power of Movement in Plants,* J. Murray, London, 1880.

Hall, A. B., R. D. Firn and J. Digby. Auxins and shoot tropisms —A tenuous connection? Journal of Biological Education **14**, 195-199, 1980.

Letham, D. S., P. B. Goodwin and T. J. V. Higgins (Eds.). *Phytohormones and Related Compounds*: *A Comprehensive Treatise,* Vols. I and II, Elsevier-North Holland, Amsterdam, 1978.

Wilkins, M. B. (Ed.). *Physiology of Plant Growth and Development,* McGraw-Hill, London, 1969.

Wilkins, M. B. Gravity-sensing guidance systems in plants, *Sci. Prog., Oxf.* **63**, 187-217, 1976.

More Advanced Reading

Audus, L. J. Geotropism. In: M. B. Wilkins (ed.), *The Physiology of Plant Growth and Development,* McGraw-Hill, London, pp. 205-42, 1969.

Audus, L. J. Linkage between detection and the mechanisms establishing differential growth factor concentrations. In: S. A. Gordon and M. J. Cohen (eds.), *Proc. Symp. on Gravity and the Organism,* Tuxedo, New York, University of Chicago Press, Chicago, pp. 137-50, 1971.

Audus, J. L. Geotropism in roots. In: *The Development and Function of Roots,* (ed. J. Torrey and D. J. Clarkson), Academic Press, London, pp. 327-63, 1975.

Audus, L. J. The mechanism of the perception of gravity by plants, *Symposia Soc. Exp. Biol.* **31**, 197-227, 1977.

Curry, G. M. Phototropism. In: M. B. Wilkins (ed.), *The Physiology of Plant Growth Development,* McGraw-Hill, London, pp. 243-73, 1969.

Dennison, D. S. Phototropism. In: W. Haupt and M. E. Feinleb (eds.), *Physiology of Movements, Encyclopedia of Plant Physiology New Series,* Vol. 7, Springer-Verlag, Berlin-Heidelberg, pp. 506-566, 1979.

Digby, J. and R. D. Firn. A critical assessment of the Cholodny —Went theory of shoot geotropism, *Current Advances in Plant Science,* **25**, 1976.

Firn, R. D. and J. Digby. The establishment of tropic curvatures in plants, *Ann. Rev. Plant Physiol.* **31**, 131-148, 1980.

Galston, A. W. Plant photobiology in the last half-century, *Plant Physiol.* **51,** 427-36, 1974.

Johnsson, A. Circumnutation. In: W. Haupt and M. E. Feinleb (eds.), *Physiology of Movements, Encyclopedia of Plant Physiology New Series,* Vol. 7, Springer-Verlag, Berlin-Heidelberg, pp. 627-646.

Juniper, B. E. Mechanisms of perception and patterns of organisation in root caps. In: M. M. Miller and C. C. Kuchnett (eds.), *The Dynamics of Meristem Cell Populations,* Plenum, New York, pp. 119-31, 1972.

Juniper, B. E. Geotropism, *Ann. Rev. Plant Physiol.* **27,** 385-406, 1976.

Kang, B. G. Epinasty. In: W. Haupt and M. E. Feinleb (eds.), *Physiology of Movements, Encyclopedia of Plant Physiology New Series,* Vol. 7, Springer-Verlag, Berlin-Heidelberg, pp. 647-667, 1979.

Reinhold, L. Phytohormones and the orientation of growth. In: D. S. Letham, P. B. Goodwin and T. J. V. Higgins (eds.), *Phytohormones and Related Compounds: A Comprehensive Treatise,* Vol. II, Elsevier-North Holland, Amsterdam, pp. 251-289, 1978.

Satter, R. L. Leaf movements and tendril curling. In: W. Haupt and M. E. Feinleb (eds.), *Physiology of Movements, Encyclopedia of Plant Physiology New Series,* Vol. 7, Springer-Verlag, Berlin-Heidelberg, pp. 442-484, 1979.

Scurfield, G. Reaction wood: Its structure and function, *Science* **179,** 647-655, 1973.

Shropshire, W., Jr. Stimulus perception. In: W. Haupt and M. E. Feinleb (eds.), *Physiology of Movements, Encyclopedia of Plant Physiology New Series,* Vol. 7, Springer-Verlag, Berlin-Heidelberg, pp. 10-41, 1979.

Seivers, A. and D. Volkmann. Gravitropism in single cells. In: W. Haupt and M. E. Feinleb (eds.), *Physiology of Movements, Encyclopedia of Plant Physiology New Series,* Vol. 7, Springer-Verlag, Berlin-Heidelberg, pp. 567-572, 1979.

Torrey, J. G. Root hormones and plant growth, *Ann. Rev. Plant Physiol.* **27,** 435-59, 1976.

Volkmann, D. and A. Sievers. Graviperception in multi-cellular organs. In: W. Haupt and M. E. Feinleb (eds.), *Physiology of Movements, Encyclopedia of Plant Physiology New Series,* Vol. 7, Springer-Verlag, Berlin-Heidelberg, pp. 573-600, 1979.

Wilkins, M. B. The role of the root cap in root geotropism, *Current Advances in Plant Science,* **6**(3), 317-28, 1975.

Wilkins, M. B. Gravity and light-sensing guidance systems in primary roots and shoots, *Symposia Soc. Exp. Biol.* **31,** 275-335, 1977.

Wilkins, M. B. Growth-control mechanisms in gravitropism. In: W. Haupt and M. E. Feinleb (eds.), *Physiology of Movements, Encyclopedia of Plant Physiology New Series,* Vol. 7, Springer-Verlag, Berlin-Heidelberg, pp. 601-626, 1979.

Wilson, B. F. and R. R. Archer. Reaction wood: induction and mechanical action, *Ann. Rev. Plant Physiol.* **28,** 24-43, 1977.

8

Photomorphogenesis

8.1 INTRODUCTION

In the preceding chapter we dealt with plant responses to the directional stimuli of light and gravity. In addition to phototropism, plants show other types of response to light signals. In the following chapter we shall consider plant responses to seasonal variations in the length-of-day (*photoperiodism*). In this latter type of response, the plant appears to have time-measuring capabilities which enable it to detect seasonal changes in the lengths of day and night. In addition to phototropism and photoperiodism, there are still other types of plant response to light signals, which are neither directional nor periodic, and which are included under the general term

photomorphogenesis which has been defined as "the developmental strategies plants adopt when growing in the light". As we shall see, photomorphogenesis includes a range of diverse phenomena which are controlled by specific photoreceptors, forming the *phytochrome* system.

8.2 THE RED/FAR-RED PHENOMENON

A major advance in our understanding of photomorphogenesis arose from the investigations of H. A. Borthwick and S. B. Hendricks of the U.S. Department of Agriculture, extending over many years and which included studies on the germination responses of a light-sensitive variety of lettuce seed.

This type of seed shows little germination in the dark at 25°C, but germinates well if exposed to a short period of illumination. By exposing seeds to various parts of the spectrum Borthwick and Hendricks were able to demonstrate that the most effective region for the promotion of germination was in the red (maximum effectiveness at 660 nm), with a subsidiary peak in the blue. Far-red (maximum at 730 nm) radiation does not promote, but earlier work had shown that it is inhibitory to germination (Fig. 11.4). Borthwick and Hendricks investigated the interaction between the effects of red (R) and far-red (FR) radiation by exposing the seeds to R and FR alternately. They made the crucial discovery that the effects of R and FR are mutually reversible and that whether germination occurs or not depends on the nature of the last radiation to which the seeds are exposed (Table 8.1). Thus, each succeeding irradiation reverses the effect of the preceding treatment.

Table 8.1. *Control of lettuce seed germination by red and far-red light (from Borthwick et al., Proc. Nat. Acad. Sci. U.S., 38, 662, 1952)*

Irradiation	Percentage germination
Red	70
Red/Far-red	6
Red/Far-red/Red	74
Red/Far-red/Red/Far-red	6
Red/Far-red/Red/Far-red/Red	76
Red/Far-red/Red/Far-red/Red/Far-red	7
Red/Far-red/Red/Far-red/Red/Far-red/Red	81
Red/Far-red/Red/Far-red/Red/Far-red/Red/Far-red	7

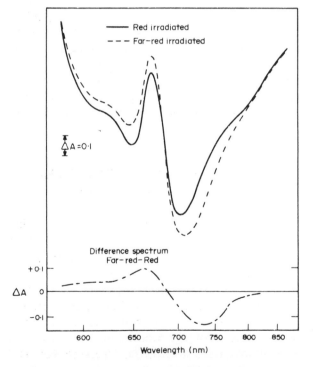

Fig. 8.1. "Difference spectrum" for phytochrome, showing absorbances of etiolated maize tissue after saturating exposures to red and far-red light (above) and the difference in absorbance (below). (From Butler, W. L., K. H. Norris, H. W. Siegelman and S. B. Hendricks, *Proc. Bat. Acad. Sci., U.S.A.* **45**, 1703, 1959.)

Now, it is clear that where a particular region of the spectrum causes a specific biological effect, the tissues of the organism must contain a photoreceptor (or "pigment") which absorbs selectively in that region. Thus, it would appear that we have to postulate the presence of two photoreceptors in lettuce-seed tissue, one of which absorbs selectively in the red region and a second which absorbs in the far-red. However, Borthwick and Hendricks made the bold suggestion that there is essentially only one photoreceptor which can exist in the two alternative forms P_r and P_{fr} and that each form is capable of being reversibly converted into the other form, an hypothesis symbolized in the equation:

$$P_r \underset{\text{Far red}}{\overset{\text{Red}}{\rightleftharpoons}} P_{fr}$$

It will be seen that this hypothesis was based upon simple experiments on the effects of light upon the germination of lettuce seed. For some years the scheme remained entirely hypothetical, but later members of the same group constructed a special dual wavelength spectrophotometer which was capable of detecting small changes in the absorp-

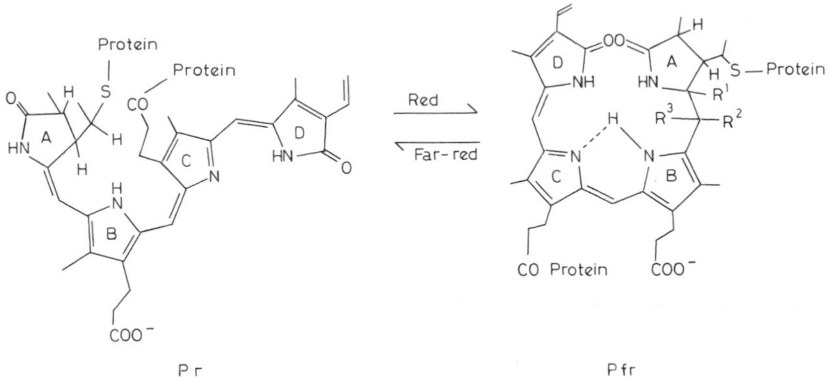

Fig. 8.2. Suggested changes within the chromophore of phytochrome occurring during photoconversion. (Courtesy of Prof. W. Rüdiger and Prof. H. Scheer).

tion spectra of etiolated plant tissues, at 660 nm and 730 nm. It was necessary to use etiolated plant tissues, since the presence of chlorophyll masks the absorption by other pigments in the red region. The instrument measured the *difference* in absorption of the tissues at 660 nm and 730 nm, during rapid alternation between irradiation at these two wavelengths, and it was found that after exposure of the dark-grown tissues to red light the absorption changed slightly, so that they absorbed more at 730 nm (Fig. 8.1), and the reverse change occurred if the tissues were now exposed to FR. Thus, the changes predicted by the hypothesis were fulfilled. In further work it was possible to demonstrate similar changes in cell-free extracts of etiolated tissues. Indeed it was possible to see a visible change in colour of the extracts with the naked eye, following exposure to R or FR. These changes in absorption properties were used to detect the presence of the photoreceptor(s) in further purification procedures and ultimately it proved possible to isolate a single protein which showed reversible changes in its absorption spectrum following exposure to R and FR, in exactly the manner originally postulated by Borthwick and Hendricks, who called this substance *Phytochrome*.

Phytochrome has subsequently been isolated in a very pure form and the protein part of the molecule apparently has a molecular weight of 120,000 daltons, while the non-protein, light-absorbing part (chromophore) has been shown to be a tetra-pyrrole compound related to the phyco-cyanins of the blue-green algae (Fig. 8.2). There is still some uncertainty as to the nature of the intra-molecular changes undergone by the chromophore when exposed to R or FR, but one suggestion is il-lustrated in Fig. 8.2. There is also some evidence that the protein part of the molecule may undergo a conformational change during photoconversion. The absorption spectra of pure P_r and P_{fr} are given in Fig. 8.3.

8.3 THE RANGE OF PHYTOCHROME-CONTROLLED RESPONSES

Since the original discovery of R/FR reversibility in lettuce seeds, similar effects have been demonstrated for a wide variety of plant responses and in all the main groups of plants from the green algae to the flowering plants (Table 8.2). It will be seen that these responses include spore germina-tion, epicotyl hook opening, leaf expansion, inter-node extension, and root initiation, as well as numerous responses at the sub-cellular and molecular level, such as chloroplast movement, enzyme synthesis and changes in membrane permeability. In all these cases it has been demonstrated that R and FR have opposite ef-

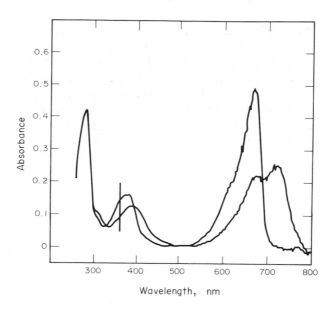

Fig. 8.3. Absorption spectra of a solution of oat phytochrome following irradiation with red and far-red light, giving the P_{fr} (broken line) and P_r (continuous line) forms, respectively. (From H. W. Siegelman and W. L. Butler, *Ann. Rev. Plant Physiol.* **16**, 383, 1965.)

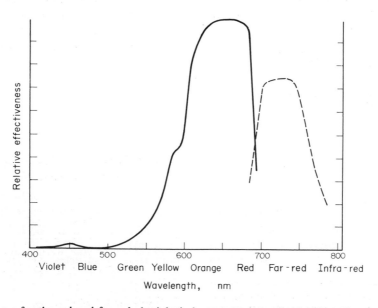

Fig. 8.4. Action spectra for the red and far-red physiological responses controlled by phytochrome. *Continuous line:* red effects; *broken line:* far-red effects. (Adapted from F. B. Salisbury, *Endeavour,* **24**, 78-80, 1965.)

fects, with similar action spectra (Fig. 8.4), and show R/FR reversibility. These R/FR effects are so characteristic of phytochrome that where they can be demonstrated for any given biological response, the latter can be assumed to involve phytochrome. It is these manifold aspects of growth and development under phytochrome control which are referred to as *photomorphogenesis*.

most general aspects of development, such as leaf expansion and stem extension in normal development of the green shoot. Everyone is familiar with the characteristic appearance of etiolated shoots which have grown in complete darkness. These symptoms of etiolation can be reduced by quite short periods (5 minutes) of daily irradiation with red light, and the effects of R can be reversed by

Table 8.2. *Some phytochrome-controlled responses*

Algae, bryophytes and pteridophytes	*Angiosperms*
Spore germination	Seed germination
Chloroplast movement	Hypocotyl hook formation
Protonema growth and differentiation	Internode extension
	Root primordia initiation
	Leaf initiation and growth
	Leaflet movement
Gymnosperms	Electrical potential
Seed germination	Membrane permeability
Hypocotyl hook formation	Phototropic sensitivity
Internode extension	Geotropic sensitivity
Bud dormancy	Anthocyanin synthesis

It is evident that phytochrome control does not apply only to special phenomena, such as light sensitivity in seeds, but is involved in some of the

FR, indicating phytochrome control (Fig. 8.5). However, the development of chlorophyll requires longer periods of irradiation and we shall see that the full development of what we regard as a "normal" green shoot requires quite high energy levels (p. 213).

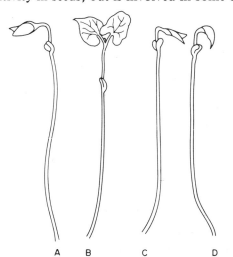

Fig. 8.5. Phytochrome control of shoot development in bean (*Phaseolus vulgaris*). Treatments: (a) grown in continuous darkness; (b) exposed to 2 minutes red light: (c) 2 minutes red and 5 minutes far-red; (d) 5 minutes far-red. (From R. J. Downs, *Plant Physiol.* **30**, 468, 1955.)

8.4 DETECTION AND MEASUREMENT OF PHYTOCHROME *IN VIVO*

The detection and measurement of phytochrome in plant tissues is based upon spectrophotometric measurements of the differences in light-absorption spectra at 660 nm and 730 nm following irradiation with R and FR, using similar methods to those employed for the original isolation of phytochrome. The tissue is first irradiated with red light to convert the phytochrome to the

P_{fr} form and the difference between the absorbance at 660 nm and at 730 nm determined (ΔA_r). The pigment is then converted to the P_r form by exposure to FR, and the difference in absorbance at the two wavelengths again determined (ΔA_{fr}).

The overall difference between the two measurements (Fig. 8.6) will be related to the total amount of phytochrome present.

That is:

$$\Delta(\Delta A) = \Delta A_{fr} - \Delta A_r$$

where ΔA is the difference in absorbance at 660 and 730 nm.

Etiolated tissues have been found to have a high content of phytochrome, the highest concentrations being found in meristematic tissues, including shoot and root tips and cambial tissue.

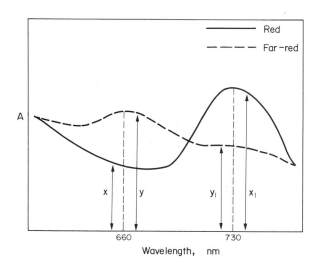

Fig. 8.6. Diagram of absorbance changes measured in the phytochrome assay. Note: $\Delta_r = x - x_1$ (after red), $\Delta_{fr} = y - y_1$ (after far-red). (From R. E. Kendrick and B. Frankland, *Phytochrome and Plant Growth*, Edward Arnold, London, 1976.)

The phytochrome content of green tissue is too low to detect spectrophotometrically, but low amounts have been found in extracts of leaves of a number of species.

8.5 THE INTRACELLULAR LOCALIZATION OF PHYTOCHROME

Attempts have been made to identify the sites at which phytochrome occurs within the cell. The most direct approach has been through the use of immunocytochemical techniques. Rabbit antibodies against purified phytochrome were produced and applied to sections of plant tissues. As a result, rabbit antibody molecules will become attached to phytochrome molecules within the cell. The section is then treated with sheep antiserum to rabbit antibody and this is followed by a rabbit antiperoxidase-peroxidase complex so that each phytochrome molecule is "tagged" with the enzyme peroxidase, the location of which can then be detected by histochemical methods (Fig. 8.7). Such studies have indicated that in the P_r form phytochrome is not strictly localized within the cell, and occurs in mitochondria and plastids, as well as in the cytoplasm generally, but not in nuclei or vacuoles. However, on conversion to the P_{fr} form by exposure to R light, phytochrome rapidly becomes localized in discrete areas which have not yet been identified.

A different approach to the problem of locating phytochrome within the cell has been used in studies on movements of the chloroplasts of the filamentous green alga, *Mougeotia*. Each cell of this alga contains a single, plate-like chloroplast which turns so that it is edgewise to the direction of the incident light at high intensities but at right angles to the light at low intensities. By using microbeams of R and FR light, it has been shown that these chloroplast movements involve phytochrome, and that the response can be obtained when only the outer layers of the cytoplasm are irradiated. Moreover, using microbeams of polarized R and FR, it was shown that R is only effective when the plane of the electric vector is *parallel* to the cell surface and that FR is only effective when the plane of the electric vector is at *right angles* to the cell surface (Fig. 8.8). This finding indicates that the phytochrome molecules are arranged in a regular manner at or near the cell surface and that photoconversion involves a change in orientation of the molecules (Fig. 8.9).

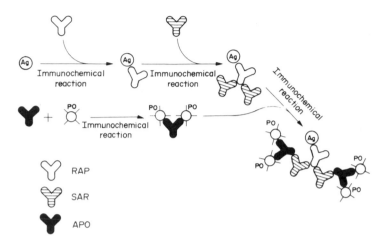

Fig. 8.7. A double indirect technique for the visualization of an antigen (see text). Components of the labelling procedure are: Ag, phytochrome; RAP, rabbit anti-phytochrome serum; SAR, sheep anti-rabbit immunoglobulin serum; APO, rabbit anti-peroxidase immunoglobulins; PO, peroxidase. (From L. H. Pratt, R. A. Coleman and J. M. Mackenzie, Jr., in *Light and Plant Development,* Ed. H. Smith, Butterworths, London, 1976.)

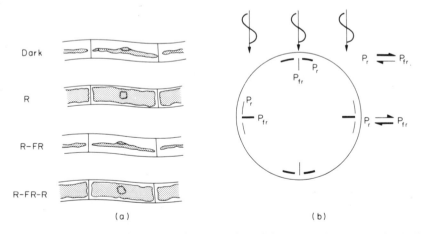

Fig. 8.8. (a) Induction of chloroplast orientation in *Mougeotia* by red light (R) and reversion of red effect by far-red (FR), starting from the profile orientation (above). (b) Absorption of polarized light by phytochrome molecules which change dichroic orientation with conversion $P_r \rightarrow P_{fr}$. Cross-section of a *Mougeotia* cell with P_r (surface parallel dashes) and P_{fr} (surface normal dashes). Molecules which are in favourable geometric position to absorb polarized light are heavy lined. As a consequence, the photostationary state is shifted, according to the position, to the left or right side (heavy-lined arrows). (Note that the electric vector is at right angles to the direction of the incident polarized light, as indicated by the sinusoidal curves above.) (From W. Haupt, in *Phytochrome,* Ed. K. Mitrakos and K. Shropshire, Academic Press, London, 1972.)

We shall see later that other evidence suggests that phytochrome acts by affecting the permeability of cells, again suggesting that it acts at the cell membrane.

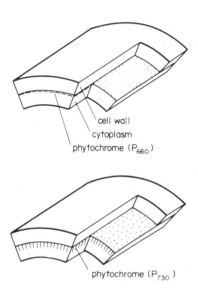

Fig. 8.9. Suggested model showing the orientation of phytochrome molecules in the plasma membrane (top) for the P_r form and (below) for the P_{fr} form. (From W. Haupt, *Phytochrome*, Ed. K. Mitrakos and K. Shropshire, Academic Press, London, 1972.)

8.6 PHOTOCONVERSION OF PHYTOCHROME IN THE CELL

We have seen that the P_r and P_{fr} forms of phytochrome can readily be interconverted by exposure to R and FR respectively. When the R/FR reversibility phenomenon was first discovered it was believed that after conversion of P_r to P_{fr} by exposure to red light, reversion of P_{fr} to P_r could take place spontaneously in the dark, without exposure to FR. However, such "dark reversion" has only been demonstrated in certain dicotyledons (e.g., in cauliflower tissue) and has not yet

been shown to occur in monocotyledons. Nevertheless, the level of P_{fr} generally declines quite rapidly in most plant tissues and concomitantly the total phytochrome (P) in the tissues also declines. This decline in P_{fr} can be arrested by exposure to FR, i.e. by conversion to P_r. Thus, whereas P_r is relatively stable, P_{fr} is highly labile and is evidently destroyed quite rapidly *in vivo*. It is not understood how such destruction of P_{fr} occurs, but evidently P_{fr} is the biologically active form and it must be involved in the initiation of the chain of processes which are ultimately manifested in an observable biological response.

In most experiments involving R/FR reversibility, the FR treatment follows immediately after exposure to R, but if exposure to FR is delayed for increasing periods a point is reached when FR no longer reverses the effect of R, indicating that the events initiated by conversion to P_{fr} have already been set in progress. This period beyond which reversibility is lost is referred to as the "escape" time and varies from 1·5 minutes to 9 hours in different tissues.

If plant tissues are allowed to remain in the dark following a single exposure to R, the initial decline in total phytochrome is later followed by an increase, suggesting that synthesis of new phytochrome in the P_r form takes place. This synthesis has been demonstrated by density-labelling (see p. 96) by demonstrating the incorporation of deuterium into P_r when plant tissues are incubated in D_2O.

As a result of the various processes described above, it is clear that the phytochrome in active plant tissues undergoes continuous "turnover", involving synthesis, interconversion, destruction and reversion, which may be symbolized as follows:

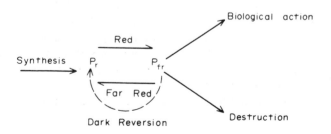

8.7 MODE OF ACTION OF PHYTOCHROME

Many phytochrome-controlled responses appear to involve the action of genes which previously were not expressed. For example, the development of epidermal hairs and of anthocyanin in mustard hypocotyls is under phytochrome control and would appear to involve the expression of genes which are not expressed in dark-grown seedlings. Moreover, gene expression involves both enzyme synthesis and enzyme activity, and phytochrome control of both these processes has been demonstrated. These observations might seem to indicate that phytochrome acts by controlling either enzyme synthesis at the transcription or translation level, or by controlling the activation of pre-existing enzymes. However, any attempt to formulate a general hypothesis to account for the very varied range of phytochrome-controlled responses must take into account various effects which appear to involve rapid changes in membrane permeability.

For example, the leaves of certain plant species, such as *Mimosa pudica* and *Albizzia julibrissin,* show "sleep movements" involving upward or downward folding of the leaflets. These movements, which involve changes in the turgor pressure in cells of the pulvini at the point of attachment of the leaflet to the midrib, are evidently under some degree of phytochrome control, since they are inhibited by exposure to FR at the end of the light period and this effect can be reversed by R. These effects of FR can be observed within about 10 minutes. The changes in turgor have been shown to be accompanied by the movement of potassium ions into and out of the cell, thereby affecting its osmotic properties. An even more rapid response is seen in root tips of barley and mung bean, which, under certain conditions, will adhere to a negatively charged glass surface following exposure to R light and this effect can be reversed by FR. This electrostatic effect indicates that the root surface becomes positively charged on exposure to R light and that this effect is reversed by FR.

Exposure to red light has also been shown to result in rapid changes in electrical potential in several organs. For example, exposure of etiolated coleoptiles to R light causes the tip to become more electropositive with respect to the base within 15 seconds, and similar changes have been observed in mung bean root tips. If coleoptiles which have been exposed to red light are returned to darkness, the electrical potential returns to the original level in a few minutes, after which far-red light causes a further decrease in potential, again with a lag of only 15 seconds.

It seems clear that these various types of rapid effect following phytochrome conversion involve changes in the permeability properties of the cell membranes to electrolytes such as potassium ions. Since these rapid effects may occur almost instantaneously and, in the case of electrical changes in coleoptiles, this applies to the opposite effects of both red and far-red, it would appear that the membrane changes involved are very close in time to the primary action of phytochrome. Hence, it is generally held that the primary mode of action of phytochrome must involve changes in cell membrane properties, a conclusion which is consistent with the evidence of the intracellular localization of phytochrome (p. 208).

However, although the *primary* mode of action may involve changes in membrane permeability, there is good evidence that the secondary effects of phytochrome conversion involve enzyme-controlled processes. Thus, in *Avena* seedlings irradiation with R light results in a rapid conversion of ADP to ATP, and in other tissues there are changes in the levels of NADP which can be prevented by FR irradiation. Moreover, exposure to R light leads to increased activity of a wide range of enzymes. Thus, R light increases the level of phenylalanine ammonia lyase (PAL) (Fig. 8.10) which catalyses the removal of ammonia from phenylalanine to give cinnamic acid, in pea seedlings. In this instance R appears to lead to activation of pre-existing enzyme rather than to new enzyme synthesis, but in mustard continuous far-red appears to stimulate the synthesis of ascorbic acid oxidase. Irradiation of etiolated bean leaves with R light leads to marked increases in polyribosomes

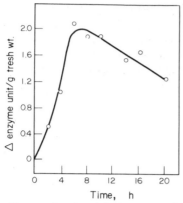

Fig. 8.10. Changes in phenylalanine ammonia-lyase activity extractable from the terminal buds after irradiation of intact, etiolated pea seedlings with 200 K ergs cm^{-2} red light. (From T. H. Attridge and H. Smith, *Biochim. Biophys. Acta,* **148,** 805, 1967.)

within 1 hour, suggesting that the treatment results in increased availability of messenger RNA.

The nature of the processes connecting the primary changes in membrane permeability with

the diverse metabolic events involving phytochrome remain obscure. However, there is now considerable evidence that exposure to R light results in rapid increases in extractable gibberellins and cytokinins, in etiolated and green leaves and in certain seeds. Thus, exposure to 5 minutes of R light causes marked increases in gibberellins in etiolated barley and wheat leaves (Fig. 8.11), and in cytokinin levels in *Rumex* seeds and leaves of poplar (*Populus robusta*). Thus, it is possible that these hormones act as "second messengers" and provide a link between the primary and secondary effects of phytochrome conversion. For example, etiolated barley and wheat leaves grown in complete darkness are tightly rolled, but they can be induced to unroll by exposure to a short period of red light, the effect of which can be nullified by far red. However, sections of etiolated barley leaves can be induced to unroll in darkness by application of exogenous GA$_3$ suggesting that the red-light effects may be mediated by the increases in endogenous gibberellins. Further studies have

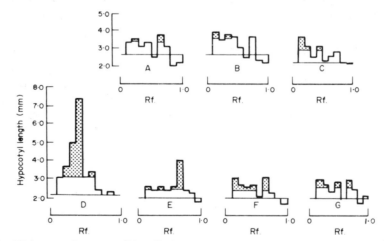

Fig. 8.11. Effect of red light on endogenous gibberellin levels in etiolated wheat leaves. Each separate histogram represents the gibberellin activity present in extracts of leaves after separation into ten fractions by paper chromatography. Each fraction was tested by its promoting effect on the growth of lettuce hypocotyls. The solid parts of the histograms represent gibberellin activity significantly above control

A = dark-grown leaves,
B = 5 minutes red light,
C = 5 minutes red light + 5 minutes dark,
D = 5 minutes red light + 10 minutes dark,
E = 5 minutes red light + 20 minutes dark,
F = 5 minutes red light + 30 minutes dark,
G = 5 minutes red light + 60 minutes dark.

(From L. Beevers, L. B. Loveys, J. A. Pearson and P. F. Wareing, *Planta (Berl.),* **90,** 286, 1970.)

shown that cell-free preparations from barley leaves containing etioplasts show similar changes in gibberellin levels in response to red light to those found in intact leaves, indicating that the etioplasts are the site of the phytochrome-controlled processes affecting gibberellin levels.

8.8 THE "HIGH IRRADIANCE REACTION"

As we have seen, the characteristic features of etiolated shoots grown in complete darkness (viz. pronounced internode elongation, reduced leaf development, lack of chlorophyll, etc.) can be greatly reduced by exposing them to quite short periods (e.g. 5 minutes) of R light each day, and the effect of R can be reserved by FR, clearly indicating involvement of phytochrome. However, the appearance of shoots exposed to only short periods of low-intensity R light in this way is still far from "normal" and in order to obtain the typical appearance of shoots grown in natural daylight it is necessary to expose the shoots to several hours of daylight at high intensity each day. Now, work on white mustard seedlings showed that whereas short periods of FR at low intensity reversed the effect of R, longer periods of FR produced the *same* effects as R, i.e. they

reduce the characteristic symptoms of etiolation by causing expansion of the cotyledons, reduction of hypocotyl extension and the development of anthocyanin. Moreover, whereas the effects of short periods of FR reached "saturation" at quite low intensities, the effects of longer periods of FR continued to increase as the intensity was increased to high levels. Hence these effects were said to involve a "High Irradiance Reaction" (HIR).

Action spectra for the HIR show major peaks in the FR at 710-730 nm, and in the blue region (Fig. 8.13). At first it was thought that the HIR must involve a separate photoreceptor, other than phytochrome. However, later studies have indicated that HIR effects in the FR region are almost certainly mediated via phytochrome. The evidence for this conclusion is based on the observation that when lettuce hypocotyl tissue are exposed to FR, 3 per cent of the total phytochrome remains in the P_{fr} form. This is because the absorption spectra of P_r and P_{fr} overlap (Fig. 8.12), so that when tissue is exposed to FR, a "photostationary state" is reached, representing an equilibrium between photoconversion of P_{fr} to P_r and the reverse reaction. It was shown that the effects of the HIR could be simulated by irradiating tissues with mixtures of R and FR at various intensities and in varying proportions, such that 3 per cent of the total phytochrome was present in the P_{fr} form. Thus, it appears that phytochrome is the

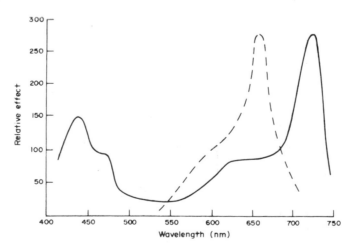

Fig. 8.12. Generalized action spectrum for the high-irradiance reaction (solid line), together with that of the low-irradiance phytochrome reaction (broken line). (From H. Mohr, *Biol. Rev.* **39**, 87, 1974.)

photoreceptor for the high irradiance FR effects. Similar results have been obtained with other responses, but the critical percentage for P_{fr} varies according to the type of tissue. Thus, it appears that phytochrome is the photoreceptor for the high irradiance FR effects. However, there is evidence that the effects in the blue region of the action spectrum for the HIR involve a different photoreceptor. Thus, the action spectrum in the blue region of the HIR is identical with that for phototropism in both higher plants and in fungi, which do not contain phytochrome. Again, gherkin seedlings transferred from dark to blue light show a reduced rate of hypocotyl extension after a lag of 30 seconds, whereas with far red the lag is 40 minutes. The photoreceptor involved in the blue region of the HIR is probably a flavin, responsible for many photoresponses in both lower and higher plants.

Under natural conditions plants are not, of course, subjected to monochromatic R and FR but normal daylight will induce a "photostationary state" in which a significant proportion of the total phytochrome will be in the P_{fr} form. From the evidence presented above it would appear that the HIR of phytochrome is involved in normal development of green shoots, under natural conditions.

8.9 SOME ECOLOGICAL ROLES OF PHYTOCHROME

Through its property of photochromicity (i.e. its property of photoreversible isomerization) the phytochrome system confers on plants the ability to sense the *quality* of light in their environment and to respond to differing light conditions. Where plants occur at high density, both in natural vegetation and in planted crops, they tend to shade each other and hence to modify the light environment for the shaded plants. Natural daylight contains approximately equal quantum flux densities of R and FR, but because chlorophyll absorbs strongly in the red region, the

light within a leaf canopy will contain a much higher proportion of FR (Fig. 8.13).

The effects of such changes in the spectral energy distribution of light within a canopy have been studied in detail by Smith and his co-workers. The ratio of the quantum flux in 10 nm wide bands centred at 660 nm and 730 nm, respectively, is used as a measure of the spectral energy distribution for the main regions which affect the phytochrome system and this ratio is denoted by the symbol ζ. The value of ζ varies from about 1.15 in full daylight to about 0.05 under dense vegetation.

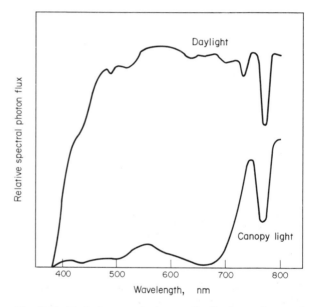

Fig. 8.13. Typical spectral photon distributions of natural daylight and of daylight filtered through vegetation canopy. (From M. G. Holmes and H. Smith, 1977.)

The value of ζ will affect the proportion of the total phytochrome which is in the P_{fr} state, and which is denoted by the symbol ϕ. The highest proportion of phytochrome existing as P_{fr} occurs in natural daylight, where the equilibrium state of P_{fr}/P_{total} $(\phi) = 0.54$. By using combinations of artificial light sources (incandescent and fluorescent lamps) it is possible to vary the value of ζ and to measure the corresponding values of ϕ in etiolated tissue, and so to plot a curve showing the relationship between ϕ and ζ for a wide range of values of the latter (Fig. 8.14). It is seen that the proportion of P_{fr} shows the greatest sensitivity to spectral changes over the range $\zeta = O$ to $\zeta = 1.0$, which is the range found in the natural environment. This

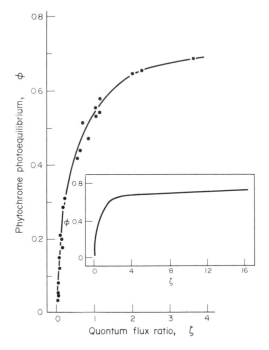

Fig. 8.14. Relationship between phytochrome equilibrium (ϕ) and quantum flux ratio (ζ) for various light regimes. (From H. Smith and M. G. Holmes, 1977.)

implies that the photoequilibrium state of phytochrome (given by ϕ) will respond markedly to changes in the spectral composition of the light environment found within the leaf canopies of crops or natural vegetation. In further studies, plants of *Chenopodium album* were grown under various light sources which provided varying proportions of red and far red, so that the value of ϕ could be varied, with parallel measurements made upon internode elongation on the plants. It was found that there was a linear relationship between the logarithm of stem extension rate and the value of ϕ (Fig. 8.15), strongly suggesting that stem elongation is controlled by the phytochrome system and that phytochrome, through the establishment of various values of ϕ allows plants

to monitor the spectral distribution of natural radiation within a leaf canopy.

The capacity of a plant to elongate its stem and so to overtop its neighbours confers on it an advantage in the fierce competition for light which frequently occurs in dense natural vegetation or crop stands.

The role of phytochrome in the dormancy of light-sensitive seeds and in their responses under natural conditions are discussed in Chapter 11.

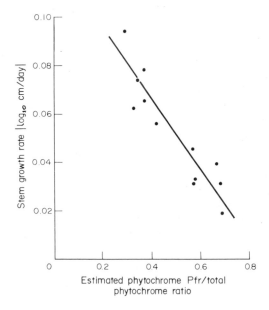

Fig. 8.15. Relationship between the stem extension rate and phytochrome equilibrium (ϕ) for seedlings of *Chenopodium album* L. (From D. C. Morgan and H. Smith, 1978.)

FURTHER READING

General

Kendrew, R. E. and B. Frankland. *Phytochrome and Plant Growth,* Edward Arnold, London, 1976.

Smith, H. *Phytochrome and Photomorphogenesis,* McGraw-Hill Book Co. (U.K.) Ltd., Maidenhead, 1975.

More Advanced Reading

Briggs, W. R. and H. V. Rice. Phytochrome: chemical and physical properties and mode of action, *Ann. Rev. Plant Physiol.* **23**, 293, 1972.

Furuya, M. Biochemistry and physiology of phytochrome, *Prog. Phytochem.* **1**, 347, 1968.

Holmes, M. G. and H. Smith. The function of phytochrome in the natural environment II, *Photochem. Photobiol.* **25**, 539, 1977.

Marmé, D. Phytochromes: membranes as possible sites of primary action, *Ann. Rev. Plant Physiol.* **28**, 173, 1977.

Morgan, D. C. and H. Smith. The relationship between phytochrome photoequilibrium and development in light grown *Chenopodium album* L., *Planta* **142**, 187, 1978.

Schoffer, P. Phytochrome control of enzymes, *Ann. Rev. Plant Physiol.* **28**, 233, 1977.

Smith, H. (Ed.). *Light and Plant Development,* Butterworths, London, 1976.

Smith, H. and M. G. Holmes. The function of phytochrome in the natural environment III, *Photochem. Photobiol.* **25**, 547, 1977.

Smith, H. and R. E. Kendrick. The structure and properties of phytochrome.

Chapters by: R. E. Kendrew, H. Smith, R. L. Salter and A. W. Galston. In: *Chemistry and Biochemistry of Plant Pigments* (ed. T. W. Goodwin), 2nd ed., Academic Press, London and New York, 1976.

Chapters by: H. A. Borthwick, B. Frankland, W. Haupt and C. J. P. Spruit, In: *Phytochrome* (ed. K. Mitrakos and W. Shropshire, Jr.), Academic Press, London and New York, 1972.

9

The Physiology of
Flowering —
I. Photoperiodism

9.1 INTRODUCTION

In the "typical" life history of a herbaceous flowering plant, there is usually an initial phase of vegetative growth, which sooner or later is followed by the reproductive phase. However, there is a good deal of variation with respect to the distinctness of these two phases; in some plants there is a fairly sharp transition from vegetative growth to reproduction, as in wheat or sunflower, while in others vegetative growth and flowering occur concurrently, as in runner beans (*Phaseolus multiflorus*) or tomato (*Lycopersicum esculentum*). In the first type of plant, growth is usually *determinate,* the axis being terminated by an inflorescence, whereas in the second group the flowers are axillary and borne on lateral shoots, while the main axis continues vegetative growth. In plants of both groups, however, there is nearly always a certain minimum period of purely vegetative growth—only very exceptionally can flowers be formed immediately following germination, e.g. in *Chenopodium rubrum,* under shortday conditions. The duration of this vegetative phase is very variable. Usually there is a certain period of growth during which a succession of new leaves is formed at the shoot apex. In some perennial species, however, the number of leaves is already predetermined in the stage of dormancy and the vegetative phase involves only the expansion of leaf primordia already laid down in the previous year, as in many bulbous plants and woody species.

9.1.1 Factors determining the onset of flowering

What causes the transition from the vegetative to the reproductive phase? Is it due to some internal control mechanism, determined by the genetical make-up of the species, or is it dependent upon a change in external conditions? Now, plants differ very greatly in their sensitivity to external conditions, the development of some species being relatively insensitive, so that provided that the environmental conditions are not so unfavourable that growth is completely prevented, they will ultimately flower under a wide range of conditions. In other species, however, the initiation of flowering is very sensitive to external conditions and will not occur under certain conditions, e.g. of temperature or daylength, even though these may be quite favourable for growth; that is to say, the requirements for flowering are not necessarily the same as for vegetative growth.

Since our knowledge of the physiology of flowering is much more complete for species which are sensitive to environmental conditions, we shall first deal with this group, and then describe briefly what is known regarding the control of flowering in the "insensitive" group.

9.1.2 Quantitative measurement of flowering responses

The difference between the vegetative and the flowering condition is a qualitative one. It is important, however, that in studying the physiology of flowering, we should have an exact measure of the flowering responses to various treatments. Various indices of flowering have been used, such as (1) the percentage of flowering plants in the total group receiving a particular treatment; (2) the total number of flowers or total number of

flowering nodes; (3) the time to the first appearance of flowers (the shorter the time the greater the flowering response); (4) the number of leaves formed before flower initiation; (5) the use of a scale of "scores" depending upon the stage of development reached by the flowers. This latter method is used where the flowers formed are still microscopic and incompletely developed. An arbitrary scale of "scores" is assigned to various stages of development (Fig. 9.1) and the total "score" of a given batch of plants is determined.

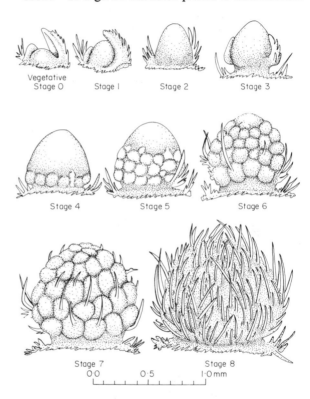

Fig. 9.1. Stages of development of the staminate inflorescence primordium of *Xanthium*. (From F. B. Salisbury, *Plant Physiol.* **30**, 327, 1955.)

9.2 PHOTOPERIODISM

The fact that seasonal changes in daylength conditions profoundly affect the life cycle of many plants was first clearly demonstrated by Garner and Allard, two American plant breeders, in 1920.

Garner and Allard were originally concerned with the peculiar seasonal flowering behaviour of certain varieties of tobacco (*Nicotiana tabacum*) and soybeans (*Glycine max*). A newly developed variety of tobacco, Maryland Mammoth, was found to grow vigorously throughout the summer, but did not flower. When grown in a greenhouse during the winter, this same variety flowered and fruited abundantly. With certain varieties of soybeans, plantings at successive intervals during the spring and summer all tended to flower at the same date in the late summer, the vegetative period being progressively shortened the later the date of sowing. Garner and Allard attempted to regulate the flowering of the tobacco and soybeans by varying the temperature, nutrition and soil moisture, but none of these factors was found to affect the date of flowering very markedly. They then investigated the effect of shortening the daily light period by a few hours by placing the plants in a dark chamber, and found that under the shortened daylight period the plants quickly initiated flowers. After this exciting discovery, they proceeded to test the effect of various daylength conditions on a wide variety of plant species. Variation of daylength was achieved in two ways: (1)

during the summer months by shortening the natural daylength, and (2) in winter by extending the natural daylength by artificial illumination.

It was found that for many plant species the daylength (i.e. the lengths of daily light and dark periods) is a very important factor in growth and development, particularly in the control of flowering, and the phenomenon is known as *photoperiodism*. On the other hand, in other species, flowering was not materially affected by length of day. The group in which daylength has a marked effect may be separated into two subdivisions, viz (1) those in which flowering is readily induced by exposure to short days, and known as *"short-day"* plants (SDP), and (2), a second group in which flowering is favoured by long days, and which are known as *"long-day"* plants (LDP). Species in which daylength does not markedly affect flowering are known as *indeterminate* or *"day-neutral"* plants.

Some species remain permanently vegetative if kept under unfavourable daylength conditions, and hence may be called obligate photoperiodic plants. They include both SDP, such as *Xanthium pennsylvanicum,* and LDP such as *Hyoscyamus niger.* Other species may show hastened flowering

Fig. 9.2. Plants of *Kalanchoë blossfeldiana* (left) and of *Rudbeckia bicolor* (right), flowering in response to short days and long days, respectively.

under SD or LD, but they will ultimately flower even under unfavourable daylength conditions, and hence show a *quantitative* photoperiodic response. Such species include the SDP *Salvia splendens,* rice (*Oryza sativa*) and cotton (*Gossypium hirsutum*) and the LDP wheat (*Triticum*) and flax (*Linum usitatissimum*). Obligate photoperiodic plants show a well-marked *critical daylength,* below or above which flowering will not occur in LDP and SDP, respectively, whereas facultative photoperiodic species show only a graded response to daylength, with no sharp cut-off point. Examples of SDP and LDP are shown in Fig. 9.2, and are listed in Table 9.1.

The short-day group includes many plants which are indigenous to regions of low latitude, north or south of the equator, such as rice, sugar cane, hemp, millet and maize, where the daylength never exceeds more than 14 hours at any season of the year. SDP of temperate regions, where the days are long in the summer, usually initiate flowers only in the late summer as the days shorten, e.g. the cultivated "Michaelmas daisies" (*Aster* spp.) and Chrysanthemums. The typical LDP are native to the temperate regions, and flower under the naturally long days of summer. They include many of the grasses and cereals, and other common cultivated plants such as spinach

Table 9.1. *Some examples of short-day and long-day plants*

Short-day plants
    A. *Species with an Absolute or Qualitative Short-day Requirement*

Amaranthus caudatus (Love-lies-bleeding)	*Ipomoea hederacea* (Morning glory)
Chenopodium album (Pigweed)	*Kalanchoë blossfeldiana*
Chrysantheum morifolium	*Lemna perpusilla* (Duckweed)
Coffea arabica (Coffee)	*Nicotiana tabacum* (Tobacco, var. Maryland
Euphorbia pulcherrima (Poinsettia)	Mammoth)
Fragaria (Strawberry)	*Perilla ocymoides*
Glycine max (Soybean)	*Xanthium strumarium* (Cockleburr)

    B. *Species with Quantitative Short-day Requirement*

Cannabis sativa (Hemp)	*Oryza sativa* (Rice)
Cosmos bipinnatus (Cosmos)	*Saccharum officinarum* (Sugar cane)
Gossypium hirsutum (Cotton)	*Salvia splendens*

Long-day plants
    A. *Species with an Absolute or Qualitative Long-day Requirement*

Alopecurus pratensis (Foxtail grass)	*Melilotus alba* (Sweet clover)
Anagallis arvensis (Pimpernel)	*Mentha piperita* (Peppermint)
Anethum graveolens (Dill)	*Phleum pratensis* (Timothy grass)
Avena sativa (Oat)	*Raphanus sativus* (Radish)
Dianthus superbus (Carnation)	*Rudbeckia bicolor* (Coneflower)
Festuca elatior (Fescue grass)	*Sedum spectabile* (Sedum)
Hyoscyamus niger (Henbane)	*Spinacia oleracea* (Spinach)
Lolium temulentum (Rye-grass)	*Trifolium* spp. (Clover)

    B. *Species with a Quantitative Long-day Requirement*

Antirrhinum majus (Snapdragon)	*Petunia hybrida* (Petunia)
Beta vulgaris (Garden beet)	*Pisum sativum* (Garden pea)
Brassica rapa (Turnip)	*Poa pratensis* (Kentucky blue-grass)
Hordeum vulgare (Spring barley)	*Secale cereale* (Spring rye)
Lactuca sativa (Lettuce)	*Triticum aestivum* (Spring wheat)
Oenothera spp. (Evening primrose)	

(*Spinacia oleracea*), lettuce (*Latuca sativa*), beet (*Beta vulgaris*), flax (*Linum usitatissimum*) and clover (*Trifolium* spp.), as well as many wild species. In addition to the two main groups of SDP and LDP, there is a smaller number of species with dual daylength requirements. Thus, certain species require to be exposed first to LD and then to SD for flower initiation to occur and hence are called "long-short-day" plants (LSDP); examples of this type of response are provided by *Bryophyllum crenatum* and *Cestrum nocturnum*. Other species, such as *Scabiosa succisa*, *Campanula medium* and *Trifolium repens,* require to be exposed first to SD and then to LD, and hence are called "short-long-day" plants (SLDP).

Where a species has a wide distribution, so that there is a considerable difference in latitude between its northern and southern limits, it is found that it is differentiated into a number of races or ecotypes, differing in their daylength responses, e.g. golden rod (*Solidago sempervirens*), which has a wide distribution along the western coast of North America, and perennial ryegrass (*Lolium temulentum*), which has a wide distribution in Europe and North Africa. These different forms within a given species usually show a closely graded series, from typical SDP at one end to LD-tolerant types at the other, or from typical LDP to SD-tolerant types.

9.2.1 Other responses affected by daylength

In addition to the onset of flowering, daylength may affect certain other purely vegetative processes in the plant. Thus, it is frequently found that the length of the internode may be much reduced under SD as compared with LD. This effect is seen at its extreme form in certain long-day species, which assume a "rosette" habit under SD, e.g. henbane (*Hyoscyamus niger*). Runner formation in strawberry (*Fragaria*) plants occurs only under LD. The strawberry has a rosette habit which is not affected by daylength, but the axillaries have extended internodes (thus forming runners) under LD.

Tuber formation is also markedly affected by daylength, and is favoured by SD, as in the Jerusalem artichoke (*Helianthus tuberosus*) and in many wild species of potato (e.g. *Solanum andigena*). (The cultivated European varieties of potato are not very sensitive to daylength, and they can form tubers even under LD.) The onion (*Allium cepa*), on the other hand, requires LD for bulb-formation.

Daylength has a marked effect on the growth and leaf-fall of many wooded plants, as will be described below.

9.2.2 Sensitivity to daylength conditions

Many plants are extremely sensitive to changes in daylength as short as 15—20 minutes, so that the flowering time of some, such as certain varieties of rice (*Oryza sativa*), may be profoundly affected by even the relatively small seasonal changes in daylength found in the tropics. Similar sensitivity is shown by cockleburr (*Xanthium strumarium*) which only flowers under daylengths of 15¾ hours or less. Evidently the mechanism whereby such plants detect changes in daylengths is very sensitive.

Variation in sensitivity to daylength is shown in the number of photoperiodic cycles required to induce flowering. Thus, plants of *Xanthium* require exposure to only *one* SD cycle for flowering, and once induced to flower in this way they will continue to produce flowers for a further 12 months. The LD grass *Lolium italicum* will also flower in response to one LD cycle. However, these are extreme cases, and the majority of photoperiodic plants require more than one cycle. Moreover, even in *Xanthium,* the rate of floral development is more rapid after exposure to 2 or 3 SDs. In soybeans, the number of nodes at which flowers are found increases linearly with the number of SD cycles, up to at least 7. In many species, the daylength requirements vary considerably with the age of the plant, and frequently the minimum number of inductive cycles is found to decrease with age. Moreover, some SDP, such as soybeans, which will flower only under SDs when they are young, ultimately become capable of flowering even under LD.

9.3 LIGHT AND DARK PROCESSES IN SHORT-DAY PLANTS

A considerable amount of work has been carried out to learn more about the light and dark responses of SDP species, and our knowledge of this type of plant is considerably greater than for LDP.

It is important to ascertain, first, which part of the plant is concerned with the "detection" of daylength conditions. Clearly, in the transition from the vegetative to the flowering condition, the response is at the shoot apices, but it does not follow that "perception" of daylength conditions occurs in these parts of the plant. Indeed, it was shown by the Russian worker, Chailachjan, that the responses of SDP, such as Chrysanthemum, are determined by the daylength condition to which the *leaves* are exposed and that the shoot-apices appear to be relatively insensitive to daylength conditions. He was able to show this by exposing the leaves and shoot-apical regions independently to LD or SD conditions (Fig. 9.3); he

found that it was only when the mature leaves were maintained under SD that flowering occurred. Thus, detection of the daylength conditions is effected by the leaves, although the response occurs at the shoot-apex. Chailachjan was quick to see that these observations imply that some "signal" must be transmitted from the leaves, which causes a response in the apices. We shall discuss the possible nature of this "signal" in a later section (p. 246).

In most plants the peak of photoperiodic sensitivity in the leaf appears to be reached when it has just attained its maximum size. At this stage quite small amounts of leaf tissue are sufficient to bring about flowering. Thus, in *Xanthium* 2 cm^2 are sufficient to bring about flowering under SD.

Although leaf tissue is the most sensitive to daylength conditions, the stem tissues of some species also show some sensitivity, a good example being seen in *Plumbago indica,* a SDP. Indeed, isolated internode sections of this species may be induced to flower under SD conditions, in aseptic culture (Fig. 9.4), whereas under LD they remain vegetative.

The next problem which arises is whether the responses of SDP are determined by (1) the length of the daily light period, (2) the length of the dark period, or (3) the *relative* lengths of the light and

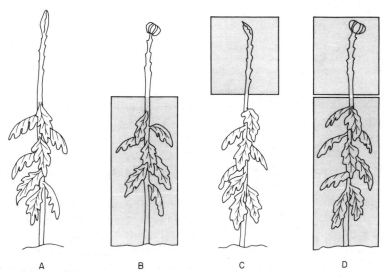

Fig. 9.3. Flowering occurs in Chrysanthemum when the *leaves* are exposed to short days (B and D), irrespective of whether the shoot apex is exposed to long days (B) or short days (D).

dark periods. Now it can easily be shown that the responses of SDP are not controlled primarily by the total duration of light which they receive each day. This is shown by the fact that soybeans will flower if exposed to three cycles consisting of 12 hours light/12 hours dark, but will not do so if exposed to 36 hours of light, followed by 36 hours of dark, or if exposed to cycles consisting of 6 hours of light alternating with 6 hours of dark (Fig. 9.10), although the total hours of light and dark are the same in all three régimes.

If we conduct experiments based upon the natural 24-hour cycles, it is impossible to vary the lengths of the light and dark periods independently; this can be done, however, if we use artificial sources of illumination, such as fluorescent lamps, since we can then choose any combination of light and dark periods we wish. K. C. Hamner was the first to take advantage of this fact in a series of experiments with *Xanthium* and soybean, which have now become "classical". Using soybean Hamner first investigated the effect of varying the length of the dark period, keeping a constant duration of light period of either (a) 4 hours, or (b) 16 hours. The results (Fig. 9.5) showed quite clearly that soybeans will not flower until the length of the daily dark periods exceeds about 10 hours, with either 4-hour or 16-hour photoperiods. Thus, the *critical dark period* for

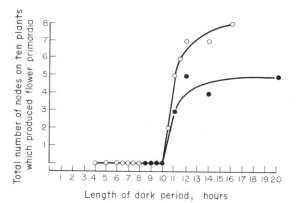

Fig. 9.5. Effect of various lengths of dark period, in association with constant 16-hour (—O—) or 4-hour (—●—) photoperiods, on flowering in soybean (*Glycine max*). (From K. C. Hamner, *Bot. Gaz.* **101**, 658, 1940.)

soybeans is about 10 hours, but the maximum flowering response is reached with dark periods of 16—20 hours. In similar experiments with *Xanthium*, Hamner showed that this species has a critical dark period of 8¼ hours to flower.

Since SDP require a certain minimum period of darkness for flowering it may be asked whether they flower most readily in continuous darkness and whether they require any light. It is found, however, that although certain species, especially those with a storage organ such as a tuberous rootstock, will flower in continuous darkness, other species, such as soybeans, will not do so, but require a regular alternation of light and dark.

Fig. 9.4. Induction of flowering *in vitro*. Flower of *Plumbago indica* L. "Angkor" produced on a segment of stem internode (7 mm in length) excised from a vegetative plant and planted aseptically on nutrient agar. Flower formation occurs only if the culture is placed under short days (10 hours of light). Under long days (16 hours of light), vegetative buds are produced. (Experiment by J. P. and C. Nitsch. Photo: Mlle B. Norreel, kindly supplied by Dr. J. P. Nitsch.)

Having investigated the dark requirements of soybeans, Hamner then proceeded to study their light requirements. Using a constant dark period of 16 hours, he varied the length of the light period and found that the flowering response increased as the length of the daily light period was increased to about 12 hours (Fig. 9.6), but with longer light periods fewer flowers were formed and there was no flowering when the light period reached 20 hours, even though these light periods were associated with long (16-hour) dark periods. Thus, the conditions for flowering in soybean are (1) that the daily dark period must exceed 10 hours and (2) that the length of the light must not exceed a certain duration. Of course, under natural conditions, dark periods of 10 hours or more can only be accompanied by light periods of 14 hours or less, and in nature the flowering response is likely to be controlled by the length of the dark period,

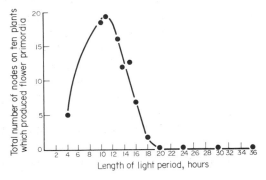

Fig. 9.6. Effect of varying length of photoperiod, with constant daily dark period of 16 hours, on flowering of soybeans. (From K. C. Hamner, *Bot. Gaz,* **101,** 658, 1940.)

rather than the duration of the light period. For this reason, it would be more appropriate to refer to short-day plants as "long-night" plants. In some SDP, e.g. *Xanthium,* long photoperiods are not inhibitory to flowering, which is determined solely by the length of the dark period.

Having thus established that flowering in short-day plants is favoured by an alternation of light and dark, the question arises as to whether the light must precede the dark or vice versa. By means of ingenious experiments with *Xanthium,* which, as we have seen, requires only one SD for flowering, Hamner was able to show that a long

dark period must be *preceded* by an adequate period of high-intensity light, and that a period of high-intensity illumination following the dark period also promotes flowering.

9.3.1 The nature of the "high-intensity" light reaction

The light requirements of SDP during the photoperiod have been investigated quantitatively by Hamner and others. In general, it appears that the light requirements are relatively high. Thus the flowering response in soybeans increases as the period of exposure to daylight is increased from 1 to 8 hours. Moreover, with photoperiods of constant duration (5 to 10 hours), the number of flowers increases steadily with increasing light intensity.

These relatively high light requirements suggest that the process involved during the main photoperiod is photosynthesis. This conclusion is confirmed by the following facts: (1) carbon dioxide is necessary during the photoperiod, if flowering is to occur; (2) if sugar is sprayed on the leaves, then certain short-day plants can flower in complete darkness. Thus, there seems little doubt that the requirement for light in SDP is for adequate photosynthesis. This requirement might be expected, since an adequate supply of energy in the form of carbohydrates will clearly be necessary to permit the other metabolic processes, more specifically related to flowering, to occur in the leaves.

9.3.2 The "dark" reactions of short-day plants

As we have seen, SDP require to be exposed to daily cycles which include a certain minimum period of darkness, the so-called "critical" dark period. This critical dark period appears to be relatively constant and independent of the length of the photoperiods over quite a wide range of the latter. At first sight it might be thought that the dark period plays a passive role, in the sense that the effects of darkness are simply due to the absence of light effects. Thus, it might be postulated that light has an inhibitory effect upon flowering, and that the flower-promoting effects of darkness are primarily due to the absence of such inhibitory effects. However, there are several pieces of evidence which suggest that certain positive flower-promoting processes occur during the dark period. Thus, it is known that the effectiveness of the dark period increases with temperature within certain limits, suggesting the occurrence of flower-promoting processes with positive temperature coefficients during the dark period.

A very remarkable feature of the dark period is that it must be *uninterrupted* if it is to be effective in promoting flowering—interruption by only a few minutes of light during the dark period may completely nullify its effect, so that flowering is inhibited. Thus, with *Xanthium,* 1 minute of light at an intensity of 150 foot-candles (fc) (= 1500 lux) during a 9-hour dark period suppresses flowering. This effect has been investigated in some detail in *Xanthium,* in which it was found that a "night-break" was most effective (maximum inhibition of flowering) when given 8 hours after the beginning of the dark period, regardless of the length of the latter, at least over the range 10—20 hours (Fig. 9.7). Thus we have the apparently paradoxical situation that light during the main photoperiod preceding a long dark period

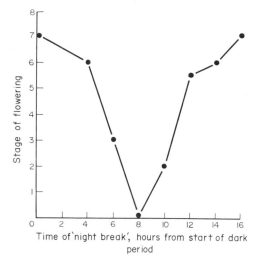

Fig. 9.7. Effect of night interruption, given at various times during dark period, on flowering in *Xanthium.* (From F. Salisbury and J. Bonner, *Plant Physiol.* **31**, 141, 1956.)

promotes flowering, whereas light given during the dark period inhibits flowering. We have seen that the requirement for light in the main photoperiod is probably for photosynthesis, to allow the formation of photosynthates necessary for metabolism of the leaf. However, it is now clear that the effects of a short interruption ("night-break") during a long dark period are not primarily due to photosynthesis but involve phytochrome. Detailed action spectra were determined for the night-break effect in soybeans and

Table 9.2. *Effect of daily interruptions of the dark period with several consecutive irradiations with red and far-red in sequence on flower initiation of* **Xanthium** *and Soybean*

Treatment	Mean stage of floral development in *Xanthium**	Mean no. of flowering nodes in *Biloxi* soybean
Dark control	6·0	4·0
R	0·0	0·0
R, FR	5·6	1·6
R, FR, R	0·0	0·0
R, FR, R, FR	4·2	1·0
R, FR, R, FR, R	0·0	—
R, FR, R, FR, R, FR	2·4	0·6

* See p. 218 for method of scoring.

in *Xanthium* by placing seedlings with single leaves in different parts of the spectrum produced by a large spectrograph, and these action spectra proved to be almost identical to those which were later demonstrated for other phytochrome-controlled responses, such as seed germination and "de-etiolation" of shoots (Fig. 8.4). Moreover, R/FR reversibility was demonstrated for the night-break effect (Table 9.2), thus showing conclusively that this is a phytochrome-controlled response.

Although flowering in the majority of species studied is inhibited by a relatively short night-break of a few minutes, other species require a longer night-break, e.g. the garden chrysanthemum requires a night-break of 4 hours.

The "escape-time" also varies considerably among different species; for example, in *Xanthium* the effect of R can be reversed if the application of FR is delayed for up to 40 minutes whereas in *Chenopodium album, Pharbitis nil* and *Kalanchoë blossfeldiana* FR must be given almost simultaneously after R to be effective.

The effects of R and FR given as night-breaks clearly indicate that the flower-promoting effects of a long inductive dark period are profoundly affected by the levels of P_{fr} in the leaf. At the end of the main photoperiod a high proportion of the phytochrome will be present in the P_{fr} form but, as we have seen, the levels of P_{fr} decline fairly rapidly during darkness, due to degradation or reversion. However, if a short exposure to R light is given after several hours of darkness high levels of P_{fr} will be restored, with inhibitory effects on the flower-promoting processes (Fig. 9.8), which apparently require low P_{fr} levels.

Although, with short night-breaks FR reverses the inhibitory effect of R, with longer periods (e.g. 60 minutes) FR acts in the same manner as R, and is inhibitory to flowering, especially after short photoperiods. This effect seems to indicate that although the low values of P_{fr} caused by FR are not effective over short periods, they can be effective if maintained over longer periods.

9.3.3 Flower promoting effects of P_{fr} in short-day plants

There is evidence that although high levels of P_{fr} are inhibitory to flowering when they occur during a long inductive dark period, as a result of a night-break, at other times of the daily cycle flowering may be promoted by high P_{fr} levels. Thus, flowering in seedlings of *Pharbitis nil* is reduced by FR (which will reduce the level of P_{fr}) given at the end of the main photoperiod. It would appear, therefore, that at that stage in the daily cycle flowering is favoured by high P_{fr} levels, whereas after several hours of darkness it is inhibited, as indicated by the effects of a R night-break. Thus, flowering in SDP is apparently promoted by high P_{fr} at the end of the main photoperiod, and inhibited by high P_{fr} later during the dark period, indicating that phytochrome has a dual action in the photoperiodic control of flowering.

9.4 RESPONSES OF LDP

As in SDP, the effect of a long dark period on LDP is nullified by a short night-break, and hence *promotes* flowering in plants of this type.

Studies on the action spectrum for the night-break effect in barley (*Hordeum*) and *Hyoscyamus* indicated, once again that it is the red region (660 nm) which is effective and that the effects of R are reversed by FR in some, but not

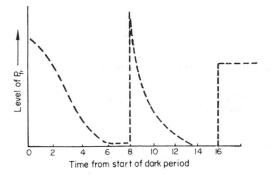

Fig. 9.8. Hypothetical changes in the P_{fr} form of phytochrome in leaves during a dark period which has been interrupted by a short exposure ("night-break") to red light.

all, LD species. However, LDP are less sensitive to night-breaks than SDP, and although a night-break of 1—2 hours is effective in some LDP, the effect is quantitative and the maximum effect may require still longer periods.

When supplementary illumination is used to extend a short photoperiod given as daylight, it is found that a mixture of R and FR is more effective in promoting flowering in LDP than R alone (Fig. 9.9). The R and FR need not be given simultaneously. Thus, in *Lolium temulentum*, when R alone was used to extend 8 hours of daylight it was ineffective in inducing flowering, but if this R was interrupted by a few hours of FR, flowering was induced. It was found that there was an optimum time for exposure to FR, and it appears that following the end of an 8-hour photoperiod of sunlight, high P_{fr} levels (caused by R) *reduce* the flowering response, whereas later in the dark period high P_{fr} tends to favour flowering and at this time R promotes. FR appears to promote flowering from about the sixth hour from the beginning of the daily photoperiod, but after about the eighteenth hour FR has little effect and flowering is now promoted by R. However, a diurnal change in P_{fr} level is not essential for LDP as it is in SDP. Thus, it appears that, as in SDP, high P_{fr} promotes flowering at certain times and inhibits at others, but the sequence of promotion and inhibition is opposite in LDP and SDP. Thus, we see that there are close analogies between the behaviour of SDP and LDP, but in certain respects their phytochrome-controlled responses are opposite, so that long dark periods promote flowering in SDP but inhibit it in LDP (Fig. 9.10). In both types of plant, moreover, the effect of the

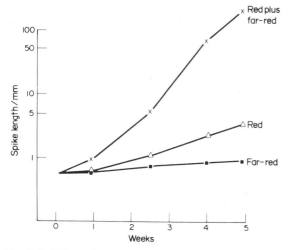

Fig. 9.9. Effect of red, far-red and mixed red plus far-red light on flowering in *Lolium temulentum*. Plants received 8 hours of daylight plus a further 8 hours of supplementary light of restricted wavelengths. (From D. Vince-Prue, *Photoperiodism in Plants,* McGraw Hill, London, 1975.)

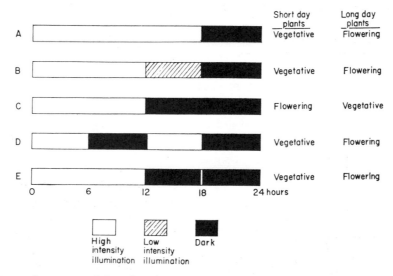

Fig. 9.10. Summary of responses of short-day plants and long-day plants to various photoperiodic régimes.

dark period is nullified by a night-break which results in high P_{fr} levels. During the earlier part of the daily cycle, flowering in SDP is promoted by high P_{fr} levels, so that they require a sequence of high P_{fr} during the main photoperiod and low P_{fr} during an ensuing long dark period (Fig. 9.11). By contrast, LDP do not require such a diurnal fluctuation in P_{fr}, since they flower even in continuous light (under incandescent lamps which produce low P_{fr} values), but under suitable experimental conditions a diurnal fluctuation in sensitivity to P_{fr} levels can be demonstrated.

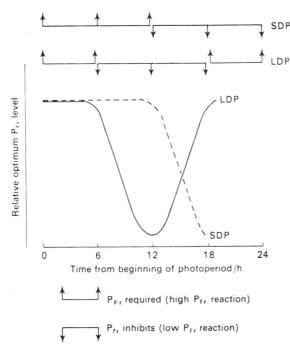

Fig. 9.11. Schematic representation of the "high P_{fr}" and "low P_{fr}" reactions in the flowering of long-day plants and short-day plants. In the SDP the P_{fr}-requiring process occurs early in the photoperiod and is followed by a period when P_{fr} inhibits flowering. In LDP similar reactions are seen, but their timing is markedly different. (From D. Vince-Prue, *Photoperiodism in Plants*, McGraw Hill, London, 1975.)

9.5 THE FLOWERING STIMULUS

We have seen that the response of the plant to daylength depends upon the conditions to which the leaves are exposed, although the response occurs at the shoot meristem. This observation immediately suggests that under favourable daylength conditions some flower-promoting "stimulus" is formed in the leaves and conveyed from there to the meristems. This hypothesis was put forward by Chailachjan very shortly after his discovery that the perceptive organs in photoperiodism are the leaves, and he postulated that a flower hormone (which he called "florigen") is synthesized in the leaves under favourable daylength conditions and transmitted to the growing points. As we shall now see, there is a great deal of evidence to support this hypothesis, but some 45 years after Chailachjan first postulated the existence of "florigen", it still remains to be isolated and characterized chemically. A considerable number of highly interesting experiments on the transmission of the flower hormone have been carried out and these will now be described.

Hamner confirmed Chailachjan's experiments, using *Xanthium*. He showed that flowering of *Xanthium* plants under SD does not occur if all the leaves are removed, but only one-eighth of one leaf is sufficient to result in flowering (Fig. 9.12). In experiments with "two-shoot" plants of *Xanthium*, it was found that the stimulus arising from a single leaf is sufficient to cause flowering not only in the shoot on which it is borne, but also on the second shoot from which all leaves have been removed.

A large number of experiments have shown that the flowering-stimulus can be transmitted across a graft union. Thus, Hamner grafted together the stems of pairs of plants of *Xanthium;* when one of the plants of a pair was exposed to SD, not only did this plant flower, but also its grafted partner which had not been exposed to SD (Fig. 9.12). In further experiments plants of *Xanthium* were exposed to SD, their tops were then decapitated and scions from other *Xanthium* plants which had been maintained under LD were grafted on to them. In due course, it was found that the "LD" scions flowered, although they had not themselves ever been exposed to SD.

In still more striking experiments carried out with various species, including *Xanthium* and *Perilla*, single leaves were removed from "donor"

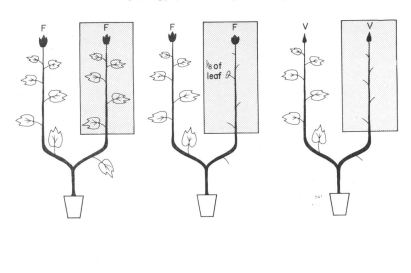

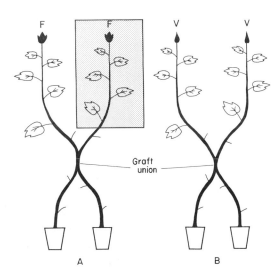

Fig. 9.12. *Above:* Two-branched plants of *Xanthium,* one branch of which was exposed to short days, the rest of the plant being exposed to long days. Both branches have flowered, provided that the "short-day" shoot has at least one-eighth of a leaf, but not if it is completely defoliated (*right*).

  Below: A. Two *Xanthium* plants were approach grafted, and the top of one plant was exposed to short days while the other plant was maintained under long days. Both plants have flowered. B. Both plants exposed to long days; neither have flowered. (From K. C. Hamner, *Cold Spring Harbor Symp.* **10,** 49, 1942.)

plants which had been exposed to SD and grafted on to plants maintained under LD. In due course the LD "receptor" plants flowered, showing that the stimulus may be transmitted from a single leaf.

Grafting experiments with LDP have been very much fewer than with SDP, but the results have proved essentially the same.

Experiments of the type just described leave little doubt that some "stimulus", presumably a substance of a hormonal nature, is formed in the leaves under favourable daylength conditions and is transmitted through the stem to the apical meristems.

Although the flower hormone has not yet been isolated from SDP, we know quite a lot about the time required for its synthesis in the leaf and about its transport in the plant. If *Xanthium* plants, each with a single leaf, are exposed to one SD cycle, and the leaves are removed at various times after the end of the long dark period, it is found that flowering does not occur unless they are allowed to remain on the plant for at least 2—4 hours following the end of the dark period, but greater flowering responses are obtained if the leaves are left for 1—2 days before removal, indicating that movement out of the leaf is a slow process. There seems no doubt that transport of the hormone occurs through living tissues, presumably by the phloem, since transport is stopped by steaming a zone of stem between the leaf and the apical region, but it may be transported through other living tissues.

Attempts have been made to estimate the rate of transport of the hormone. One method was to use two-shoot plants of *Pharbitis* in which a single donor leaf on one branch was exposed to SD, and the differences in time of flower initiation on the second branch (kept under LD) were determined for varying distances between the donor leaf and the receptor bud. Methods of this type, which admittedly are very indirect, suggest that the rate of movement of the flower hormone is much slower (2—4 mm/hour) than the normal rate of translocation of sugars in the phloem. An experiment with *Pharbitis* has given evidence of much more rapid transport, however (Fig. 9.13).

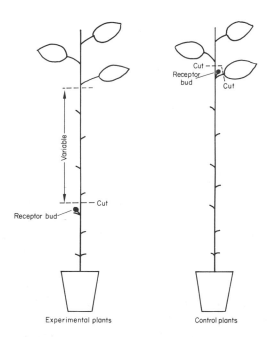

Fig. 9.13. *Pharbitis* plants of various heights were decapitated above the uppermost fully expanded leaf and all leaves except for the three upper ones were removed. The plants were divided into two groups. In the experimental group all axillary buds were removed except for one receptor bud at the base of the stem, and in the control group all buds were removed except for the one in the axil of the lowest donor leaf. The plants were given a single dark period which ranged in duration from 14 to 16 hours in different batches of plants, and at the end of the dark period all leaves and stems were cut off just above the receptor buds. No flower buds were formed in either group when the leaves were removed 14 hours after the start of the dark period. Flower buds were, however, initiated in many plants whose donor leaves were removed 16 hours after the start of the dark period, irrespective of the length of stem between the donor leaf and the receptor bud. The tallest plant had a stem 102 cm long between the lowest donor leaf and the receptor bud at the base, and initiated two flowers. As there was no flowering in the lowest donor leaf in control plants given a 14-hour dark period, the flowering stimulus must have moved through 102 cm of stem within the last 2 hours of the 16-hour dark period, yielding a rate of movement of more than 51 cm/hour. (From G. Takeba and A. Takimoto, *Bot. Mag., Tokyo,* **79,** 811—14, 1966.)

Evidence was obtained that the flowering stimulus had moved over a distance of 102 cm in 2 hours, i.e. a rate of 51 cm/hour, which is commensurate with the rate of movement of photosynthates in the phloem.

9.6 PHOTOPERIODIC "AFTER-EFFECT" AND THE INDUCTION OF LEAVES

As we have seen, a short-day plant does not have to be maintained continuously in short days in order to flower. After a certain number of favourable photoperiodic cycles a SD plant will flower even though it may subsequently be transferred to LD. Thus, as we have seen, *Xanthium* will ultimately flower in response to a single SD cycle, although a considerable number of LD cycles may intervene between the SD and the ultimate appearance of flowers. Thus, certain species showed a marked photoperiodic "after-effect", and a plant is said to become "induced" after exposure to SD.

This after-effect of favourable photoperiodic cycles is evidently a property of the leaves, which apparently continue to produce "flower-hormone" even after they have been transferred from favourable to unfavourable daylength conditions. Thus, the inductive response of the whole plant reflects the inductive changes occurring in the leaves. This conclusion is supported by many experiments. For example, Lona subjected a single leaf of a *Perilla* plant to SD, the other leaves remaining under LD conditions. After this period of SD treatment, the leaf was allowed to remain under LD for 4 weeks, when it was removed and grafted on to a vegetative *Perilla* plant, which in due course flowered. Thus, the leaf "remembered" the previous SD treatment although 4 weeks of LD treatment intervened between the end of the last SD cycle and the grafting on to the receptor plant. What is the basis of this after-effect shown by leaves?

Two theories have been put forward to account for the after-effect. Chailachjan postulated that a store of flower hormone is accumulated in the leaf under favourable conditions and is gradually exported from it for a long period even under favourable conditions. On the other hand, another Russian worker, Moshkov, postulated that the metabolism of the leaf somehow becomes *permanently* changed in response to favourable daylength conditions, so that it continues actively to *produce* flower hormone even if it is subsequently transferred to unfavourable daylengths. The available evidence seems strongly to support Moshkov's theory. Thus, it would seem unlikely that in the experiment of Lona described above, sufficient flower hormone would still remain in the SD-treated leaf after 4 weeks of LD treatment to induce flowering when it was grafted on to a vegetative plant.

Even stronger evidence that leaves may become permanently changed under favourable daylength conditions is provided by certain experiments of Zeevaart. In one experiment, *Perilla* plants were first exposed to forty SD cycles and the leaves were then grafted on to vegetative plants growing under LD. Every 14 days, ten of these leaves were removed and regrafted on to a new group of vegetative plants. It was found that the leaves continued to induce flowering in each new group of vegetative stocks on to which they were grafted (Fig. 9.4). There was no diminution in the flowering response even with the fifth group of leaves, which had been maintained under LDs for 10 weeks. In a second experiment there was no diminution in the flower-inducing effect after seven such graftings, although more than 3 months had elapsed since the leaves had received the SD treatment. Thus, in *Perilla,* the leaves appear to become permanently induced.

It is found that the state of induction is strictly localized within the *Perilla* plant and even within the individual leaves. Thus, if single pairs of leaves of *Perilla* are exposed to a series of SD cycles and the rest of the plant is kept under LD, it is found that only those leaves which directly received SDs are capable of inducing the plants to flower when grafted on to them (Fig. 9.14). In other experiments by Lona, only one half of single leaves were exposed to SD, and the other half of each leaf to LD. The leaves were then divided longitudinally and grafted separately on to

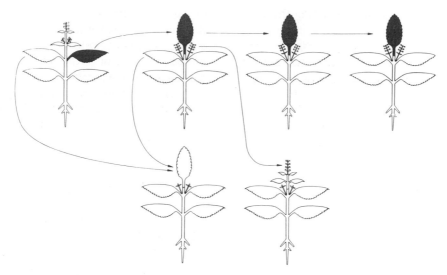

Fig. 9.14. Permanency of the photo-induced state in *Perilla* (see text). Black: leaves exposed to short days; in outline: plants on long days. Leaves which have directly received SD treatment will induce flowering when grafted successively to negative receptor plants, whereas non-induced leaves from the original or from the receptor plants, and flowering shoots from the latter, will not do so. (From A. Lang, *Encycl. Plant Physiol.* **15**(1), 1416, 1965, after data from Zeevaart.)

vegetative receptor plants. Only those half leaves which had been directly exposed to SD were capable of inducing flowering.

The situation is different in *Xanthium,* however. If a single leaf from a flowering *Xanthium* plant (A) is grafted on to a vegetative plant (B) growing under LD, this will flower, as has already been described. If other young leaves, *which have never themselves been exposed to SD,* are taken from plant B and grafted on to further vegetative plants (C) these will also flower. Thus, the leaves of plant B have themselves become induced by the grafting of an induced leaf from plant A, although these leaves of B have never directly been exposed to SD. This effect is desribed as *secondary induction,* and it probably explains why *Xanthium* does not revert to the vegetative condition when it is transferred from SD to LD conditions, since all the new leaves formed will become secondarily induced. On the other hand, the new leaves formed by a *Perilla* plant which is transferred from SD to LD will not be induced and they appear to inhibit the flower-promoting effects of the older leaves which

became induced by the previous SD treatment (see below), and hence the plant reverts to the vegetative condition.

Although *Xanthium* and *Perilla* show marked persistence of photoperiodic induction of leaves, it is not known how general are these effects. Reversion of whole plants to the vegetative condition on exposing to a minimal number of SDs and then returning them to LD certainly occurs in soybean and a number of other species, but whether this is due to "de-induction" of the leaves exposed to SD, or to the production of new, non-induced leaves, as in *Perilla,* is not known.

As Zeevaart has pointed out, the permanent induction of leaves makes it necessary to distinguish between two distinct phenomena, viz.

(1) the *induced state* (i.e. the ability to produce the floral stimulus), gradually built up under the influence of favourable daylengths, which is irreversible and strictly localized in some plants;

(2) the *floral stimulus,* which is transmissible from induced leaves to the growing points where it exerts its morphogenetic effect.

9.7 NATURAL FLOWER-INHIBITING EFFECTS

Several lines of evidence suggest that not only flower-promoting processes, but also flower-inhibiting effects occur in photoperiodic plants. Thus, if one branch of a two-shoot plant of soybean or *Perilla* is exposed to SD and the other to LD, the latter will not flower unless its own leaves are removed. That is to say, the flowering stimulus from the "donor" branch does not produce any effect in the "receptor" branch in the presence of LD leaves. Similarly, if a scion from a vegetative plant of *Perilla* is grafted on to a stock which has previously been maintained under SD, the scion will only flower if its leaves are removed. Thus, in some SD species, LD leaves exert an inhibitory effect on flowering.

This inhibitory effect of LD leaves is only manifested if they occur between the shoot apex and the source of the flowering stimulus. Thus, if a single leaf of *Kalanchoë* is exposed to SD and the remainder of the plant is maintained under LD, the presence of LD leaves between the SD leaf and the shoot apex prevents flowering, but if all leaves are removed *above* the SD leaf and only LD leaves below are allowed to remain the plant will flower. A LD leaf is particularly inhibitory if it is immediately above the SD leaf (i.e. on the same orthostichy). LD leaves on the opposite side of the stem from the SD leaf have less inhibitory effect.

It has been suggested that these inhibitory effects of non-induced leaves can be interpreted in terms of interference with the translocation of the flower hormone, which, it is assumed, is carried with the mains stream of photosynthates. The leaves which supply the greater part of the photosynthates to the shoot apical region are the uppermost mature leaves. If the latter are SD leaves, then they will supply flower hormone, as well as photosynthates, to the shoot apex, but if LD leaves are interposed between the apex and the SD leaves, then the supply of both photosynthates and flower hormone from the SD leaves to the apex will be reduced. However, more recent work by Lang has produced strong evidence for the production of a transmissible flower inhibitor in non-induced leaves of some species. Thus, when scions of the long-day plants *Nicotiana silvestris* and *Hyoscyamus niger* were grafted sideways into the stem of plants of the day-neutral tobacco cultivar *Trapezond,* the latter flowered when the grafted plants were maintained under long days, but flowering of both scion and stock was inhibited when they were maintained under short days (Fig. 9.15), although control plants of *Trapezond* flowered under short days.

Fig. 9.15. Flowering response of day neutral tobacco "Trapezond" after grafting with LDP *Nicotiana silvestris* under LD (left) or under SD (right) Trapezond receptor shoots defoliated). (From A. Lang, M. C. Chailakyan and I. A. Frolova, *Proc. Nat. Acad. Sci.* U.S.A., **74,** 2412, 1977.)

In these latter experiments we evidently have a situation in which a transmissible inhibitor counteracts the action of the floral stimulus in the shoot apex. A different situation exists where there is inhibition of formation of the stimulus in the leaf when it is exposed to alternating inductive and non-inductive photoperiods. Thus, when the SDP *Kalanchoë* is exposed alternately to two SD cycles and one LD cycle, flowering is completely inhibited, and analysis of the data indicates that the effect of the intercalated LD cycle is not merely "neutral" but positively inhibitory to flowering. In this type of experiment it would seem that we are dealing with inhibitory effects of LDs within the leaf itself, and not with the export of a flowering inhibitor to the shoot apex.

Other flower-inhibiting effects are seen in experiments in which SDP such as *Kalanchoë* are exposed to SD cycles between which are intercalated one or more LD cycles. It can be shown that if the plants are exposed alternately to two SD cycles and one LD cycle, flowering is completely inhibited, and a careful analysis of the data indicates that the effect of the intercalated LD cycle is not merely "neutral" but positively inhibitory to flowering. In this type of experiment it would seem that we are dealing with inhibitory effects of LDs within the leaf itself, and not with the export of a flower-inhibitor to the shoot apex.

Flower-inhibiting effects also occur in LDP, since, as we have seen, in these plants long dark periods appear to have an inhibitory effect. It seems likely, however, that the mechanism of inhibition in this case is different from that occurring in the leaves of SDP when they are kept under LD.

It might be asked whether, if there are flower-inhibiting processes in both SDP and LDP, it is necessary also to postulate active flower-promoting processes? Might not the flowering of SDP when transferred to SD conditions be due primarily to the removal of flower-inhibitory processes occurring under LD? If so, we may not need to postulate the existence of the elusive "flower hormone". However, most workers in this field consider that other experimental evidence points strongly to the existence of both flower-promoting

and flower-inhibiting processes. Thus, the observation that a single leaf taken from a flowering plant of *Xanthium* will induce flowering in a vegetative plant maintained under LD is difficult to explain in terms simply of the removal of flower-inhibition in the receptor plant. It would seem, therefore, that the regulation of flowering in both SDP and LDP may involve the interplay of both flower-promoting and flower-inhibiting processes.

9.8 TIME MEASUREMENT IN PHOTOPERIODISM

We have seen that flowering in SDP occurs when they are kept under daylength conditions in which the length of the night exceeds the critical dark period and that some plants can detect differences in the length of the dark period of as little as 15 minutes. Thus, *Xanthium* has a critical dark period of 8¼ hours at 25°C. A difference of only 15 minutes in the length of the dark period can determine whether or not sugar cane will flower. It is clear, therefore, that these species have rather accurate time-measuring mechanisms. Several suggestions have been made as to the nature of this mechanism.

Thus, the "clock" might be of the "hour-glass" type, in which the time taken for a particular substance to accumulate or be depleted to a certain threshold value may be the time-measuring process. Since flowering in SDP requires that the phytochrome in the leaves shall be in the P_r form, and that there is a gradual conversion of P_{fr} to P_r during the first hours of darkness (p. 226), it has been suggested that the critical dark period may represent the time taken for P_{fr} to decline to a certain level. However, it is found that the reaction $P_{fr} \rightarrow P_r$ is effectively complete after 2—3 hours of darkness, whereas the critical dark period of *Xanthium* is 8¼ hours.

Again, if the length of the critical dark period is determined by the time taken for the conversion of P_{fr} to P_r, then irradiation with far-red light at the beginning of the dark period should greatly reduce the length of the critical dark period for flowering in SDP, by hastening the conversion of P_{fr} to P_r, but this is found not to be the case for all but one of the species so tested. On these and other grounds it seems unlikely that the rate of conversion of P_{fr} to P_r is the factor determining the length of the critical dark period of SDP. We are therefore forced to consider other hypotheses regarding the time-measuring mechanism, and among these is the "Endogenous Rhythm" hypothesis of Bünning.

9.9 ENDOGENOUS RHYTHMS IN PHOTOPERIODISM

It is many years since Bünning first drew attention to the existence of persistent rhythms in plants. He investigated the diurnal movement of leaves in the runner bean (*Phaseolus multiflorus*) in which the primary leaves rise during the early part of the day and later fall towards evening and reach a minimum position during the night; they then start rising again towards the morning. These movements are regulated by turgor changes in the "pulvini" of the leaves.

Now, if a bean seedling is exposed to a period of daylight and then kept in continuous darkness for several days, it will continue to show the typical diurnal movements, rising and falling on a 24-hour cycle even though it is itself not being exposed to the natural alternation of light and dark. That is to say, the bean plant shows a persistent endogenous rhythm in its leaf movements. If the position of the leaf is plotted against time then we obtain a sinusoidal curve. As the period of darkness is extended, the amplitude of the movements gradually declines to zero. The plant then needs a further exposure to light to set the rhythm in train again.

Since these early observations on leaf movements, it has been shown that many other processes in plants show a regular diurnal rhythmicity, including the following:

(1) Opening and closing of flowers.
(2) Root pressure.
(3) Growth rate.
(4) Respiration and other metabolic processes.
(5) Activity of certain enzymes.
(6) Mitosis and size of nucleus.
(7) Discharge of fungal spores, e.g. *Pilobolus.*

The length of a single cycle of an endogenous rhythm when "free-running" in darkness is approximately, but not exactly, 24 hours. It is important that under natural conditions the endogenous rhythm of the plant should be synchronized with the diurnal cycle of day and night, and this is achieved by the *entrainment* of the endogenous rhythm by the natural day and night. Although the rhythms are endogenous in the sense that, once started, they can be maintained for several days even in continuous darkness, nevertheless, many such rhythms require the stimulus of the change from dark to light or vice versa, to start them off. Thus, synchronization between endogenous rhythm and natural day and night is achieved by daily setting of the rhythm by a "light-on" signal at dawn, or a "light-off" signal at dusk.

Thus, there can be no doubt as to the existence of endogenous rhythms in plants. Now Bünning postulated that there is also an endogenous rhythm in photoperiodic sensitivity, and he put forward a theory of photoperiodism based upon this endogenous rhythm. He suggested that during the day SDP are in a "photophile" phase, when light is favourable for flowering, and that during the night they are in a "skotophile" phase, when darkness is favourable and light is inhibitory to flowering. Thus we have to envisage a regular rhythm in photoperiodic sensitivity, as the plants enter first the photophile and then the skotophile phase. It was postulated that SDP will flower when the daily light and dark periods correspond with the photophile and skotophile phases of the plants, i.e. when they are under SD. Under LD light will extend into the skotophile phase and will inhibit flowering. The responses of LDP have proved more difficult to account for on this theory, and these responses may be the reverse of those of SDP.

Bünning's theory has been subjected to certain criticisms and put to various experimental tests, some of which have not given the results predicted by the theory. Nevertheless, there now seems no doubt that there is an endogenous rhythm in photoperiodic sensitivity in at least some species. This rhythm has been demonstrated in various ways.

Thus, Hamner grew soybean plants on a wide range of different cycle lengths; each plant received 8 hours of light followed by a dark period the length of which varied from 8 to 62 hours. It was found that the plants flowered maximally when the total cycle length was 24 hours or a whole multiple thereof, i.e. 48 hours, 72 hours, but remained vegetative when the cycle length amounted to about 36 or 60 hours (Fig. 9.16). These results strongly suggest that the light and dark processes in soybean are geared to a 24-hour cycle, so that when the cycle length is 24, 48 or 72 hours in length, the endogenous "photophile" and "skotophile" phases will correspond with the environmental light and dark periods, but that when the latter are on 36- or 60-hour cycles they will be out of phase with the endogenous rhythm of the

plants, which will therefore not flower. It is also found that the vegetative growth of tomato plants is best when the cycle length is 24 hours or a whole multiple thereof, but is poor on cycle lengths of 36 or 60 hours. However, other species, including *Kalanchoë* and *Xanthium,* show no marked responses to variations in cycle length.

A second type of approach to the problem of endogenous rhythms in photoperiodism has involved the introduction of short light interruptions ("night-breaks") at different stages of long dark periods, to determine whether there is any evidence of rhythmic variations in photoperiodic sensitivity, as judged by the flowering response. Experiments have been carried out with plants maintained on 48-hour or 72-hour cycles in which, following a short photoperiod (of say 10 hours), night-breaks are given at various stages during the following 38- or 62-hour dark periods. In soybean and in *Chenopodium rubrum,* such night-breaks are found to be inhibitory to flowering during the early and late part of a 38-hour dark period, corresponding to the skotophile phases postulated by

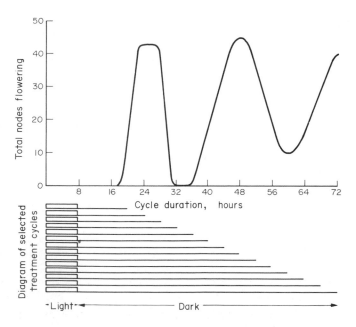

Fig. 9.16. Flowering responses of Biloxi soybeans to cycles by various lengths. Plants were exposed to seven cycles, each cycle consisting of 8 hours of high-intensity light (1000—1500 fc) and associated dark periods of various lengths. Total nodes flowering per 10 plants is plotted against cycle length. (After K. C. Hamner, Chapter 13 in *Environmental Control of Plant Growth,* Academic Press, New York, 1963.)

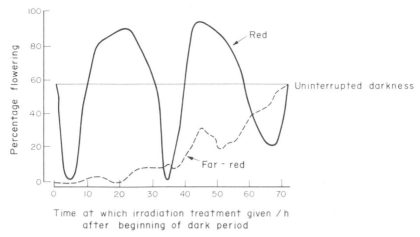

Fig. 9.17. Flowering of *Chenopodium rubrum* with a single 72-hour dark period interrupted at different times by 4 minutes of red light or by 10 seconds of intense far-red light. (From D. Vince-Prue (1975) adapted from B. G. Cumming, S. B. Hendricks and H. A. Borthwick, *Can. J. Bot.* **43**, 825, 1965.)

Bünning. With 62-hour dark periods *three* inhibitory phases are found, corresponding again to the predicted three skotophile phases in a 72-hour cycle (Fig. 9.17).

Thus, two different types of approach have given results, with certain species, which appear to indicate that there is, indeed, an endogenous rhythm in photoperiodic sensitivity, similar to that postulated originally by Bünning. However, other species have given different results. Thus, although "night-breaks" given during long dark periods show rhythmicity in *Kalanchoë,* flowering in this species is little affected by the cycle length. Again, night-break experiments with *Pharbitis nil* show very little evidence of rhythmicity, and the experimental data can be interpreted in terms of an "hour-glass" model.

Bünning originally postulated that the photoperiodic cycle is set off by the onset of dawn, which starts the oscillation, and that the skotophile phase follows on 12 hours after this "light-on" signal. However, later experimental data seem to indicate that in some species the oscillation is established by the transfer from light to dark, i.e. by a "light-off" signal. There is some evidence that phytochrome may be involved in the "light-off" response. For example, in *Lemna per-*

pusilla, an endogenous rhythm in carbon dioxide output is established by a light-off signal at the end of a period of red light, and R/FR reversibility has been demonstrated for the light phase.

From the evidence presented there seems little doubt that some plants do show an endogenous periodicity in photoperiodic sensitivity. It still remains an open question, however, whether such rhythms occur in all species showing photoperiodic responses. Thus, endogenous rhythms may modify the photoperiodic responses in some species but may not be a universal factor in the photoperiodism of all plant species. Moreover, the demonstration of rhythmicity by the type of experiment described above does not imply the correctness of Bünning's hypothesis regarding photophile and skotophile phases.

It is clear that occurrence of endogenous rhythms implies the existence of some sort of "oscillator" mechanism in the plant, but the nature of this oscillator is still completely unknown. However, such an oscillator could serve as a time-measuring mechanism or physiological clock, but whether this is the actual time-measuring mechanism involved in photoperiodism must remain an open question for the present.

9.10 THE SEQUENCE OF PROCESSES LEADING TO HORMONE SYNTHESIS

We have seen that it has been possible to recognize a number of partial processes leading to the production of the flowering stimulus in SDP, and we are now in a position to summarize briefly the present state of knowledge regarding the sequence and interrelations of these processes.

Firstly, there is a requirement for a high-intensity light reaction (photosynthesis) which is met during the photoperiod, and which evidently provides the energy and substrates necessary for the processes occurring during the ensuing dark period. However, there is reason to believe that phytochrome effects are also involved during the photoperiod, since the flowering of SDP, such as *Kalanchoë blossfeldiana,* when maintained in continuous darkness for long periods, is *promoted* by a short daily period of red light, suggesting that a minimal level of P_{fr} during the photoperiod is required for flowering. Normally, at the end of the photoperiod, the phytochrome will be present in approximately equal concentrations of P_r and P_{fr}. During the first few hours of the dark period, P_{fr} decays, so that P_r now predominates. Flower hormone synthesis does not commence until a certain minimum period of darkness (critical dark period) has elapsed and it then proceeds rather rapidly in the next few hours. It is apparently necessary for phytochrome to be present in the P_r form in order for flower hormone synthesis to proceed, since we know that a short interruption with red light (which converts P_r to P_{fr}) inhibits flowering. However, the commencement of hormone synthesis is apparently not directly controlled by the decay of P_{fr}, since the duration of the critical dark period is apparently considerably longer than the time required for P_{fr} decay. Hence it has been suggested that there must be some "time-measuring" mechanism, which more directly controls the commencement of hormone synthesis. We have seen that it is possible that the time-measuring mechanism involves an endogenous rhythm in certain unknown processes, and that the role of phytochrome in flowering may lie in its effects on the time-measuring processes, rather than directly on the process of flower hormone synthesis.

These ideas are summarized as follows:

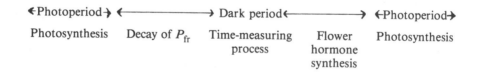

←Photoperiod→	←————————→	Dark period	←————————→	←Photoperiod→
Photosynthesis	Decay of P_{fr}	Time-measuring process	Flower hormone synthesis	Photosynthesis

FURTHER READING

General

Bünning, E. *The Physiological Clock,* Academic Press, New York, 1967.
Evans, L. T. (Ed.). *Induction of Flowering,* MacMillan, New York and London, 1969.
Vince-Prue, D. *Photoperiodism in Plants,* McGraw-Hill, London, 1975.
La Physiologie de la Floraison, Editions du Centre Nationale de la Recherche Scientifique, Paris, 1979.

More Advanced Reading

Bernier, G. *Cellular and Molecular Aspects of Floral Induction,* Longmans, London, 1970.
Evans, L. T. Flower induction and the florigen concept, *Ann. Rev. Plant Physiol.* **22,** 365, 1971.
Hillman, W. S. Biological rhythms and physiological timing, *Ann. Rev. Plant Physiol.* **27,** 159, 1976.
Hoskikazi, T. and K. C. Hamner. Interaction between light and circadian rhythms in plant photoperiodism, *Photochem. Photobiol.* **10,** 87, 1969.
Lang, A. Physiology of flower initiation, *Encycl. Plant Physiol.* **15**(1), 1380, 1965.
Naylor, A. W. The photoperiodic control of plant behaviour, *Encycl. Plant Physiol.* **16,** 331, 1961.
Zeevaart, J. A. D. Physiology of flower formation, *Ann. Rev. Plant Physiol.* **27,** 321, 1976.

10

The Physiology of Flowering — II. Temperature and Other Factors

10.1 VERNALIZATION

As we have seen, photoperiodic responses to seasonal variation in daylength conditions will account for the periodicity in flowering behaviour of many plant species, both temperate and tropical. It will be noticed, however, that among the examples of temperate species which show photoperiodic responses there were relatively few spring-flowering species, although it is a matter of everyday observation that there is a considerable number of "flowers that bloom in the spring", and many of these spring-flowering forms, such as celandine (*Ficaria verna*), primroses (*Primula vulgaris*), violets (*Viola* spp.), etc., show marked seasonal behaviour, and remain vegetative for the remainder of the year after the spring flush of flowers. It might have been expected that spring-flowering is a response to the short days of winter, but this does not appear to be the case in many species.

Daylength is not, of course, the only environmental factor showing an annual variation, and clearly temperature also shows well-marked seasonal changes, especially in temperate regions, although there is considerable variations from day to day and from year to year. We now know that seasonal variations in temperature, as well as in daylength, have a profound effect upon the flowering behaviour of many species of plant.

The clue to the importance of temperature as a regulator of flowering came first from the studies of Gassner in 1918 on the flowering of cultivated varieties of cereals. Cereals such as wheat (*Triticum*) and rye (*Secale*) can be grouped into two classes depending upon whether they are to be sown in the autumn ("winter" varieties) or in the spring ("spring" varieties). Winter wheat sown in the autumn, or spring wheat sown in the spring, both flower and mature in the following summer. If the sowing of the winter wheat is delayed until the following spring, however, it fails to ear and remains vegetative throughout the growing season. It seems unlikely that the necessity for sowing winter wheat in the autumn is simply to secure a longer growing season as such, since autumn-sown plants make relatively little growth during the winter, and winter wheat plants from spring sowings certainly seem to make adequate leaf development, yet they will not flower.

Gassner therefore investigated the effects of different temperature régimes during the germination and early growth of winter and spring rye. He sowed winter and spring rye in sand at different dates between 10th January and 3rd July and kept them at the following temperatures during germination: 1—2°C, 5—8°C, 12°C and 24°C. They were later planted out-of-doors. He found that the temperature during germination had no influence on the subsequent flowering of spring rye, and all seedlings planted out on the same date flowered at approximately the same time, irrespective of the temperature treatment during germination. In winter rye, however, only the plants which had been germinated at 1—2°C flowered regardless of the planting date out-of-doors. Seedlings that were germinated at temperatures above 1—2°C were found to flower only if they were planted not later than March or early April, so that they would have been exposed to some actual chilling out-of-doors (in Central European climatic conditions). Gassner concluded that whereas the temperature conditions during early growth do not affect the

flowering of spring rye, the flowering of winter rye depends on its passing through a cold period, either during germination or later (Fig. 10.1).

This work of Gassner was later followed up in the U.S.S.R. and a great deal of work was carried out there, particularly in relation to possible economic applications. The severity of the Russian winter in many regions does not permit autumn sowing of winter wheats, which, however, usually have a higher yield than spring varieties. A technique was devised by Lysenko, whereby the cold treatment required by winter wheat was applied to the seed before sowing in the spring. The method used was to allow partial soaking of the seed, so that there was sufficient imbibition of water to allow slight germination and growth of the embryo, but not sufficient for complete germination. Seed of winter wheat in this condition, which was exposed to cold treatment by burying in the snow, attained flowering and maturity in the same season if sown in the spring. The technique came to be known as *vernalization,* and this term

Fig. 10.1. Effect of vernalization of flowering in "Petkus" winter rye (*Secale cereale*). *Left:* maintained for several weeks at 1°C after germination; *right:* seed unvernalized. (From O. N. Purvis, *Ann. Bot.* **48**, 919, 1934.)

has subsequently been extended to other treatments involving exposure to winter chillings, not only at the seed stage, but also at later stages of development of the plant.

10.1.1 Types of plant showing chilling requirements for flowering

We have seen that chilling of winter wheat is effective in stimulating subsequent flowering whether given during the early stages of germination or later, when considerable development of leaves has occurred. Later work has shown that many species have a chilling requirement for flowering, including winter annuals, biennials and perennial herbaceous plants. Winter annuals are species which normally germinate in the fall and flower in the early spring. They include such species as *Aira praecox, Erophila verna, Myosotis discolor* and *Veronica agrestis*.

It now appears that winter annuals and biennials are effectively monocarpic* plants which have a vernalization requirement—they remain vegetative during the first season of growth and flower the following spring or early summer, in response to a period of chilling received during the winter. The need for biennials to receive a period of chilling before flowering can occur has been demonstrated experimentally for a number of species, including beet (*Beta vulgaris*), celery (*Apium graveolens*), cabbages (and other cultivated forms of *Brassica*), Canterbury bells (*Campanula medium*), honesty (*Lunaria biennis*), foxglove (*Digitalis purpurea*) and others. If plants of foxglove which normally behave as biennials, flowering in the second year after germination, are maintained in a warm greenhouse they may remain vegetative for several years. In regions with a mild winter climate cabbages may grow for several years in the open without "bolting" (i.e. flowering) in the spring, as they do in areas with cold winters. Such species have an obligate requirement for vernalization, but there is a number of other species in which flowering is hastened by chilling

but will occur even in unvernalized plants; such species showing a facultative cold requirement include lettuce (*Lactuca sativa*), spinach (*Spinacia oleracea*) and late-flowering varieties of pea (*Pisum sativum*).

As well as biennial species, many perennial species show a chilling requirement and will not flower unless they are exposed each winter to cold conditions. Among common perennial plants which are known to have such a chilling requirement are primrose (*Primula vulgaris*), violets (*Viola* spp.), wallflowers (*Cheiranthus cheirii* and *C. allionii*), Brompton stocks (*Mathiola incarna*), certain varieties of garden chrysanthemum (*Chrysanthemum morifolium*), Michaelmas daisies (*Aster* spp.), Sweet William (*Dianthus*), rye-grass (*Lolium perenne*). Perennial species require revernalizing every winter.

It seems very probable that many other spring-flowering perennials will prove to have a chilling requirement when they are investigated. Bulbous spring-flowering plants such as daffodil (*Narcissus*), hyacinth, bluebell (*Endymion nonscriptus*), crocus, etc., do *not* have chilling requirements for flower initiation, the flower primordia being laid down within the bulb during the previous summer, but their growth is markedly affected by temperature conditions. For example, in tulip flower initiation is favoured by relatively high temperatures (20°C), but the optimum temperature for stem elongation and leaf growth is initially 8-9°C, rising to 13°C, 17°C and 23°C at successively later stages. Similar temperature responses are shown by hyacinth and daffodil (*Narcissus*).

In many species flower initiation does not occur during the chilling period, but only after the plants are exposed to higher temperatures following chilling. However, some plants, such as Brussels sprouts, have to remain at low temperatures until flower primordia have actually been formed. It appears that all the species with a chilling requirement for flowering can be vernalized at the "plant" stage, i.e. as leafy plants, but not all species can be vernalized at the "seed" stage, as can winter cereals. Among the other species which can be vernalized at the seed stage are mustard

*See p. 281 for an explanation of this term.

(*Sinapis alba*) and beet (*Beta*). On the other hand, cultivated varieties of *Brassica* (cabbages; Brussels sprouts) and celery cannot be vernalized in the seed stage but the seedlings must attain a certain minimum size before they become sensitive to chilling, and thus show a "juvenile" phase. Generally, those species which can be vernalized at the seed stage are facultative cold-requiring plants, whereas those which can only be vernalized at the plant stage show an obligate chilling requirement.

Many plants with a chilling requirement resemble long-day plants, in that they have a rosette habit in the vegetative phase, and show marked internode elongation associated with flowering.

The requirement for chilling for flower initiation must not be confused with a chilling requirement for the removal of bud dormancy (see p. 264). Thus, many woody plants flower in the spring, but the flower initials are laid down within the bud during the preceding summer and they have a chilling requirement, not for flower initiation, but for the removal of bud dormancy.

10.1.2 Species showing both chilling and photoperiodic responses

Interactions between vernalization and photoperiodic responses have been studied in a number of species. Henbane (*Hyoscyamus niger*) exists in annual and biennial forms, corresponding to the spring and winter forms of rye. The annual form does not require vernalization, but is a long-day plant, which flowers in the summer. The biennial form requires vernalization, followed by long days, for flowering to take place.

As an example of a perennial plant which shows both vernalization and photoperiodic responses, we may mention perennial rye-grass (*Lolium perenne*). In this species, flowers are initiated in response to winter chilling, but long days are required for emergence of the inflorescence, so that elongation of the flowering stem does not com-

mence until the daylength exceeds 12 hours in March. The new tillers (lateral shoots) which emerge during the spring and summer are unvernalized and remain vegetative throughout the growing season, until the following winter. Consequently, flowering of perennial rye-grass is seasonal and restricted to the spring and early summer.

Vernalization requirements are less common among short-day plants, but the garden chrysanthemum constitutes an example which has been studied intensively. Certain varieties of chrysanthemum have a vernalization requirement which must be met before they will respond to short days. After the parent plant has flowered in the autumn a number of horizontally growing rhizomes emerge from the base of the plant and grow just beneath the surface of the soil. Under outdoor conditions these new shoots will become vernalized by natural winter chilling, and in the spring they grow into normal, upright leafy shoots, which grow vegetatively under the long days of summer and initiate flowers in response to short days in the autumn. If, however, the plants are not exposed to chilling during the winter, but are grown under warm conditions in a greenhouse, the new shoots do not become vernalized, and although they grow actively throughout the summer, they are incapable of forming flowers in the short days of autumn. Thus, the chrysanthemum provides another example in which the new shoots arising each year need to be vernalized, since the vernalized condition is not transmitted from the parent shoots, as it is in rye (p. 243).

Genetical aspects of vernalization responses have been studied in several species. In rye, the difference between spring and winter forms is found to be controlled by a single major gene, the requirement for vernalization of the winter forms being recessive to the non-requirement of spring rye. The opposite situation exists in henbane, where the biennial habit (vernalization required) is dominant to the annual habit (no chilling requirement). In other species, however, the inheritance of chilling responses is more complex and in rye-grass (*Lolium perenne*) several genes appear to be involved.

10.1.3 Physiological aspects of vernalization

Intensive studies on the physiological changes underlying vernalization have been carried out on relatively few species and our knowledge of the subject is based largely on the work of Gregory and Purvis on winter rye, of Melchers and Lang on henbane (*Hyoscyamus niger*) and of Wellensiek on several other species.

The work of Gregory and Purvis has established a number of important characteristics of the processes occurring during vernalization of rye. Firstly, they showed that the changes occur in the embryo itself and not in the endosperm, as had been suggested. This was done by removing the embryos from the grain and cultivating them on a sterile medium containing sugar; such embryos were given a period of cold treatment and showed the typical hastened flowering responses given by vernalized grain, when planted. It was even possible to vernalize isolated shoot apices, which had been removed from the embryos and cultivated under sterile conditions. Such apices developed roots and regenerated into seedlings which ultimately flowered in response to the earlier chilling treatment. Moreover, it was shown that vernalization may even be effected whilst the young embryos are still developing in the ear of the mother plant. This was done by enclosing the developing ears in vacuum flasks packed with ice, or by removing developing ears and keeping them in a refrigerator until mature. In this way it was shown that vernalization is effective in embryos, even if commenced only 5 days after fertilization.

It has been shown that in older plants it is the shoot apical region which must be chilled. This was shown for celery, beet and chrysanthemum by placing a cooling coil around the shoot apical region. Thus, whereas in photoperiodism it is the *leaves* which are the sensitive organs to daylength, in vernalization the shoot apex itself is sensitive to the chilling temperatures. Wellensiek has shown that in *Lunaria* young leaves are also capable of being vernalized, but older leaves, which have ceased growth, do not respond, unless they show some cell division at the base of the petiole.

Wellensiek maintains that only tissues which have dividing cells are capable of being vernalized.

For the majority of species, the most effective temperatures are just above freezing, viz. 1-2°C, but temperatures ranging from −1°C to 9°C are almost equally effective. Hence freezing of the cells is not necessary to bring about the changes occurring during vernalization, suggesting that physiological, rather than purely physical processes are involved. This conclusion is confirmed by the observation that cold treatment of rye grain is ineffective under anaerobic conditions, indicating that probably aerobic respiration is essential. By cultivating excised embryos on media with and without sugar it was shown that a supply of carbohydrate is necessary during cold treatment. Thus, although the metabolism of most plants is considerably retarded at cool temperatures, there seems little doubt that vernalization involves active physiological processes, but the nature of these processes is still quite unknown.

It is found that the degree of hastening of flowering in rye varies with the duration of exposure to cold, the longer the cold treatment the shorter the period from sowing to flowering, up to a certain limit, beyond which further cold treatment has no further hastening effect (Fig. 10.2A). Quite short exposures to cold, such as 7-11 days, have a noticeable vernalizing effect, and this effect increases progressively with the duration of treatment.

The vernalization process can be reversed by exposing the grains to relatively high temperatures (25—40°C) for periods of up to 4 days. Seeds treated in this way show a reduced flowering response, and are said to be "devernalized". As the period of chilling increases, it becomes increasingly difficult to reverse the effect, and when vernalization is complete, high temperature is ineffective. Devernalized grains can again be vernalized by a further period of chilling.

Once the rye plant has been vernalized it appears that the condition is transmitted to all new tissues formed subsequently, so that all new laterals are also vernalized. Indeed, if the main shoot apex is removed, so that the laterals are

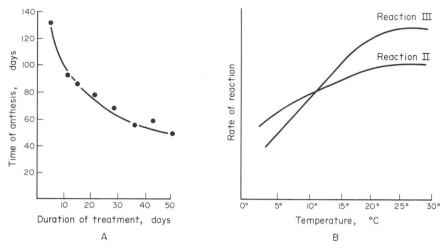

Fig. 10.2. A. Effect of the duration of chilling of seed, on subsequent flowering behaviour of Petkus winter rye. Curve indicates time to anthesis from planting, for various periods of vernalization. (From O. N. Purvis and F. G. Gregory, *Ann. Bot.* **1**, N.S. 569, 1937.)

B. Hypothetical scheme, illustrating two reactions with different temperature coefficients (see text).

stimulated and then these are decapitated to stimulate secondary laterals and so on, it is found that even fourth-order laterals are still fully vernalized although the apices of these laterals were not present at the time of the chilling treatment. Thus, the vernalized condition is transmitted from a parent cell to its daughter cells in cell division and it does not appear to be "diluted" in the process.

10.1.4 The nature of the changes occurring during vernalization

One of the striking features of vernalization is that, at first sight, it appears to involve processes which go on more rapidly at lower than at higher temperatures. This effect is most unusual for chemical processes and yet we must assume that the changes occurring during vernalization are essentially enzyme-controlled reactions showing the usual characteristics of such reactions. How then are we to explain the apparent "negative temperature coefficient" of vernalization? One very simple hypothesis which has been put forward postulates the occurrence of two separate processes competing for a common substrate, each having positive (though different) temperature coefficients:

$$\text{Precursor (A)} \xrightarrow{\text{I}} \underset{\displaystyle\downarrow\text{III}}{\underset{\displaystyle\text{D}}{\overset{\displaystyle\text{product (B)}}{\text{Intermediate}}}} \xrightarrow{\text{II}} \text{End product (C)}$$

In this scheme reaction II and III compete for the common "Intermediate product" B. Suppose reaction III has a higher temperature coefficient than reactions I and II (Fig. 10.2B). This means that high temperatures will favour reaction III and more of B will be diverted into this reaction, so that little C will be formed. When, however, the temperature is markedly reduced, this will reduce the rate of reaction III more than that of II (since by definition reaction III shows greater response to changes in temperature). Consequently, reaction II will be favoured at the reduced temperature and C will accumulate. C will, therefore, be formed at lower temperatures but not at higher temperatures. Thus, the overall production of C will appear to have a "negative temperature coefficient", although each of the three involved reactions has a positive temperature coefficient.

There is no direct evidence to support this hypothesis, but it is of value as indicating how an overall process may proceed more rapidly at lower temperatures, without contravening the normal laws of chemical reactions.

We have already seen that vernalization appears to involve relatively stable changes, so that once the fully vernalized state has been attained by meristematic tissue, it appears to be transmitted by cell-lineage without "dilution". This conclusion in turn may imply that the vernalized state is transmitted through some self-replicating cytoplasmic organelle, but it is equally possible that certain genes become activated during vernalization and that once this has occurred, this change is transmitted to the daughter nuclei during division.

10.1.5 The flowering stimulus in vernalization

Although so much study has been devoted to vernalization, our understanding of the nature of the physiological and biochemical processes involved is still very fragmentary. It is not yet clear whether a specific transmissible flower hormone is formed as a result of vernalization, although there is evidence that this may be the case, at least in some species. We have seen that in photoperiodism some of the strongest evidence for the existence of a flower hormone is provided by grafting experiments, and somewhat similar results have been obtained with certain species showing a vernalization requirement.

A notable example of flower induction by grafting in a vernalized plant is provided by henbane. If a leaf from a vernalized plant of the biennial variety of henbane is grafted on to an unvernalized stock of the same variety, the latter is induced to flower without chilling. A similar response can be obtained by grafting on to the unvernalized biennial variety a leaf from any of the following: (1) the annual variety of henbane (a LDP, with no vernalization requirement); (2) *Petunia hybrida*, an annual LDP; (3) tobacco, day-neutral variety;(4) tobacco, variety "Maryland Mammoth" (SDP), under either SD or LD. Thus,

transmission of a flowering stimulus can take place between cold-requiring and non-cold-requiring plants, even of different genera. Similar results have been obtained in the biennial species beet (*Beta vulgaris*), cabbage (*Brassica oleracea*), carrot (*Daucus carota*) and *Lunaria biennis*. On the basis of these results, Melchers and Lang suggested that a transmissible flowering stimulus, which they called *vernalin*, is formed as a result of chilling in biennial plants.

On the other hand, in some species, including *Chrysanthemum*, it has not proved possible to obtain transmission of a flowering stimulus from vernalized to non-vernalized plants by grafting. Moreover, if the tip region of the plant was given localized cold-treatment and hence flowered, the other buds not directly chilled remained vegetative. Similarly, when the tip of an unvernalized radish plant was grafted on to a vernalized one, flowering did not occur. These latter results are consistent with the conclusion that the *vernalized condition* (as opposed to a flowering stimulus) is only transmitted through cell division, i.e. by "cell lineage". It would seem, therefore, that we must make a distinction between the vernalized ("thermo-induced") state and the formation of a flowering hormone, just as we saw that in photoperiodism we have to distinguish between the induced state of a leaf, and the transmissible stimulus formed in it.

The question then remains as to whether there is a specific flowering stimulus, vernalin, formed in plants with a chilling requirement. In most of the successful grafting experiments, referred to above, the donors were from species which have a requirement for both chilling and LDs or for LDs alone, and the experiments were usually carried out under LD conditions. Moreover, it is generally found necessary that the donor shoots should have leaves. Thus, it seems possible that "vernalin" is identical with the flower hormone produced in the leaves of LDP. Melchers and Lang have argued against this last conclusion, on the ground that leaves of Maryland Mammoth tobacco will induce flowering in stocks of biennial henbane, both under SD and under LD, although "Maryland Mammoth" itself will not flower under the latter conditions (Fig. 10.3). On the other hand, biennial *Hyoscyanus* causes flower formation in Maryland Mammoth tobacco only if the donor has been vernalized. These results suggest that there is a distinct transmissible agent which in cold-requiring plants arises only after chilling, whereas in non-cold-requiring ones its formation is not

Fig. 10.3. Flower induction in non-vernalized biennial *Hyoscyamus niger* by Maryland Mammoth tobacco donors in short days (*left*) and long days (*right*). Note the flowering response caused by the non-photoinduced donor. (From Lang, 1965—see Further Reading.)

dependent on photoperiodic induction, and hence it cannot be identical with florigen. It has therefore been suggested that vernalin and florigen are interdependent, and that vernalin must be present if florigen is to be formed. Thus, the sequence* of events for cold-requiring plants may be summarized as follows:

Low temperature →vernalized state →vernalin→ florigen.

More generally, the sequence* can be written as follows:

──────▶ Vernalin ──────▶ Florigen ──────▶ Flower formation

Cold requiring plants:
  only after cold
Others: without cold

Photoperiodic plants:
  only in inductive
  photoperiods
Day neutral plants:
  independent of
  daylength

*From Lang, 1965 (see Further Reading).

10.2 THE NATURE OF THE FLOWERING STIMULUS

10.2.1 Attempts to isolate the flowering stimulus

As we have seen, the evidence from the various types of grafting experiment described in Chapter 9 strongly suggests that a flowering stimulus arises in the leaves of both SDP and LDP under favourable daylength conditions and is transported to the meristems where it causes a vegetative apex to change to the flowering condition. Moreover, it would seem that the same type of flowering stimulus is produced in both SDP and LDP, as the following experiment suggests. *Kalanchoë blossfeldiana* is a SDP, whereas the species of *Sedum ellacombianum* and *S. spectabile* of the same family (Crassulaceae) are LDP. If vegetative shoots of *Sedum* are taken from plants growing under SD and grafted on to *Kalanchoë* under SD, not only do the stocks of the latter flower, but also the *Sedum,* which will not itself flower under SD (Fig. 10.4). Thus, a stimulus is produced in the SDP *Kalanchoë* which will cause flowering in the SDP *Sedum.* Conversely, if vegetative shoots of *Kalanchoë* are taken from plants growing under LD and grafted on to *Sedum* plants under these conditions, again both stock and scion will flower. Thus, the LDP *Sedum* produces a flowering stimulus under LD which is capable of causing the SDP *Kalanchoë* to flower. These experiments suggest that the flowering stimulus is identical in both SDP and LDP.

A range of experimental results of the type described are consistent with the hypothesis that a specific flower hormone arises in the leaves under inductive daylength conditions, and is transmissible by grafting. As we have seen, grafting experiments with species having a chilling requirement have also given evidence of a transmissible flowering stimulus.

Although the circumstantial evidence for the existence of flower hormones is very strong, repeated attempts to extract a specific flower hormone from SDP over the past 40 years have nearly always given negative results. However, recently

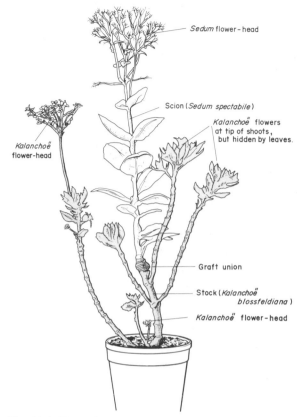

Fig. 10.4. Transmission of flowering stimulus from short-day plant, *Kalanchoë blossfeldiana*, to long-day plant, *Sedum spectabile*, by grafting. A scion from a vegetative plant of *Sedum* growing under short days was grafted on to a flowering plant of *Kalanchoë*, also under short days. The scion of *Sedum*, as well as the *Kalanchoë* stock, has flowered under these conditions. (Original by I. D. J. Phillips from plant supplied by J. Hillman.)

Chailachjan and his co-workers in Moscow have reported that they have obtained extracts of induced leaves of a short-day variety of Maryland tobacco which will induce flowering when applied to vegetative seedlings of *Chenopodium album* (SDP) maintained under long days. Thus, some progress may now be made towards the isolation and identification of the flowering stimulus in SDP.

It remains to be seen, however, whether there is a specific "flower hormone", since it has been argued that flowering may be controlled not by a specific substance, but by an interaction between two or more growth substances, each of which may have a wide spectrum of effects. We have already seen that dramatic morphogenetic changes can be induced in tissue cultures by varying the relative levels of substances such as auxins and cytokinins in the culture medium (p. 157). Thus, it is argued, a transition from the vegetative to the flowering condition may be controlled by changes in the levels of, for example, certain growth substances, such as auxins, gibberellins and cytokinins, or by the balance between these substances. We shall now consider this possibility.

10.2.2 Effects of gibberellins and other growth hormones on flowering

Although attempts to regulate flowering through application of auxins have almost all been unsuccessful (apart from the exceptional case of the pineapple, p. 143) it has been found that a number of LDP can be induced to flower under SD by application of GA_3 (Table 10.1). LDP which respond to GA_3 are typically species which form a pronounced rosette under SD and which show marked internode elongation ("bolting") under LD. When GA_3 is applied to such species growing under SD there is a very marked stimulation of internode elongation and this process is accompanied by flower initiation.

Gibberellin is also effective in stimulating flowering the "long-short-day" plants *Bryophyllum crenatum* and *B. daigremontianum*, which normally require to be exposed first to LD and then to SD for flower initiation but which will flower under continuous SD if treated with GA_3. Thus, GA_3 substitutes for the LD requirement in these species.

A number of species which normally require vernalizing before flower-initiation will occur can also be induced to flower by external application of GA_3 (Table 10.1, Fig. 10.5). Thus, in these species GA_3 apparently replaces the chilling requirement. However, GA_3 will apparently not stimulate flowering of unvernalized rye and certain other species, although it will stimulate stem elongation in these species. In general, treatment of seeds with GA_3 is not effective in stimulating flowering, even in species which respond to seed vernalization.

Fig. 10.5. Effect of vernalization and gibberellic acid on flowering of carrot. *Left:* untreated control plant; *right:* plant chilled for 8 weeks; *centre:* plant unchilled but treated with 10g GA$_3$ per day. (From A. Lang, *Proc. Nat. Acad. Sci., U.S.A.* **43**, 709, 1957.)

Table 10.1. *Species showing flowering under non-inductive conditions in response to applied gibberellic acid (GA$_3$)*

A. *Long-day Plants*
    Cichorium endivia (Chicory)
    Hyoscyamus niger (Henbane, annual)
    Lactuca sativa (Lettuce)
    Papaver somniferum (Poppy)
    Petunia hybrida (Petunia)
    Raphanus sativus (Radish)
    Rudbeckia bicolor (Coneflower)
    Silene armeria
    Spinacia oleracea (Spinach)

B. *Plants with Chilling Requirement*
    Apium graveoleus (Celery)
    Avena sativa (Oat)
    Beta vulgaris (Sugar-beet)
    Bellis perennis (Daisy)
    Brassica oleracea (Cabbage)
    Daucus carota (Carrot)
    Digitalis purpurea (Foxglove)
    Hyoscyamus niger (Henbane, biennial)
    Matthiola incana (Stock)
    Myosotis alpestris (Forget-me-not)
    Solidago virgaurea (Golden rod)

These observations raise the question of whether the endogenous gibberellins are not the "flower hormone" in LDP and species showing vernalization responses. Thus, it might be postulated that in LDP growing under SD the level of endogenous gibberellins is too low for flowering, and that the effect of LD is to raise the level of endogenous gibberellins to the threshold necessary for flowering. Indeed it has been shown for certain species, including spinach and henbane, that there is a marked rise in the levels of endogenous gibberellin when the plants are transferred from SD to LD conditions. Moreover, extracts of gibberellins of the LDP *Rudbeckia,* growing under LD conditions, will induce flowering in plants of the same species growing under SD. Similarly, during vernalization of the biennial species, hollyhock (*Althaea rosea*), there is an increase in the levels of a gibberellin-like substance which will stimulate flowering in *Rudbeckia,* although it will not stimulate flowering in unvernalized hollyhock.

Although these latter observations suggest that changes in endogenous gibberellins may be important in flowering, there is other evidence against the hypothesis that flowering in LDP and plants which require vernalization is regulated primarily by gibberellins, including the following:

(1) As we have seen, there is evidence that the flowering stimulus is identical in both LDP and SDP, and yet gibberellins are quite ineffective in inducing flowering in most SDP.

(2) Not all LDP and species with a chilling requirement can be induced to flower in response to applied GA_3. (However, certain of these latter species can be induced to flower when other types of gibberellin are applied. For example, GA_4 and GA_7 are effective in promoting flowering in *Myosotis alpestris,* which does not respond to GA_3. Thus, in some species there appear to be rather precise requirements with respect to the nature of the gibberellin which will stimulate flowering, and these requirements differ between one species and another.)

(3) Nearly all rosette plants respond to GA_3 by elongation of the internodes, including those species which do not flower in response to this treatment. Moreover, in henbane, when flowering is induced by LD treatment, the formation of flower primordia *precedes* the elongation of the internodes, whereas in response to GA_3 treatment plants of this species kept under SD begin to show internode elongation before the flower primordia appear. Again, stem growth in *Silene armeria* can be completely suppressed with AMO-1618 while flower formation proceeds normally, thus showing that stem growth and flower formation are independent processes controlled by different hormones. Furthermore, the genetic analysis of *Silene* clearly shows that there are separate genes for stem growth and for flower formation. These facts suggest that internode elongation and flowering initiation are distinct processes and that the primary effect of GA_3 is on internode elongation, with flower-initiation occurring as a secondary effect of stem elongation in certain species.

(4) Exogenous GA_3 will not stimulate flowering of "normal" genotypes of red clover (*Trifolium pratense*), but in certain non-flowering genotypes of this species both LD and GA_3 are necessary for flowering. Thus, in these non-flowering genotypes, GA_3 does not substitute for LD, suggesting that some other flower-promoting factor is also normally involved in this species.

Thus, although there is good evidence that changes in the levels of endogenous gibberellins may play an important role in the flowering responses of LDP, it would seem that this is not "the whole story" in LDP and certainly the responses of SDP cannot be accounted for solely in these terms.

In addition to the stimulation of flowering by gibberellins, a number of other growth-regulating substances, both natural and unnatural, have been found to promote flowering in some species under certain conditions. Thus, flowering can be induced in pineapples not only by synthetic auxins (p. 143) but also by ethylene. Under certain conditions kinetin and adenine will promote flowering in *Perilla* and zeatin does so in the aquatic plant *Wolffia microscopica*. Similarly, the naturally occurring growth inhibitor, abscisic acid (p. 70) promotes flowering in several species, while the synthetic growth retardants CCC and B.9 promote flowering in a number of species, including apple and pear trees. A number of other substances, including tri-iodobenzoic acid, maleic hydrazide, vitamin E and even sugars, have been reported to promote flowering in a few species.

Nevertheless, in the majority of SDP and in some LDP, flowering cannot be induced by any combination of the known naturally occurring hormones. Hence we cannot at present account for the flowering behaviour of the majority of species in terms of interaction between known growth hormones.

The conclusion that flowering behaviour cannot be accounted for in terms of the known growth hormones, and yet all attempts to extract a specific flower hormone have been unsuccessful, may indicate that we have adopted an over-simplified approach to the problem, in assuming that flowering is controlled by a single, specific hormone. The isolation and identification of the flowering stimulus remains one of the most challenging problems in developmental plant physiology.

10.3 SEX EXPRESSION AND GROWTH HORMONES

There is some evidence that growth hormones may be involved in the determination of sex in some plants. This has come from studies of dioecious species (male and female flowers on *separate* plants) such as hemp (*Cannabis sativa*), and of those monoecious species in which the male and female organs, stamens and ovaries are borne in separate flowers on the same plant (e.g., some varieties of cucumber (*Cucumis sativus*)). Treatment of genetically male plants of hemp with an auxin spray causes female flowers to be produced.

In monoecious cucumber varieties, it is normally the case that male flowers develop during the earlier stages of growth and that female flowers form only later on. However, application of an auxin to the leaves of young cucumber plants results in an acceleration of the transition from production of male flowers to production of female flowers. It has been suggested, therefore, that female flowers or female parts of flowers tend to differentiate under conditions of higher auxin concentration than do male flowers or parts. This conclusion is supported by the finding that genetically determined forms of cucumber which bear only male flowers contain lower levels of endogenous auxin than the normal hermaphrodite forms. Femaleness in cucumber is also enhanced by treatment with ethylene, or ethrel (a commercial preparation of 2-chloroethane-phosphonic acid, which is converted to ethylene in plant tissues, p. 147).

Gibberellin treatment of monoecious cucumber plants, in contrast to auxin treatment, increases the number of male flowers formed. Treatment of gynoecious cucumber (i.e. a dioecious variety which normally produces only female flowers) with gibberellin results in the formation of male as well as female flowers. Moreover, endogenous gibberellin levels are lower in the gynoecious types than in the normal hermaphrodite forms. It is possible, therefore, that sex expression in plants is effected by a balance between endogenous auxins and gibberellins. The effect of auxin on flower sexuality may involve the participation of ethylene (see p. 68).

10.4 CHANGES OCCURRING IN THE SHOOT APEX DURING FLOWER INITIATION

We have seen that the transition from the vegetative to the reproductive condition involves drastic changes in the structure of the shoot apical meristems (p. 38). The earliest steps in this transition have been studied in a number of SDP and LDP. Indeed, SDPs such as *Xanthium* or *Chenopodium,* and LDPs such as *Lolium temulentum,* in which flowering may be induced by exposure to a single inductive cycle, provide very favourable material for studying the transition, since the timing of the latter can be rigorously controlled to within a few hours. Studies on *Xanthium* have revealed that the earliest changes leading to flower initiation can be first detected about 4 days after exposure to a single SD cycle, but when the plants are exposed to two SD cycles changes can already be detected at the end of this treatment.

As we have seen, in *Xanthium* and *Pharbitis* the first detectable changes occur in the region between the central zone and the rib-meristem and involve cell division. However, in *Sinapis* and *Anagallis* activation occurs first in the peripheral zone and then in the central zone. The changes associated with flower evocation include an increase in the percentage of mitoses and nuclei synthesizing DNA and an increase in nucleolus diameter, especially in the central zone; there is

frequently also a mounding up of the meristem and an increase in its size, accompanied by vacuolation and elongation of cell of the pith-rib meristem. A stage is soon reached in which a "mantle" of small, densely staining and actively dividing cells overlies a central core of more vacuolated cells (Fig. 2.19).

It is clear that when a plant meristem changes from the vegetative to the flowering phase, there is a drastic change in gene expression, many "flowering" genes which remained unexpressed in the vegetative phase now being brought into action. We know very little concerning the manner in which gene expression is controlled in plants, but this subject is discussed in Chapter 13. Since the control of development must involve the synthesis of specific enzymes, through a process involving various types of RNA coded for by the DNA, it is evident that nucleic acid metabolism is likely to be very intimately involved in the events

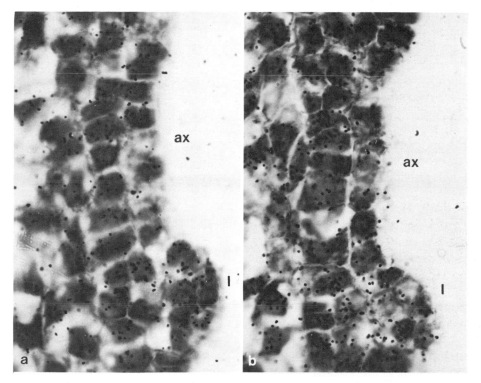

Fig. 10.6. Autoradiographs of vegetative and induced shoot apices of *Lolium temulentum,* labelled with ̇ ;H· orotic acid. A. Axillary bud site (Ax) and leaf primordium (L) in a vegetative (short-day) apex. B. Axillary bud site of plant which has been exposed to long day, in which there has been active incorporation of orotic acid (as shown by the density of silver grains), indicating active nucleic acid metabolism. (From R. B. Knox and L. T. Evans, *Austr. J. Biol. Sci.* **21,** 1083, 1968.)

occurring at the shoot apex during flower initiation (evocation). The flowering stimulus, formed in the leaves of photoperiodic plants, must therefore act at the shoot apex by promoting the expression of specific genes and thus involve nucleic acid metabolism.

Nucleic acid changes occurring at the shoot apex during flower evocation have been studied in two main ways: (1) by following the incorporation of radioactive precursors into RNA, and (2) by the use of inhibitors of RNA synthesis.

In the LDP, *Lolium temulentum,* there is a marked incorporation of radioactive precursors into RNA (thus indicating active RNA synthesis) at the shoot apices on the morning of the day following a single LD cycle, which is precisely the time at which the flowering stimulus is estimated to arrive at the apex. These changes in RNA and protein synthesis are most prominent in the cells on the flanks of the meristem which are destined to give rise to the spikelets (Fig. 10.6). Similarly, in the LDP, *Sinapis alba* (mustard), which will also flower in response to a single LD cycle, there is a marked increase in RNA synthesis in the central and peripheral zones of the apical meristem about 17 hours following the beginning of the LD cycle. Later, active DNA synthesis occurs, and this is followed by mitosis.

Although work with inhibitors has suggested that early RNA and protein synthesis are essential components of evocation in several species, these studies have not provided evidence for the appearance of RNA or protein molecules specifically related to evocation. However, using immunodiffusion techniques it has been shown that in *Sinapis* there is a qualitative change in the protein complement of the meristem during evocation, two new proteins appearing and one disappearing.

Bernier and his colleagues have attempted to obtain information on the nature of the flowering stimulus by detailed studies on the kind of early changes that occur during evocation. They have found that an application of exogenous cytokinin to vegetative plants of *Sinapis* results in an increase in mitosis in the central zone of the meristem in a manner analogous to that observed in response to an inductive photoperiod, but further development into a flowering apex does not occur. Moreover, it was found that the contents of soluble sugar and active cytokinins show early increases in plants of *Sinapis* induced by exposure to long photoperiods. It was concluded that both sugars and cytokinins may be components of the floral stimulus and that they act independently and sequentially at the meristem.

10.5 NUTRITION AND FLOWERING

Farmers and gardeners have long known that fertilizers and manures have a marked effect on the balance between vegetative and reproductive growth. Indeed, for many practical purposes there seems to be an antagonism between vegetative growth and reproduction, and manurial treatment which favour strong vegetative growth may be unfavourable for flowering and fruiting. Experimental studies give some support for this view. Thus, it is found that low levels of nitrogen tend to result in earlier flowering in certain long-day plants. It has been shown that high nitrogen and carbohydrate nutrition delay flower initiation in pea (*Pisum sativum*). However, the number of cases in which a clear effect of mineral nutrition on the onset of flowering has been demonstrated is rather small. Mineral nutrition appears to have an important effect on flower initiation in fruit trees, high levels of nitrogen tending to promote vegetative growth and reduce flowering.

10.6 FLOWERING IN "NEUTRAL" SPECIES

In a large number of species the onset of flowering is a daylength or chilling response, and the discovery of photoperiodism and vernalization represents a very real advance in our understanding of the physiology of flowering. However, it must be remembered that in many other species, probably equally numerous, flowering is not greatly affected by daylength or winter-chilling (Table 10.2). This is the group referred to earlier (p. 218) in which flowering is relatively insensitive to external conditions, and although in these species the length of the vegetative phase may be *modified* by environmental conditions, the effects of the latter are not completely overriding. However, such species, which will be referred to as "neutral" species, are not sharply distinguished from those showing a "quantitative" response to daylength, which have been referred to as "facultative" LDP or SDP (p. 220).

environmental changes and will occur even under constant conditions.

The changes in leaf shape appear to reflect progressive changes in the size and shape of the shoot apex. For example, in the sunflower plant the diameter of the sub-apical region increases progressively as the plant grows and ultimately these changes are terminated by the formation of an in-

Table 10.2. *Some examples of species showing "neutral" flowering responses.*

Cucumis sativus (Cucumber)	*Phaseolus vulgaris* (Dwarf bean)
Fagopyrum tataricum (Buckwheat)	*Poa annua* (annual meadow grass)
Fuchsia hybrida (Fuchsia)	*Rosa* spp. (Rose)
Helianthus annuus (some varieties)	*Senecio vulgaris* (Groundsel)
Lathyrus odoratus (Sweet pea)	*Solanum tuberosum* (Potato)
Lycopersicum esculentum (Tomato)	*Vicia faba* (Broad bean)
Nicotiana tabacum (Tobacco, certain varieties)	

In these neutral species, flowering appears to be determined primarily by some *internal* mechanism, since even when grown under constant environmental conditions a species such as sunflower will grow vegetatively for a certain period and then become reproductive; thus the transition cannot be caused by any change in external conditions and must be regulated by some "internal" mechanism. Moreover, it would appear that the transition from the vegetative to the flowering condition in such species is but one manifestation of a more general phenomenon, since progressive changes during development are very common, as shown by the development of morphological differences in successive organs such as leaves. It is commonly found that there are changes in the size and shape of successive leaves, the seedling or primary leaves being smaller than the later ones, and in species with deeply indented or compound mature leaves the primary leaves are usually much simpler in form (Fig. 10.7) with a progressive series of later formed leaves showing increasing segmentation. Plants showing such changes are said to exhibit *heteroblastic development*. These changes may be affected by environmental factors, such as light intensity and mineral nutrition, but they are not *dependent* on

florescence. It seems likely that the changes in leaf shape result from these changes in the shoot apex, leaf primordia produced on a larger apex apparently being capable of continuing their development for a longer period and hence producing a more mature leaf form.

This heteroblastic development appears to be an indication of the progress of the plant towards

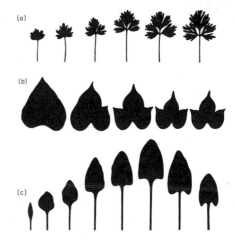

Fig. 10.7. Heteroblastic development, as shown by successive leaves of (a) *Delphinium ajacis*, (b) Morning glory (*Ipomoea hederacea*), (c) sugar beet (*Beta vulgaris*). (From E. Ashby, *New Phytologist*, **47**, 153, 1938.)

maturity and the attainment of the reproductive condition. For example, early flowering varieties of the cotton plant (*Gossypium*) show a steep gradient in leaf shape changes, whereas in later-flowering types the rate of change in leaf shape is less steep.

Evidence of progressive changes towards the flowering condition has come from experiments in which internode segments of tobacco were taken and grown in sterile culture. These segments produced callus and regenerated buds. Internode segments from young plants or from the lower region of the stem of older plants produced vegetative buds, whereas stem segments from the upper part of a flowering plant produced flower buds, with few leaves or bracts. Segments taken from an intermediate region of the stem first produced leaves and bracts and then a flower. Thus, there appeared to be a gradient in the propensity to produce flowers, from the lower to the upper part of the plant.

These latter observations suggest that the intrinsic capacity of the tissues to initiate flowers increases during ontogeny. On the other hand, other observations suggest that the progress towards "ripeness-to-flower" does not involve changes in the intrinsic properties of the cells, but rather reflects a change in the conditions to which the new tissues are exposed as the plant increases in size. Thus, experiments were carried out with sunflower plants in which seedling tips were grafted on to stocks of various ages and tips from older plants were grafted on to seedlings. In the first type of graft it was found that seedlings formed fewer nodes before flowering than they would have done on their parent plants. On the other hand, grafts or tips from older plants on to seedlings formed a greater number of nodes than they would have done on the parent plants. These results would seem to indicate that the conditions to which the short apices are subjected are determined by influences from other parts of the plant and that these change with age and/or size.

The possibility that this "size factor" may involve influences from the roots is suggested by the results of further experiments with a day-neutral variety of tobacco, "Wisconsin 38". This variety normally flowers after producing 30 to 40 nodes, but when adventitious roots were induced on the stems, the number of nodes produced before flower initiation was greatly increased. In further experiments plants were allowed to form 6 to 10 leaves which were 10 cm or greater in length and were then decapitated and the tops placed in a rooting medium. When the rooted plants had again developed 6-10 leaves the process was repeated. Such treatment was found to delay flowering indefinitely. Thus, it would appear that flowering in this plant is affected by the number of nodes which are present between the roots and the shoot apices.

It will be remembered that among plants which are sensitive to daylength or chilling, there are some species which will not respond until they have reached a certain minimum size ("ripeness-to-flower"), and before they have reached this stage seedling plants may be described as in a "juvenile phase". This phenomenon would seem to correspond to the necessity for neutral species to reach a certain size before flowering. Indeed, photoperiodic and vernalizable species will behave like neutral species if they are maintained from germination under constant conditions favourable to flowering, in that they will undergo a certain period of vegetative growth and then commence to initiate flowers. Thus, it would appear that the physiology of flowering is not qualitatively different in the "sensitive" and "neutral" groups of plant, but rather that they differ in *degree* of sensitiveness to external conditions.

If we are correct in assuming that the difference between the flowering responses of "sensitive" and "neutral" species is primarily one of degree, then it might be assumed that a flower hormone is involved in neutral as well as in sensitive species, but there is little evidence as yet for the occurrence of flower hormones in neutral species.

10.7 THE DIVERSITY OF FLOWER-CONTROLLING FACTORS

As we have now seen, flowering is affected and controlled, in various species, by a range of factors, some of which are external to the plant, such as daylength, temperature and nutrition, and others which arise within the plant itself, as seen in day-neutral species. There is a bewildering mass of facts on the effects of these various factors and it is difficult to see how these parts of the "jigsaw puzzle" fit together to give an overall, unified picture. It might be asked whether the fact that quite different environmental factors may control flowering implies a different control mechanism for each factor. In attempting to answer this question it is useful to approach the problem from a different viewpoint and to ask not, what makes a plant flower under a given set of conditions, but rather, why they *fail* to flower under a different set of conditions. Thus, flowering may be prevented by a number of different factors in various species, although the flower-promoting processes might be the same in these species.

For example, in apple trees, flower initiation is rather sensitive to mineral nutrition, but not to photoperiod. We do not need to postulate that flower hormone synthesis depends upon low nitrogen levels—it may well be the case that the "flower hormone" is always present in apple shoots, but that under conditions of high nitrogen nutrition there are high levels of, for example, endogenous gibberellins, which are known to inhibit flowering in this plant. On the other hand, in *Xanthium* it would appear that synthesis of the flower hormone is blocked under long days. In plants requiring vernalization, yet another step in the flowering processes may be blocked in unchilled plants. From this viewpoint, under favourable conditions the flowering processes may be blocked in different ways and at different steps in plants of various response types. Similarly, as we have seen, flowering may be blocked by different factors in the same plant at different stages of its life cycle. In the seedling stages flowering may be prevented because the plant is still in a juvenile phase and is not yet capable of responding to favourable environmental conditions. But even when it has attained ripeness to flower, it may be prevented from flowering by unfavourable environmental conditions.

It is not difficult to envisage that in some cases flowering does not occur because the stimulus is not being synthesized in the leaves, while in other cases the limiting step may be the inability of the shoot apices to respond to the stimulus, possibly because of inappropriate levels of gibberellins.

10.8 FLOWERING IN WOODY PLANTS

So far, our consideration of the physiology of flowering has been restricted to herbaceous plants. Flowering in woody plants presents a number of characteristic features which will be briefly described.

The first important feature to note is that there is considerable variation between species in the length of time elapsing between flower initiation and complete development of the flower. In some species, such as sweet chestnut and many other late-summer-flowering trees and shrubs, such as *Buddleia, Fuchsia, Hypericum, Hibiscus syriacus, Caryopteris,* etc., the flowers are formed on the current year's shoots and flower initiation is followed immediately by the further development of the flower, as in herbaceous species, i.e. there is no gap between the early and later stages of flower development. In many other woody plants, however, especially temperate species, flower initiation occurs during the summer within resting buds formed earlier in the same year, but the development of the flower parts becomes arrested at an early stage and the further development and emergence of the flower does not occur until the following spring. This is the situation in a large number of common European woody plants, e.g. oak (*Quercus*), ash (*Fraxinus*), sycamore (*Acer*), elm (*Ulmus*), pine (*Pinus*), apple, plum, peach, black currants, gooseberries, etc. In such species the buds containing flower primordia become dormant in the late summer or autumn and require a period of exposure to chilling temperatures to break the dormancy (p. 264). Once the dormancy has been broken by winter chilling they become capable of emerging as the temperatures rise in the spring. In some winter- or early spring-flowering trees and shrubs the buds containing flowers are able to grow at lower temperatures than the vegetative buds, so that the flowers may actually emerge before the leaves, e.g. hazel (*Corylus*), willow (*Salix*), jasmine (*Jasminium nudiflorum*), almond (*Prunus persica*), elm, ash, oak, etc.

We have very little information regarding the effect of environmental factors on flower initiation in woody plants. This is mainly due to the technical difficulties in experimenting with mature trees. It is not possible to work with seedling trees since these are still in the juvenile phase and not capable of flowering (see below). However, it is possible to study the effects of environmental conditions on flowering of trees by taking scions from mature trees and grafting them on to seedling stocks. In this way it is possible to obtain small trees which are potentially capable of flowering. In a few woody plants, flower initiation appears to be controlled by daylength conditions. Thus, long days are necessary for flower initiation in birch (*Betula*), *Erica* and *Calluna*. On the other hand, short days promote flower initiation in coffee (*Coffea arabica*), black currant (*Ribes nigrum*), *Poinsettia* and *Hibiscus*. Flower initiation in other woody plants, e.g., pine, larch, beech, apples, cherries, plums, appears to be unaffected by photoperiodic conditions, which, nevertheless, have a profound effect on vegetative growth of some of these species (p. 262).

In general, vernalization does not appear to play an important role in flower initiation in woody plants, with a possible exception of the olive tree (*Olea europaea*), which apparently initiates flowers in response to cool temperatures in the winter.

On the other hand, there is no doubt that temperature, rainfall and soil nutrients may have a profound effect on the flowering of many woody plants. It is well known that a hot, sunny summer is frequently followed by abundant flowering of many tree species in the following spring, and the flowering of beech has been shown to be particularly affected in this way. Evidently high temperatures and possibly high light intensity (and hence the formation of abundant carbohydrate reserves) favour flower initiation in many tree species. The effect of mineral nutrition on flower initiation in fruit trees was mentioned earlier (p. 252).

A number of instances have been reported in which flower initiation in woody plants has been influenced by gibberellins and by synthetic growth retardants. Thus, "flowering" has been stimulated by gibberellic acid in a number of conifers, including *Cupressus, Chamaecyparis, Juniperus* and *Thuja*. By contrast, gibberellic acid inhibits flowering in apple, pear, black currant, grape (*Vitis vinifera*), *Syringa, Fuchsia,* and a number of other woody plants. On the other hand, growth retardants such as "CCC" (chlorocholine chloride) and "Phosfon D", which apparently inhibit gibberellin biosynthesis in plant tissues promote flowering in apples, pears and azaleas (*Rhododendron* spp.).

10.9 PHASE CHANGE IN WOODY PLANTS

We have seen that in "day-neutral" annual species the onset of flowering is apparently regulated by some endogenous mechanism, the nature of which is unknown, but that the effect of this regulatory mechanism is that the plant does not flower until it has attained a certain size. This "size effect" is seen also in species showing daylength or chilling responses, so that there is a juvenile phase during which flowering cannot be induced. An apparently analogous phenomenon is seen in woody plants, the seedlings of which normally show a juvenile phase during which they make active growth but remain vegetative. The transition to the flowering condition occurs only after a delay which varies greatly from species to species, ranging from 1 year in certain shrubs to 30-40 years in forest trees such as beech (Table 10.3). Once a given tree commences flowering, it normally continues to do so every year, although, as we have already seen, in species such as beech, flowering is sensitive to weather conditions and may occur irregularly. Thus, on the basis of the flowering behaviour we may distinguish between a *juvenile* and an *adult* (or *mature*) phase in the life history of the tree.

Table 10.3. *Duration of juvenile period in forest trees*

	years
Pinus sylvestris (Scots pine)	5-10
Larix decidua (European larch)	10-15
Pseudotsuga taxifolia (Douglas fir)	15-20
Picea abies (Norway spruce)	20-25
Abies alba (Silver fir)	25-30
Betula pubescens (Birch)	5-10
Fraxinus excelsior (Ash)	15-20
Acer pseudoplatanus (Sycamore, Maple)	15-20
Quercus robur (English oak)	25-30
Fagus sylvatica (Beech)	30-40

Differences between juvenile and adult stages are seen not only in flowering behaviour but also in various vegetative characters. Thus, in certain species, such as ivy (*Hedera helix*), mulberry (*Morus*), *Acacia, Eucalyptus* and juniper, the leaf shape in the juvenile phase is very different from that of the adult stage. In oak and beech there is a marked tendency for the dead leaves to be retained on the shoots of juvenile trees during the winter, whereas in the adult stages they are shed normally. In some species the juvenile and adult stages are distinguished by marked differences in phyllotaxis, as in ivy (see below). Another morphological character which changes during ontogeny is the development of thorns—for example, in lemon trees the juvenile stages are commonly more thorny than the adult.

Among the physiological differences observed between juvenile and adult stages is the rooting ability of cuttings; it is very commonly found that whereas cuttings from young trees root readily, after the parent trees attain a certain age this rooting ability is greatly diminished or entirely lost.

An interesting feature of these phase differences is that the lower parts of the tree retain the juvenile stage after the upper parts have developed adult characters. This can be seen very well in ivy, where the lower regions of an erect-growing vine show the palmate type of leaf and are purely vegetative, whereas the upper parts show ovate leaves and normally produce abundant flowering shoots.

Not only is there apparent stability of the juvenile and adult phases in the same plant body, but cuttings taken from different regions and rooted retain their juvenile or adult characters for a long period. This is well known in ivy, for example, where cuttings taken from the juvenile part of the vine had palmate leaves with opposite phyllotaxis, the shoots are trailing and are coloured with anthocyanin and produce abundant adventitious roots, but do not flower; cuttings of adult shoots, on the other hand, produce plants with ovate leaves, spiral phyllotaxis, and the shoots are erect, green and produce few, if any, adventitious roots, but flower readily (Fig. 10.8). Cuttings from adult ivy shoots may grow for many years and produce shrubs known to gardeners as "tree ivy".

Fig. 10.8. Cuttings taken from adult part of parent vine of ivy (*Hedera helix*). These cuttings will continue to retain for several years the "adult" characters, such as leaf shape, phyllotaxis, and the capacity to flower, although high temperatures favour reversion to the juvenile condition. (Print supplied by Dr. L. W. Robinson.)

A similar retention of juvenile and adult characters is seen in grafting experiments. Thus, scions from flowering regions of mature trees continue to flower when grafted on to quite small seedling stocks. This is a common observation in forest tree breeding and is known for species such as birch, larch and pine. Likewise, scions from mature fruit trees grafted on to suitable root stocks flower readily, whereas scions from young seedlings treated in the same way show delayed flowering.

The phenomenon of phase change, involving stable, non-genetic changes which can be transmitted through many cell-generations, shows certain interesting parallels with the changes occurring during vernalization. This subject will be discussed further in Chapter 13.

FURTHER READING

General

Doorenbos, J. Juvenile and adult phases in woody plants, *Encycl. Plant Physiol.* **15**(1), 1222, 1965.
Evans, L. T. (Ed.). *Induction of Flowering,* MacMillan, New York and London, 1969.
Lang, A. Physiology of flower initiation, *Encycl. Plant Physiol.* **15**(1), 1380, 1965.
Purvis, O. N. The physiological analysis of vernalization, *Encycl. Plant Physiol.* **16**, 76, 1961.
Zimmerman, R. H. (Ed.). Juvenility in woody perennials, *Acta Horticulturae,* **56**, 1976.
La Physiologie de la Floraison, Editions du Centre Nationale de la Recherche Scientifique Paris, 1979.

More Advanced Reading

Zimmerman, R. H. (Ed.). *Symposium on Juvenility in Woody Perennials,* Acta Horticulturae, 56, 1976.

11

Dormancy

11.1 THE BIOLOGICAL SIGNIFICANCE OF DORMANCY

Outside the equatorial regions there are seasonal variations in climatic conditions, which are most notable in the temperate zones. These variations are especially marked with respect to light intensity, daylength, temperature and frequently also to rainfall. As a result there is a regular alternation of seasons favourable and unfavourable for growth and this alternation has had a marked effect on the pattern of the life cycles evolved by the higher plants. The necessity to withstand low temperatures during the winter and, in some regions, hot dry conditions during the summer, poses special problems for the plant, and we now have to consider some of the ways in which these problems have been met.

Plant cells normally contain a large amount of water which is liable to freeze at low temperatures, with grave risk of damage to the protoplasm. Tropical plants are very easily killed by freezing, but it is evident that plants of temperate and arctic regions must have become adapted to survive the period of winter frost—they have developed *cold-resistance*. Although cold-resistance has been studied for many years, our understanding of its biochemical basis is still far from complete, and a discussion of this subject here would take us too far afield.

In many cold-resistant species the general morphological appearance of the plant during winter is not essentially different from that in the summer—the growth rate of the plant is reduced or arrested during the winter, but the growing points of the shoots are still in a potentially active condition, and may make some growth during mild periods, as in many biennial plants. In such species the whole plant, including the apical meristems, is relatively cold-resistant. Other species, of course, show distinct differences between their summer and winter states. Thus, in woody plants the shoot apices cease active growth and become enclosed in bud scales, to form winter-resting buds. They are then said to have become *dormant*. Many woody plants are much more cold-resistant in the dormant than in the actively growing condition. Thus, seedlings of forest trees, such as larch (*Larix*) and *Robinia,* which continue growing late into the autumn, are very liable to be damaged by early frosts, but if they have ceased growth and their growing points have entered the dormant condition, they then remain cold-resistant throughout the winter.

The reason why dormant buds should be more cold-resistant than actively growing tissues is not fully understood. However, it is fairly clear that the cold-resistance of dormant tissue is due to certain protoplasmic characters, and that it is not primarily due to the presence of the bud scales, the protective function of which is probably concerned with the reduction of water loss—one of the secondary effects of winter cold is to increase the difficulties of the plant in maintaining an adequate water balance. Under frosty conditions, especially when accompanied by wind, the plant continues to lose water, but is unable to take up replacement supplies if the soil is frozen. There is thus a considerable danger of damage from drought under winter conditions, but water loss is reduced by the enclosure of the growing parts within a covering of bud scales, and also, in deciduous trees, by the falling of leaves in autumn which reduces the total surface area over which evaporation can occur.

The danger of winter drought and low temperature seems to have influenced not only the evolution of woody plants, but has apparently had a profound effect on the form of many other types of plant. Many plants over-winter entirely below ground, as bulbs, corms and rhizomes; although such organs will be partly insulated against frost, they will also be protected against drying when the soil does become frozen to some depth. Some dormant organs, such as bulbs, are probably adaptations to hot, dry summer conditions, such as are found in the Mediterranean region.

Whereas perennial plants have developed special organs which resist the unfavourable conditions of winter, annual plants have pursued yet another course—they frequently over-winter in the form of seeds. The seeds of many annual plants, particularly of the common weeds of arable land, germinate almost immediately they are shed, if conditions of temperature and moisture are favourable. But the seed of many other plants does not germinate immediately (or only a proportion of the seeds do so) and remains in the soil until conditions become favourable for germination in the following spring. Now seeds are generally very much more cold-resistant than the growing plant of the same species. Dry seeds may resist freezing down to as low as -234°C. Some seeds, such as those of Leguminosae (clover (*Trifolium*), broom (*Cytisus*), *Laburnum,* etc.) do not, in fact, take up water immediately they are shed, due to the fact that they have a coat which is impermeable to water, and such seeds will be capable of withstanding very severe frost.

The majority of seeds imbibe water as soon as they fall on to moist soil, but, as already stated, they do not necessarily germinate immediately. Such imbibed seeds are less cold-resistant than in the dry state, but nevertheless many of them still retain a considerable degree of cold-resistance, and apparently certain annual species which are frost-tender in the actively growing state are able to survive the winter in the form of seed.

11.2 FORMS OF DORMANCY

Dormancy may be defined as a state in which growth is temporarily suspended. In some species the cessation of growth is directly due to unfavourable temperature and light conditions; thus many pasture grasses remain in continuous growth throughout a mild winter and cease growth only when temperatures fall to about 0-5°C. Similarly, certain annual weeds, such as groundsel (*Senecio vulgaris*), chickweed (*Cerastium* spp.) and Shepherd's purse (*Capsella bursa-pastoris*), stop growing only during the coldest part of the winter. In such cases the dormancy of plants is evidently caused by the unfavourable external conditions, and in this case we speak of *imposed* or *enforced dormancy*.

However, in many cases the unfavourable conditions are not directly the cause of dormancy. Thus, many trees form winter-resting buds during the summer and autumn, when temperatures and light conditions are still favourable, and long in advance of the onset of winter. In such woody plants the cause of dormancy appears to lie within the tissues of the buds themselves, and we then speak of *innate* or *spontaneous dormancy*. This form of dormancy also occurs in many seeds. Thus, if freshly harvested barley grains are planted under warm, moist conditions, a high percentage of them will fail to germinate. If, however, the barley is stored dry for a few months, the seed will then be found to germinate readily when planted under the same conditions as previously. Thus, the failure of freshly harvested barley grains to germinate is not due to external conditions being unfavourable for growth, but must be due to some cause within the seed itself.

Innate or spontaneous dormancy is found not only in buds and seeds, but in other types of resting organs such as rhizomes, corms and tubers.

11.3 BUD DORMANCY IN WOODY PLANTS

The majority of temperate woody plants, including both coniferous and dicotyledonous species, show a well-marked dormancy or resting phase during the annual growth cycle and this is usually accompanied by the development of resting buds. The typical resting bud involves the "telescoping" of the bud scales and leaf primordia in the apical region, due to the arrest of normal internode extension. In some genera (e.g., *Betula, Fagus, Quercus*) having stipules, this telescoping of the shoot apical region leads to the formation of a resting bud, since the overlapping stipules in this region form the bud scales. In other species the protective scales represent leaves,

which may be only slightly modified, as in *Viburnum* spp., or more highly modified, so that they frequently represent only the leaf base, as in *Acer, Fraxinus, Malus* and *Ribes* (Fig. 11.1). During the development of such buds, certain leaf primordia show greater marginal growth than occurs during normal leaf development, whereas lamina development is suppressed, and these primordia give rise to the bud scales. The younger leaf

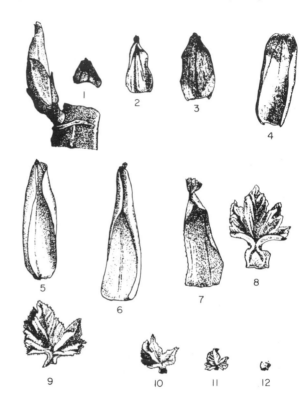

Fig. 11.1. Bud of *Ribes* and bud scales and leaves from a dissected bud. (From J. H. Priestley and L. I. Scott, *An Introduction to Botany,* 3rd ed., Longmans, Green & Co., London, 1955.)

primordia, formed within the bud scales, are arrested at an early stage in their development, and give rise to normal leaves when the buds resume growth in the following spring. In some species, such as pines, growth of a bud may continue for several months from June to September. In some trees a terminal bud is not formed, since growth of the shoot is terminated by the death and abscission of the apical region, and growth is later continued from the uppermost axillary bud. Such species (which include *Tilia, Ulmus, Castanea, Robinia* and *Ailanthus*) are said to show a *sympodial* (as opposed to a monopodial) growth habit.

When terminal buds are first formed they can frequently be induced to resume growth by various treatments including defoliation, either by hand or by insect attack. It would appear, therefore, that at this stage the terminal buds are not themselves innately dormant, but their growth is apparently inhibited by the mature leaves on the shoot. Similarly, lateral buds may be inhibited by the leaves, or by the main apical region in actively growing shoots, and are thus held in check by correlative inhibition (p. 133) rather than by innate dormancy. This phase of bud development is referred to as *summer dormancy* or *predormancy*. Later, in many species, the buds are found to have entered a state referred to as *true dormancy, winter dormancy* or *rest*. When they have entered this condition the buds will no longer resume growth if the shoots are defoliated, so that they are now innately dormant and not simply held in check by environmental conditions or inhibitory influences within the plant itself, as is the case during predormancy.

After a certain period of true dormancy the buds become capable, in the later part of the winter or early spring, of resuming growth when external conditions, particularly temperature, are favourable for growth. Thus, at this stage the buds are no longer innately dormant, but nevertheless for some time they may fail to grow because of low outdoor temperatures. This phase is referred to as *post-dormancy*. We shall now consider some of the environmental factors controlling the development and breaking of dormancy in temperate woody plants, of which daylength and temperature are the most important.

11.3.1 The development of bud dormancy

One of the most important factors affecting and controlling the induction of dormancy in woody plants is daylength. In the majority of species so far studied, long days promote vegetative growth

and short days bring about the cessation of extension growth and the formation of resting buds in seedlings of woody plants (Fig. 11.2). However, a

Fig. 11.2. Photoperiodic control of bud dormancy in seedlings of birch (*Betula pubescens*). Plants transferred to short days (left) have ceased growth and formed resting buds, whereas seedlings maintained under long days will continue to grow actively for many months.

number of common cultivated fruit trees (*Pyrus, Malus, Prunus*) and certain other species, including the family Oleaceae, appear to be relatively insensitive to daylength changes.

The seedlings of some species, e.g. black locust (*Robina pseudacacia*), birch (*Betula pubescens*) and larch (*Larix decidua*), can be maintained in continuous growth for at least 18 months under LD conditions in a warm greenhouse, whereas under SD they cease growth within 10-14 days. On the other hand, other species, such as sycamore

(*Acer pseudoplatanus*), horse chestnut (*Aesculus hippocastanum*) and sweet gum (*Liquidambar styraciflua*), show delayed dormancy under LD, but they cannot be maintained in growth indefinitely under these conditions. For those species which can be maintained in continuous growth under LD, there appears to be a certain critical daylength below which dormancy is induced and above which dormancy does not occur.

As in the flowering responses of herbaceous plants, the photoperiodic responses of woody seedlings appear to depend upon the length of the dark period, rather than of the photoperiod, and if a long dark period is interrupted by a short "light-break", the effect of the dark period is nullified and dormancy is delayed. However, the effect is less pronounced than in flowering responses. The most effective region of the spectrum for this light-break effect lies in the red, suggesting that phytochrome is involved.

The response of woody seedlings depends on the daylength conditions to which the leaves are exposed. In sycamore, as in herbaceous species, it is the young, fully expanded leaves which are the most sensitive to daylength, but in birch seedlings even quite young leaves in the apical regions show sensitivity to photoperiod.

How important are these photoperiodic responses in determining the formation of resting buds and the onset of dormancy in nature? It has been shown that the seasonal decline in daylength is important in determining the onset of dormancy in seedlings of species which normally continue active growth into the autumn, e.g. *Larix decidua*, *Populus* spp. and *Robina pseudacacia*. But it is very common to find that older trees show a very much shorter period of extension growth than do seedlings of the same species, and they frequently cease growth in June or July, when the natural photoperiods are still long. In such cases, it seems doubtful whether declining daylength is important in determining the formation of resting buds, and it seems more likely that some change, in either nutrient levels or in hormonal balance, arising within the tree itself, determines the period of growth and the onset of dormancy. However, as we have seen, resting buds are at first in a state of predormancy and only later enter a state of true dormancy; it is possible that the declining daylength in the autumn plays a role in the transition of the buds from predormancy to true dormancy.

It has been found that leaf fall in some woody plants is promoted by short days, and delayed leaf fall is sometimes observed in trees growing near street lights. However, in nature, low temperatures and possibly low light intensity are probably at least as important as daylength in determining the onset of leaf senescence and abscission.

In addition to the observable morphological changes associated with the induction of dormancy by short days there are also biochemical changes which are reflected in increased cold-resistance. Seedlings of black currant (*Ribes nigrum*) and *Robinia pseudacacia* which have been exposed to SD are markedly more cold-resistant than seedlings which have been grown throughout under LD. Cold acclimatization in *Cornus stolonifer* depends upon exposure to both SD and decreasing temperature.

It is now well established that wide-ranging woody species, such as *Pinus sylvestris* and *Picea abies,* show marked ecotypic differences in photoperiodic responses in relation to both the latitude and the altitude at which they occur naturally. Northern races are frequently found to require longer photoperiods for active extension growth than do more southern races, adapted to shorter natural photoperiods. This fact suggests that woody plants are rather closely adapted to natural daylength conditions, and that the latter probably play an important controlling role in the seasonal cycle of growth and dormancy.

11.3.2 Emergence of buds from dormancy

Usually the terminal resting buds formed during the summer or fall remain dormant until the following spring, when they expand and form new shoots. The dormancy of the buds diminishes during the course of the winter, as can be demonstrated very simply by collecting twigs of trees such as lime (*Tilia*), sycamore (*Acer pseudoplatanus*), poplar (*Populus*) and willow (*Salix*) at different times during the winter and placing them in water in a warm room or greenhouse. It is found that twigs collected in October, November and early December usually remain dormant when they are brought into warm conditions. A fairly high proportion of buds collected in January will be found to expand in 2-3 weeks, and with later dates of collection, e.g. February or March, the buds burst increasingly rapidly after they are brought into the warm.

Many woody plants require to be exposed to a period of winter chilling to overcome the dormancy of their buds, as can be shown by growing small trees of species such as poplar or sycamore in pots and, when they have become dormant in the autumn, keeping some of them out-of-doors throughout the winter and some of them in a warm room or greenhouse. In the spring, the buds of young trees which have been kept outdoors will expand in the normal way, but those which have been kept in the warm will still be dormant and may remain so until well into the summer; indeed, a certain proportion may ultimately die without ever resuming growth. Temperatures in the range 0-5°C are the most effective in overcoming bud dormancy, and chilling periods of 260 to 1000 hours are required. In regions with cold winters, the chilling requirements are normally fully met by the spring, but in warm climates, such as those of California and South Africa, where the winters are very mild, difficulties may be met in cultivating certain fruit trees, such as peaches (*Prunus persica*), since the chilling requirements of the buds may not be met and delayed and irregular bud-break may occur in the spring.

It should be noted that although chilling is necessary to remove the dormancy of the buds of many trees, *warm* temperatures are necessary for the growth of the buds after chilling. Frequently the chilling requirements are met by January, but in many regions the buds may fail to resume growth then because the temperatures are still too low, and they remain in the phase of postdormancy. Thus, the time of bud burst in the spring is normally determined by the return of warmer conditions.

In the majority of North Temperate woody species so far studied, once bud dormancy has been fully induced by SD, they cannot be induced to resume growth by transfer to LD, and the dormancy can normally be overcome by chilling. However, in a few species the unchilled buds can be induced to resume growth under LD or continuous illumination. Thus, if leafless seedlings of beech (*Fagus sylvatica*), birch (*Betula* spp.) or larch (*Larix decidua*) are placed under continuous light in a warm greenhouse in the autumn, the buds will soon expand.

At first sight it would seem that the response of dormant buds to photoperiod contradicts the rule that it is the leaves which are the organs of photoperiodic "perception", and that the apical meristematic region is insensitive to daylength. However, it should be remembered that resting buds contain well-developed leaf primordia and that therefore the differences between species, such as beech, and other species relates primarily to a difference in the age at which the leaves become sensitive to photoperiod.

It is not clear whether the photoperiodic control of bud-break is important in nature, but there is evidence that bud-break in *Fagus sylvatica* may be dependent upon lengthening daylengths in the spring, although in many regions temperature is also likely to be a limiting factor for this species. In *Rhododendron,* also, bud-break appears to be determined by daylength.

11.3.3 Dormancy in various organs

Various other types of organ, such as rhizomes, corms, bulbs, tubers and the winter-resting buds of aquatic plants, show dormancy. In the aquatic plants *Stratiotes, Hydrocharis* and *Utricularia* the dormancy of the winter-resting buds is induced by short days, in association with high temperature. Short days also promote the formation of resting buds in the insectivorous plant, *Pinguicula grandiflora,* and the dormancy of the buds is overcome by chilling.

By contrast, dormancy in bulbs of onion (*Allium cepa*) is promoted by LD so that the bulbs develop and "ripen" in the summer. The period of dormancy of onion bulbs is shortest when the bulbs are stored at cool temperatures.

The rhizomes of lily-of-the-valley (*Convallaria majalis*) normally become dormant during the summer and they require a 1-week period of chilling at 0·5-2°C, or 3 weeks at 5°C to remove this dormancy. Similarly, when *Gladiolus* corms are exposed to warm soil temperatures they do not grow, but periods as short as 24 hours at 0-5°C will break their dormancy.

Although the tubers of some varieties of cultivated potato have a chilling requirement for dormancy breaking, this is apparently not the case with all varieties.

11.3.4 Artificial means of breaking bud dormancy

A very wide range of treatments, especially using various chemicals, has been found to break the dormancy of resting organs. One of the simplest methods of breaking the dormancy of woody plants consists of immersing the shoots in warm water (at 30-35°C) for 9-12 hours; this treatment is used by florists for "forcing" the flower buds of lilac and *Forsythia,* to get blooms much earlier than normally. Exposure to ether vapour is also effective in removing the dormancy of lilac buds and of lily-of-the-valley rhizomes, the flowers of which can thus be obtained very early in the winter for sale by florists. Among the other substances which have been found very effective in breaking dormancy are thiourea and ethylene chlorhydrin, which will remove the bud dormancy of a wide range of woody plants and are also effective with potato tubers and rhubarb root stocks.

As will be shown below, various growth regulators, including gibberellins, cytokinins and ethylene, will also break the dormancy of buds and seeds of many species.

11.4 DORMANCY IN SEEDS

Morphologically, the seed consists of an embryo surrounded by one or more covering structures, of which the most important is the testa, which is usually derived from the integuments of the ovule. Some seeds contain a well-developed endosperm, which lies within the testa and may surround the embryo, or may lie to one side of it. Functionally, the seed is a "propagule" or dispersal unit, i.e. an organ of propagation. In many species the seeds are liberated from the fruit and the isolated seeds become the dispersal units. In other species, however, the fruits may contain a single seed which is retained within the fruit coat (pericarp), the fruit itself being shed as a whole and becoming the dispersal unit. Examples are provided by achenes, nuts, caryopses and so on, the precise definition of which does not concern us here. Although these latter structures are distinguishable morphologically from seeds, they perform the same biological function as seeds, i.e. they are functional propagules and hence it is convenient to refer to all such structures as seeds, although it is not strictly accurate to do so.

Although, as we shall see, dormancy in seeds shows many parallels with that in buds and other organs, the presence of enclosing coats introduces complications not found in buds, and we find several types of seed dormancy, which do not appear to correspond to any form of bud dormancy.

11.4.1 Hard seed coats

The seeds of certain families, such as the Leguminosae, Chenopodiaceae, Malvaceae and Geraniaceae, possess testas which are impermeable to water, so that such seeds are liable to lie dormant in the soil for considerable periods before germination occurs. Water uptake by these seeds can be brought about by various treatments, such as abrasion by sand, treatment for short periods with concentrated sulphuric acid, etc., which remove the impermeable outer layer of the testa and permit penetration of water to the embryo. Seedsmen treat clover seed by rotating it in a drum lined with carborundum. Probably under natural conditions the activities of micro-organisms in the soil slowly break down the outer layers of the testa and so render water uptake possible.

11.4.2 Immaturity of the embryo

In certain seeds the embryo is still immature when the seed is shed and germination cannot occur in such seeds until the embryo has undergone development. This is true of the seeds of wood anemone (*Anemone nemorosa*), lesser celandine (*Ficaria verna*), marsh marigold (*Caltha palustris*), ash (*Fraxinus excelsior*), and other species. In order for this further development to take place, the seeds must be imbibed with water and maintained under favourable temperature conditions. The time required for the embryo to complete its development may vary from about 10 days in *C. palustris* to several months in *F. excelsior*.

11.4.3 After-ripening in dry storage

The seeds of many species fail to germinate if sown immediately after harvesting, even though the embryo is fully mature. If they are stored dry at ordinary room temperatures, however, they gradually lose their dormancy and become capable of germinating when provided with suitable conditions (Fig. 11.3). This effect is called "after-ripening in dry-storage", and is found in several types of cereal, e.g. barley, wheat, oats and rice. The duration of the dormant period may range from a few weeks to several months. Other species showing this type of dormancy include many grasses, black mustard (*Brassica nigra*), evening primrose (*Oenothera* spp.) and cultivated varieties of lettuce.

It is not known what causes this type of dormancy nor what are the changes which occur during the storage period which ultimately release the seed from dormancy. It would seem that the processes involved during this period of after-ripening are not of a metabolic nature, since they occur even in the dry seed, when metabolism is at a very low level.

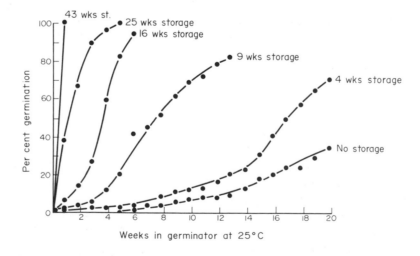

Fig. 11.3. Effect of after-ripening in dry storage at room temperature on germination rate of seeds of *Impatiens balsamina*. (From W. Crocker, *Growth of Plants,* Reinhold, New York, 1949.)

This type of dormancy is of considerable economic importance in cereals. In regions where the weather during the harvest period is liable to be wet, dormancy of the grain is an advantage, since varieties which show such dormancy are less liable to germinate in the ear under wet conditions. From this point of view, dormancy is a desirable economic "character", which is deliberately selected for by plant breeders. On the other hand, for the production of malt from barley, dormancy of the grain is often a major problem, since it may be impossible to germinate the grain for several weeks after harvesting.

11.4.4 Light-sensitive seeds

One of the most interesting forms of dormancy, and one which has received intensive study in recent years, is that shown by light-sensitive seeds. In a considerable number of species, exposure to light is necessary for germination, e.g. seeds of tobacco (*Nicotiana* spp.), foxglove (*Digitalis pururea*), hairy willow-herb (*Epilobium hirsutum*), purple loosestrife (*Lythrum salicaria*), dock (*Rumex crispus*), and many others. On the other hand, in certain other species germination is *inhibited* by light, although the number of such species is considerably smaller than for light-promoted seeds; among the known light-inhibited seeds are those of love-in-a-mist (*Nigella*), *Nemophila*, *Phacelia* and annual phlox (*Phlox Drummondii*).

Light-sensitive seeds will only respond to light after they have imbibed water. The duration of illumination required by light-promoted seeds is often very short; for example, a high percentage of germination is obtained with lettuce seeds exposed to only 1-2 minutes of light, while with seed of purple loosestrife a light-flash of only 0·1 seconds duration has a marked effect in stimulating germination. The responses of light-sensitive seeds are strongly affected by temperature and many seeds which are light-

requiring at, say, 25°C become capable of germinating in the dark at lower temperatures, e.g., certain light-requiring varieties of lettuce. If certain light-requiring seeds are exposed to daily alternations of temperature, e.g., between 15°C and 25°C, they can be induced to germinate without exposure to light. Other treatments which can replace the light requirement of seeds include treatment with certain inorganic ions, especially nitrate, and certain organic substances, e.g., thiourea.

Many seeds which are light-requiring when freshly harvested gradually lose their light-requirement during storage and ultimately give full germination in complete darkness, e.g., light-sensitive varieties of lettuce. It would seem, therefore, that the changes occurring during after-ripening in dry storage in some way remove the light requirement.

As we have already seen, studies on the responses of light-sensitive varieties of lettuce seed played a key role in the discovery of phytochrome (Chapter 8, p. 204). It was shown that the red region of the spectrum promotes and far-red inhibits the germination of lettuce seed (Fig. 11.4). If red and far-red are given alternatively, then whether germination occurs or not depends upon the nature of the last radiation to which the seeds were exposed. Thus, when phytochrome is converted to the P_{fr} form by red light, it evidently initiates a chain

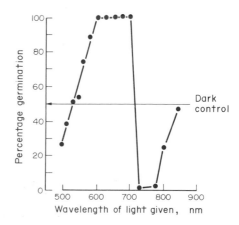

Fig. 11.4. Effect of red and far-red radiation on germination of lettuce seed. (From L. H. Flint and E. D. McAlister, *Smithsonian Inst. Misc. Collections*, **96**, 1-8, 1937.)

of processes which ultimately result in germination. It has been found that similar red/far-red responses are shown by other species of light-sensitive seeds and it seems probable that phytochrome is universally involved in light-promoted seeds.

Light-inhibited seeds have been very much less intensively studied than light-promoted ones, but it now seems probable that the same phytochrome system is involved in both types of seed, and that in the light-inhibited species the effect of far-red is enhanced so that it predominates over the effect of red. Thus, it has been shown that the light-inhibition of *Nemophila* seed is due mainly to the far-red region of the spectrum; red light, on the other hand, seems to have little or no promotive effect on this seed.

11.4.5 Dormancy removed by chilling

Gardeners have long known that the seeds of many species will not germinate if sown under warm conditions, but will lie dormant in the soil for long periods; if, however, they are sown out-of-doors in the autumn and exposed to winter conditions they will germinate in the following spring.

This behaviour led to the horticultural practice of "stratifying" the seed, i.e. placing it between layers of sand and leaving it out-of-doors during the winter. Such "stratified" seed is no longer dormant and germinates readily in the spring (Fig. 11.6). From such observations it is clear that exposure to winter cold is, in some way, necessary to break the dormancy of many seeds.

At one time it was believed that the dormancy of such seeds is due to hard and impermeable coats, and that freezing temperatures are necessary to break the coats. It is now known, however, that freezing temperatures are not required, and that, in fact, temperatures just above freezing (0-5°C) are more effective than lower temperatures (Fig. 11.5). Moreover many seeds which have a chilling requirement do not, in fact, have hard coats, e.g., apple, birch.

The range of seeds showing a chilling requirement is very wide (Table 11.1) and includes both woody and herbaceous plants. In some seeds there is an obligate requirement for chilling, e.g., ash (*Fraxinus excelsior*), whereas in others, e.g., *Pinus* spp., a period of pretreatment at chilling

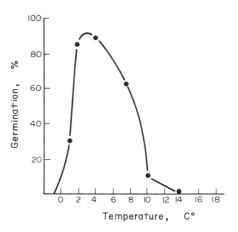

Fig. 11.5. Effect of chilling temperatures on dormancy of apple seed (germination after 85 days chilling at temperatures shown). (From P. G. de Haas and H. Scharder, *Zeitschrift fur Pflanzen-zuchtung,* **31,** 457, 1952.)

temperature, although not essential, nevertheless increases and hastens subsequent germination. It should be noted that for chilling temperatures to be effective the seeds must be imbibed with water, there being no effect with dry seeds. The minimum period of chilling necessary to remove dormancy varies from species to species, but usually amounts to several weeks. In some species the embryo itself is dormant and can only be induced to germinate with difficulty if it is unchilled, e.g., mountain ash (*Sorbus aucuparia*), whereas in other species the embryo will germinate if the testa is removed and only the intact seed has a chilling requirement, e.g., sycamore (*Acer pseudoplatanus*). Seedlings grown from unchilled embryos frequently show "dwarfism", however, making sluggish growth and having very short internodes. This dwarfism of seedlings can itself be removed by chilling or by treatment with gibberellic acid.

Certain seeds, such as acorns (*Quercus*) and those of *Viburnum,* show "epicotyl dormancy"; such seeds germinate and develop a radicle in the autumn without any prior chilling but development of the epicotyl is dependent upon chilling, i.e. the epicotyl, but not the radicle, shows dormancy.

A few species have "2-year seeds", so called because they do not normally germinate until the second spring after shedding. Certain types of 2-year seeds have hard coats, as well as a chilling requirement, e.g., hawthorn (*Crataegus*) and *Cotoneaster;* because of the hard coats, the embryos are prevented from imbibing water as soon as shed, and hence chilling during the first winter is ineffective in removing dormancy. The hard coats are rendered permeable to water during the following summer, however, as a result of the activities of soil micro-organisms. When such imbibed seeds enter the second winter, the dormancy is broken and they become capable of germinating in the following spring.

The cause of the "2-year" behaviour is different in other species. Thus, seeds of lily-of-the-valley (*Convallaria*) and Solomon's seal (*Polygonatum*) require a chilling period to bring about growth of the radicle, but development of the epicotyl does not follow until the seeds have been subjected to a second winter's chilling.

Fig. 11.6. Effect of winter chilling on dormancy of seeds of *Rhodotypos*. A. Seeds exposed to outdoor temperatures throughout the winter. B. Seeds maintained in a warm greenhouse throughout experiment. (Photograph supplied by late Dr. Lela V. Barton.)

Table 11.1. *Woody species with seeds having a chilling requirement to overcome dormancy*

Acer spp. (Maples)	*Malus* spp. (Apple and Crab apple)
Betula spp. (Birches)	*Picea* spp. (Spruce)
Cornus florida (Dogwood)	*Pinus* spp. (Pines)
Corylus avellana (Hazel)	*Prunus* spp. (including Peach)
Crataegus spp. (Hawthorns)	*Rosa* spp. (Roses)
Fagus sylvatica (Beech)	*Sequoiadendron giganteum* (Wellingtonia)
Fraxinus spp. (Ash)	*Tilia* spp. (Lime)
Hamamelis virginiana (Witch hazel)	*Thuja occidentalis* (Western red-cedar)
Juglans nigra (Walnut)	*Vitis* spp. (Grape)
Liriodendron tulipifera (Yellow poplar, Tulip tree)	

11.4.6 The role of the coats in seed dormancy

The seed coats have been found to play an important role in the dormancy of the seed of many species. It has already been mentioned that although the embryos are dormant in some seeds which have a chilling requirement, nevertheless in other species it is only the *intact* seeds which show dormancy, and the isolated embryos will germinate without chilling if the testa is removed. Similarly, certain light-requiring seeds, such as birch and lettuce, will germinate in darkness if the seed coverings are removed or even if they are only slit. Again, certain seeds which show a requirement for after-ripening in dry storage will germinate if the seed coverings are removed; for example, removal of the husks of barley, wheat, oats and rice will permit germination soon after harvesting, whereas when the coats are left intact such seeds normally require several weeks of after-ripening. It is clear, therefore, that the seed coats play an important role in at least three different types of dormancy, and, indeed, in all cases where the embryo itself is not dormant, the dormancy of the intact seed depends upon the presence of coats, which will include the testa, together with the endosperm and pericarp in some seeds. The conclusion raises the question as to the mechanism of these seed-coat effects.

It is possible that the seed coats present a physical barrier to gaseous exchange of oxygen and carbon dioxide between the embryo and the external air. It seems unlikely that seed coat effects are due to the accumulation of high internal concentrations of carbon dioxide since germination of lettuce seed is actually *stimulated* by keeping the seeds in high concentrations of this gas. On the other hand, several types of seed show higher oxygen requirements than do actively growing plants of the same species, suggesting that seed coats may present a physical barrier to oxygen uptake. The testas of marrow (*Cucurbita pepo*) seed have been shown to be much less permeable to oxygen than to carbon dioxide. Certain seeds can be induced to germinate either by slitting or removing the coats, or by maintaining the intact seed in a high concentration of oxygen, e.g., in

Betula and non-after-ripened cereals. Studies on the respiration of germinating pea seeds suggest that anaerobic conditions may occur in the initial stages of germination until the testa is ruptured, when there is a marked increase in oxygen uptake (p. 278). Thus, several kinds of evidence support the view that seed coats may limit the uptake of oxygen.

Interference with oxygen uptake, especially in association with high temperatures, appears to be important in what is known as *secondary dormancy*. Thus, non-dormant seeds of *Xanthium* can be rendered dormant by embedding them in clay (which restricts gaseous exchange) and keeping them at 30°C for several weeks. Similar secondary dormancy phenomena have been shown for a number of other species, including members of the Polygonaceae and Rosaceae, e.g., apple and pear. In all these cases, the secondary dormancy can be overcome by chilling treatment, and it appears that the development of secondary dormancy is the reverse of after-ripening.

Secondary dormancy shows many resemblances to primary dormancy and it has been suggested by Vegis and others that restricted oxygen uptake, in association with high temperature, is the cause of normal dormancy in seeds and, indeed, in buds also. Thus, Vegis has pointed out that embryos of developing seeds are liable to experience oxygen deficiency because of the surrounding seed coats and maternal tissues, and postulates that under such conditions of partial anaerobiosis normal oxidative breakdown through the tricarboxylic acid cyle and "terminal oxidation", necessary for growth in most species, does not take place. Instead, the products of glycolysis, such as phosphoglyceric acid, cannot undergo normal oxidative breakdown, but become diverted into alternate pathways, leading to the formation of fatty acids and fats, which tend to accumulate in dormant tissues.

The possible importance of oxygen deficiency within the seed as a factor in dormancy is suggested by work on rice seeds, which are dormant immediately after harvesting, but gradually emerge from dormancy during dry storage. The dormant seeds of rice can be induced to germinate by removing the husks, thus indicating the importance of coat effects in these seeds. Storage in oxygen greatly reduces the dormancy period, suggesting that some oxidation reaction may be involved in the after-ripening processes during dry storage. However, Roberts tested the effects of various respiratory inhibitors (including inhibitors of terminal oxidation, Krebs cycle and glycolysis) and obtained the unexpected result that they *stimulated* germination of dormant rice seeds. He suggested that it is necessary for some oxidation reaction to proceed before germination can take place, and that this reaction is in competition with respiratory processes involving gycolysis, the Krebs cycle and the terminal oxidase system for the low levels of oxygen present in the seeds; hence the inhibition of these respiratory processes by various substances will release greater amounts of oxygen for the other oxidation reaction, which he suggests may involve the "pentose phosphate pathway" of carbohydrate metabolism. There is indeed evidence that the loss of dormancy is accompanied by a shift from the glycolytic to the pentose phosphate pathway in several species of seed. It has been further suggested that the action of respiratory inhibitors in stimulating germination is brought about by their inhibiting the enzyme catalase, which catalyses the breakdown of hydrogen peroxide. The hydrogen peroxide so spared is postulated to enhance the activity of the pentose phosphate pathway.

There is also some evidence that the effect of seed coats may be due to mechanical resistance to the growth of the radicle. Thus, several types of dormant seed will germinate if the seed coat is removed in the radicle region, but if seeds so treated are placed in a high osmotic concentration of 0·2 M mannitol (which will reduce the ability of the seeds to take up water and hence replace the mechanical effect of the seed coat) their germination is inhibited. However, the osmotic effect of the mannitol solution can be overcome by treatments, such as exposure to light or treatment with gibberellic acid, which will stimulate germination of the intact seed. It is concluded, therefore, that the normal effect of the seed coat is a mechanical one, to overcome which the radicle

needs to develop a sufficient turgor. Whether this mechanical effect of seed coats is important in many species is not yet clear. Nevertheless, whatever the effect of the coats, it is clear that they play a very important role in many forms of seed dormancy.

11.4.7 Similarities between seed dormancy and bud dormancy

Inhibition of germination is not dependent solely upon coat effects in seeds which show embryo dormancy, where even the naked embryos are dormant. Hence we must seek some other cause of dormancy in such cases. Now the dormancy of seeds showing a light or chilling requirement shows certain features in common with that of buds and other organs, which may be summarized briefly as follows:

(1) Chilling for several weeks at 0-5°C is effective in breaking the dormancy of buds, rhizomes, corms and many types of seed.
(2) Certain substances will break the dormancy of several kinds of organ; thus, thiourea and gibberellic acid will remove the dormancy of tree buds, potato tubers and several types of seed.
(3) Certain tree buds and certain seeds may be induced to grow by exposure to long days, whereas short days are ineffective.

The close parallel between dormancy in buds and in seeds is particularly clear in instances where the buds and seeds of a single species are compared. For example, in birch (*Betula pubescens*) the dormancy of both the seeds and buds can be removed by chilling, by exposure to long days or by gibberellic acid. This parallel between seed and bud dormancy in a single species strongly suggests that the cause of dormancy is the same in both organs.

Now, bud dormancy is apparently not due to interference with gaseous exchange by the bud scales since (1) many dormant buds are not tightly enclosed by the scales, and (2) removal of the bud scales does not usually cause resumption of apical activity. Moreover, interference with gaseous exchange cannot be important in the *induction* of dormancy in buds, since until the buds are actually formed there can be no question of interference with oxygen uptake by the bud scales. On the other hand, we have seen that in many woody species resting buds are formed under short days, and the response is determined by the daylength conditions to which the *leaves* are exposed. In view of the evidence for the role of hormones in the control of bud dormancy it is pertinent to examine their possible importance in some forms of seed dormancy.

11.5 HORMONAL CONTROL OF DORMANCY

Since hormones appear to play an essential role in most aspects of growth and differentiation it is reasonable to examine their possible role in the control of dormancy of both buds and seeds. Studies on this problem involve two main types of approach, viz. (1) observations on the effects of application of exogenous hormones, and (2) investigations on endogenous hormones, especially to establish whether there is any meaningful correlation between variations in the levels of endogenous hormones and the state of dormancy of buds and seeds.

Experiments with exogenous hormones have shown that the dormancy of many seeds can be overcome by application of gibberellins, cytokinins and ethylene. Species which show a response to gibberellins include a number which normally require after-ripening in dry storage, others which are light-requiring and many which have a chilling requirement. Among the various gibberellins which have been tested, GA_4 and GA_7 have been found to be particularly active in stimulating germination of dormant seeds. In a smaller proportion of species dormancy can be overcome by cytokinins; usually a given species responds either to gibberellins or to cytokinins, but some seeds respond to both types of growth substance, e.g. lettuce, pear (*Pyrus communis*) and sugar maple (*Acer saccharum*). Ethylene has also long been known to stimulate germination in

a number of species and in some species, such as lettuce, ethylene increases the germination percentage which can be obtained with cytokinins or gibberellins alone.

Although greatly increased dark germination can be obtained in many light-requiring seeds by application of exogenous growth substances, these do not always fully replace the light requirement. Thus, the effects of red light and gibberellins have often been found to be synergistic in overcoming dormancy, suggesting that their modes of action are not identical. Similarly, in certain weed species, such as *Spergula arvensis,* the effects of red light, ethylene and carbon dioxide are markedly synergistic in promoting germination. Again application of kinetin reduces, but does not entirely replace, the light requirement of lettuce seeds.

The dormancy of resting buds of many types of woody plants and other resting organs may be overcome by the same three main types of hormone as are active with dormant seeds. In most such cases the exogenous hormone removes a chilling requirement, but in buds of beech and birch exogenous gibberellin will also substitute a requirement for long photoperiods. It has been claimed that GA_3 is only effective in hastening bud-break in *Acer pseudoplatanus* after the chilling requirement has been met, but there seem to be other well-authenticated instances in which GA_3 is effective in replacing a chilling requirement, e.g. in unchilled hazel seeds.

Studies on changes in the levels of endogenous growth-promoting hormones in plant extracts have provided a number of instances in which levels of gibberellins and cytokinins have been found to decline during the development of dormancy and to increase during emergence from dormancy. Thus, it is well established that whereas levels of endogenous gibberellins and cytokinins are very high in young developing embryos, the levels of these hormones decline drastically to almost zero during the later stages of seed development (Fig. 11.7), although these changes are not confined to species which have dormant seeds. Concomitant with the decline in free gibberellins, there is an increase in gibberellin conjugates, especially gibberellin glucosides and

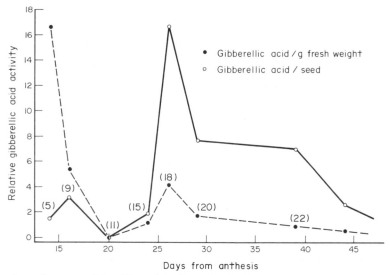

Fig. 11.7. Changes in endogenous gibberellin activity in seeds of *Phaseolus vulgaris,* during their development. Relative gibberellin activity per gramme fresh weight and per seed plotted against days from anthesis. The ordinate scale refers to gibberellic acid per seed; one unit of this scale is equivalent to 5 units of gibberellin activity per gramme fresh weight. (From K. G. M. Skene and D. J. Carr, *Austr. J. Biol. Sci.* **14**, 13-25, 1961.)

glycosyl esters, which may serve as "reserves" in the seed. A similar decline in endogenous gibberellin and cytokinin levels appears to occur in the resting buds of woody plants during the development of dormancy, both under natural conditions and in response to short days applied experimentally.

By contrast, it has been found that levels of extractable gibberellins and cytokinins increase in both seeds and buds during chilling treatments. In several species of seed, the levels of cytokinins and gibberellins rise to a peak in succession and then decline, so that by the end of the chilling period these hormones have returned to a low level (Fig. 11.8). In several light-requiring seeds rapid increases in gibberellins and cytokinins have been observed following exposure to short periods of red light (Fig. 11.9). Thus, both chilling

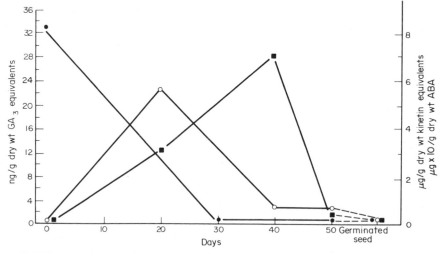

Fig. 11.8. Effects of chilling at 5°C on the levels of endogenous cytokinins, gibberellin-like substances and abscisic acid in seeds of *Acer saccharum*. Squares, acidic gibberellin-like substances. Circles, cytokinin-like substances. Solid circles, endogenous abscisic acid. (From D. P. Webb, J. van Staden and P. F. Wareing, *J. Exp. Bot.* **24**, 105-16, 1973.)

treatments and red light, which overcome the dormancy of various species of seed, result in increases in endogenous gibberellins and/or cytokinins.

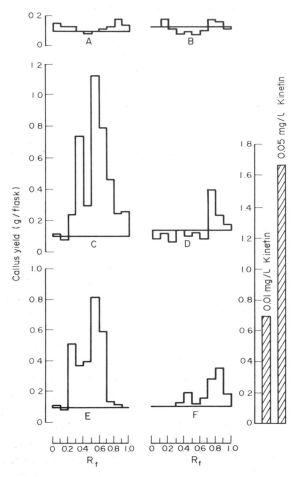

Fig. 11.9. Effects of short exposures to red (R) and far-red (FR) light on endogenous cytokinin levels in seeds of *Rumex obtusifolius*. All seeds were first imbibed in the dark for 2 days and then treated as follows before extraction of cytokinins: A, nil treatment (extracted immediately); B, 2 days in dark; C, 10 minutes R light; D, 10 minutes R light followed by 2 days in dark; E, 10 minutes in R light and 5 minutes FR light; F, 10 minutes R light and 20 minutes FR light. After paper chromatography seed extracts were assayed for cytokinin activity by the soybean callus assay. (From J. van Staden and P. F. Wareing, *Planta*, **104**, 126-33, 1972.)

We have seen that ethylene will stimulate germination in seed of a number of species, and there is increasing evidence that endogenous ethylene may play an important role in seed dormancy.

Thus, the embryonic axes of non-dormant varieties of peanut (*Arachis hypogea*) produce ethylene during germination whereas those of dormant varieties produce only low levels. In dormant *Xanthium* seeds there is a complex interaction between the effects of ethylene, carbon dioxide and oxygen in the promotion of germination. Non-dormant seeds actively produce endogenous ethylene under aerobic conditions, whereas dormant ones show only a low rate of ethylene production, so that dormancy in *Xanthium* seeds may be caused by the repression of the capacity for ethylene biogenesis. Germination of the dormant seeds can be stimulated by thiourea and the cytokinin, benzyladenine, which also increase ethylene production, but it would appear that the primary effect of these substances is to stimulate growth and that the increased production of ethylene is a *consequence* of such growth rather than its cause.

Although such general correlations between hormone levels and the state of dormancy suggest that the effects of dormancy-breaking treatments may be mediated through the variations in endogenous hormone levels, it still remains to be demonstrated unequivocally that this is the case, since (1) it remains to be shown whether the variations in the hormones are the cause or the effect of the changes in the state of dormancy, and (2) more detailed studies have sometimes shown a lack of a close correlation between levels of extractable hormones and dormancy. For example, although cytokinins increase rapidly in seeds of dock (*Rumex obtusifolius*) following exposure to red light (which stimulates their germination), there is evidence that these increases in endogenous cytokinins is *not* the primary cause of the release from dormancy. Moreover, most of the determinations of endogenous hormone levels in the past have been carried out on relatively crude plant extracts, using biological assay techniques, with all the errors inherent in these methods.

Apart from the problem as to the role of gibberellins, cytokinins and ethylene in the control of dormancy, there is good reason to believe that dormancy is not simply brought about by the absence of these growth-promoting hormones. On the contrary, there is circumstantial evidence to suggest that metabolism is actively blocked in dormant tissues, suggesting that possibly dormancy also involves natural growth-inhibiting substances. Substances which inhibit growth in various tests can be extracted from many plant tissues and hence it is possible that dormancy involves the active inhibition of growth by such substances, as was first suggested by Hemberg. He showed that extracts of dormant potato tubers and buds of ash (*Fraxinus excelsior*) contain substances which inhibit growth of *Avena* coleoptiles and that treatments, such as exposure to ethylene chlorhydrin, which overcome the dormancy of potato tubers also cause a marked reduction in the inhibitory activity of tissue extracts. It was also shown for a number of woody species that extracts of resting buds become less inhibitory during the course of the winter and this change is correlated with a gradual emergence of the buds from dormancy. However, again it is clear that such a correlation does not establish a causal relationship between the level of growth inhibitors and dormancy. Bioassays can rarely differentiate between changes in amounts of "inhibiting" compounds and of growth-promoting compounds which interact with them. Moreover, because certain substances extracted from plant tissues inhibit growth of coleoptiles it does not necessarily follow that such substances normally function as growth inhibitors in the tissues from which they were extracted. It is probable, indeed, that some of these "inhibitors" are toxic substances which are normally restricted to the vacuoles of differentiated cells and hence do not normally have access to the growing tissues of the plant.

However, considerable interest was aroused by the discovery of abscisic acid (ABA), which is a powerful growth inhibitor in many tests. It was, indeed, the search for a natural dormancy-inducing substance ("dormin") which led to the isolation of ABA from leaves of *Acer pseudoplatanus* contemporaneously with its isolation from cotton fruits.

The application of exogenous ABA inhibits growth and germination in many tests. However, to inhibit the germination of non-dormant seeds, ABA must be supplied continuously, and if the seeds are rinsed in water they germinate rapidly. On the other hand, ABA induces the formation of resting buds ("turions") in the duck weed, *Spirodela polyrrhiza,* and the experimentally induced resting buds appear to show normal dormancy which can only be overcome by chilling or by application of cytokinins. Similarly, dormant immature embryos of yew (*Taxus baccata*) can be made to germinate by soaking them in a nutrient medium, which results in leaching of the endogenous ABA from the embryos. However, these leached embryos can be rendered dormant again by treating them with exogenous ABA. Again, it is possible to induce the formation of resting buds in seedlings of sycamore (*Acer pseudoplatanus*) by application of exogenous ABA. However, it is necessary to apply relatively high concentrations of ABA for prolonged periods in order to induce dormancy, so that it is not clear whether the formation of resting buds by this treatment is of any significance for the normal process. Moreover, some unsuccessful attempts to induce bud formation by application of ABA have been reported.

The effects of ABA and of gibberellins and cytokinins are mutually antagonistic in a number of tests. For example, the inhibitory effect of ABA on the germination of lettuce seeds can be completely reversed by cytokinins such as kinetin (Fig. 11.10). Similar interactions between ABA and GA_3 have been found in other tests. These observations have led to the idea that dormancy may be regulated by an interaction between growth inhibitors such as ABA, and growth promoters, such as GA_3 and cytokinins. In a number of seed species, the effects of ABA can be overcome by cytokinins, but not by GA_3. It has been suggested that gibberellins, ABA and cytokinins have "primary", "preventive" and "permissive" roles, respectively, in the regulation of seed dormancy, i.e. it is postulated that gibberellins have a primary role in overcoming seed dormancy, that at appropriate concentrations ABA prevents germination, but that its effect can be overcome by conditions which lead to an increase in endogenous cytokinins.

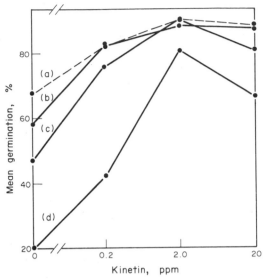

Fig. 11.10. Effect of abscisic acid and kinetin alone and in combination on the germination of lettuce seed. ABA concentration, ppm: (a) nil; (b) 0·02; (c) 2·0; (d) 20. (From P. F. Wareing, J. Good, H. Potter and J. A. Pearson, *S.C.I. Monograph* No. 31, Soc. for Chem. Ind., London, 1968.)

Although experiments with exogenous ABA seem to point to its possible role in the regulation of dormancy, determinations of levels of endogenous ABA have given some results which appear to support the hypothesis, but others which appear to be inconsistent with it. Thus, endogenous ABA levels decline during chilling in seeds of apple and sugar maple (*Acer saccharum*), but experiments with other seeds have given less clear-cut results. Similarly, the ABA levels in buds of black currant (*Ribes nigrum*) and beech (*Fagus sylvatica*) have been reported to decline during the course of the winter, accompanied by a concomitant increase in the levels of the glucosyl-ester of ABA, suggesting a possible conversion of free ABA into conjugated forms. On the other hand, similar studies on birch and sycamore buds gave less marked changes in the levels of free ABA and the ratio of free/bound ABA, during the winter.

Since the formation of resting buds in woody plants is promoted by short days it might be expected, if bud formation and dormancy involves ABA, that levels of endogenous ABA would be higher under short days than under long days, but no such differences could be detected in the leaves and shoot apices of birch, maple and other species. Again, if high levels of ABA bring about cessation of growth and bud formation in woody plants, the very high ABA levels resulting from drought stress (p. 139) would be expected to be very effective in causing the formation of resting buds, but this does not appear to be the case in birch seedlings maintained in long days under water stress, although drought does promote bud formation in other species.

Thus, the present state of knowledge regarding the role of hormones in the control of dormancy remains confused, and although it seems very likely that variations in endogenous gibberellins, cytokinins and ABA are important in bud and seed dormancy, their precise role remains uncertain.

Many studies have been carried out on the effects of growth substances on RNA and protein synthesis in seeds, but diverse and to some extent conflicting results have been reported. There have been several reports that ABA inhibits RNA and enzyme synthesis and that GA_3 or kinetin will reverse this effect in certain systems. In pear embryos ABA inhibits RNA polymerase activity and this effect is overcome by kinetin and GA_3. In barley aleurone a m-RNA for α-amylase is formed in response to GA_3 (p. 96) and this effect is inhibited by ABA. Other studies have indicated that GA_3 increases the DNA template availability for transcription in hazel seeds. The significance of these observations in relation to the hormonal control of dormancy and germination is by no means clear.

11.6 THE LONGEVITY OF SEEDS

The period during which seeds retain their viability varies greatly between species. The seeds of certain species remain viable for only short periods if kept in air at ordinary atmospheric conditions of humidity and temperature. Thus, seeds of willow (*Salix*) are viable for only a few days and must be sown very shortly after attaining maturity. The seeds of poplar and elm retain viability for a few weeks, and those of oak, beech and hazel for a few months, when stored under cool, moist conditions.

The seeds of most species, however, remain viable for considerable periods, generally for at least a year and frequently for much longer. Various types of experiments have been carried out to determine the longevity of such seeds. Thus, various workers have tested the viability of very old seeds from herbaria, and Becquerel found appreciable germination of seeds of various members of the Leguminosae ranging from 100 to 200 years old. Other families characterized by long-lived seeds are the Euphorbiaceae, Malvaceae, Convolvulaceae, Solanaceae, Labiatae and Compositae.

Several long-term experiments have been set up in the United States to determine the longevity of seeds. In 1902, the U.S. Department of Agriculture set up an experiment in which seeds of 107 species were placed in sterile soil in pots, which were then buried outside at different depths. After 20 years, some seeds of fifty-one of the species were still viable, but the seeds of most cultivated plants tested were dead. After 39 years, low germination was shown by twenty species and high germination by a further sixteen species.

A number of field observations also support the finding that seeds may retain their viability for long periods in the soil, under natural conditions. Thus, there is a well-authenticated case in which viable seeds of arable weeds were found in the soils of forests which had been planted on farm land some 20-46 years previously and had presumably lain dormant in the soil for that time. Other similar examples are known. Thus, living seeds of the Indian water lily, *Nelumbium nucifera,* were found in the bed of a former lake in Manchuria, which was estimated to have been drained at least 120 years, and more probably 200 years, previously. Seed from herbarium specimens of this species which were 237 years old have also germinated. The seeds of *Nelumbium* have very thick coats and are impervious to water, so that the embryo is not imbibed with water until the coat is rendered permeable. Similarly, the hard-coated seeds of the Leguminosae will survive in the soil without taking up water until the coats have been eroded by the activities of soil micro-organisms. Such seeds may therefore lie for long periods in the soil in the *unimbibed* condition.

Many other types of seed which survive in the soil for long periods do not have impermeable coats, however, and hence they must survive in a moist condition. This fact raises the problem as to why it is that these seeds do not germinate in the soil, where they appear to have adequate conditions of moisture and temperature for germination. This problem has not been satisfactorily solved, but it has been suggested that high carbon dioxide concentrations in the soil render the seeds dormant. However, recent studies have shown that a very high proportion of buried weeds have a light-requirement for germination. It is of interest that among the buried seeds found to be light-requiring are those of certain species which do not normally show a light-requirement, and hence it would appear that burial in some way leads to the development of a light requirement. Where such light-requiring seeds lie buried in the soil they will remain dormant until ploughing or some other disturbance bring them to the surface, when they rapidly germinate. Seeds showing this behaviour include those of *Digitalis purpurea, Juncus* spp., *Polygonum* spp., *Veronica persica, Spergula arvensis* and *Hieracium* spp.

The ultimate cause of loss of viability of seeds is not fully understood, but it is known that seedlings from old seeds show various cytological abnormalities such as chromosome breakage, disturbance of the mitotic spindle etc. Moreover, both ribosomal RNA and messenger RNA in dry seeds apparently undergo degradation with increasing age, although such RNA fractions isolated from old seeds will still support protein synthesis *in vitro* (i.e. in cell free systems). However, RNA synthesis is greatly reduced and there is only a very low rate of protein synthesis in old, intact seeds which have lost their viability.

11.7 GERMINATION

11.7.1 The conditions for germination

Since the tissues of the ripe seed are in a highly dehydrated condition, it is not surprising that water supply is frequently a limiting factor controlling the germination of seeds. Most seeds take up a relatively large amount of water when planted in a moist medium, and this is initially an imbibitional process, in which various substances present in the seed, especially proteins and starch, are involved. The imbibitional forces involved are enormous and certain seeds can take up considerable quantities of water from relatively dry soil.

Temperature is a second factor which plays an important role in controlling germination. The minimum temperature at which germination can occur varies considerably from one species to another, and the seeds of some species, such as beech, will germinate at temperatures only a little above freezing, whereas the seeds of tropical and sub-tropical species have much higher temperature requirements. Whereas most species will germinate under constant temperature conditions, other species require a daily alternation in temperature. For example, the seed of the dock, *Rumex obtusifolius,* germinates best when subjected to daily temperature alternations of 15° and 30°C. Alternating temperatures are also required by seeds of evening primrose (*Oenothera biennis*), Yorkshire fog grass (*Holcus lanatus*) and celery (*Apium graveolens*). Presumably these requirements are usually met by the normal variation between day and night temperatures under natural conditions. Little is known of the physiological basis of this phenomenon.

Although it is well known that seeds need conditions of adequate aeration for germination, there is little precise information on the oxygen requirements of seeds. It is clear, however, that the oxygen requirements of the intact seed will depend not only upon the metabolic demands of the embryo, but also on the permeability of the enclosing testa or other seed coats. As we have seen, these seed-coat effects appear to play an important role in the dormancy of many species. Whereas the oxygen content of the air is far above that needed for normal growth of plants, it is not far above that required for the germination of many seeds, no doubt due to the physical barrier to oxygen uptake presented by the seed coats.

There are marked differences between species in the ability of their seeds to germinate under water,

no doubt because the partial pressure of the oxygen dissolved in water is considerably less than in air. Other seeds will germinate well under water; these latter include both aquatic species, and certain land species, such as rice.

11.7.2 The process of germination

In non-dormant seeds, active metabolism evidently commences soon after they are placed under conditions favourable for germination. The question arises as to what are the first stages in the complex overall process known as germination. Usually, of course, we take the emergence of the radicle as the primary criterion of germination, i.e. for most practical purposes the initiation of growth is taken as the first detectable sign of germination. It appears that the initial elongation of the radicle involves cell extension rather than cell-division, but cell division starts very early in the growth of the radicle. It is probable that the commencement of radicle growth is preceded by a number of preliminary processes, but little is known regarding the nature of these processes.

In the dry condition of the resting seed, metabolism must obviously be at an extremely low level on account of lack of water, but full metabolic activity does not develop immediately water is imbibed, even in non-dormant seeds.

When a non-dormant seed is planted under conditions favourable for germination there is a rapid increase in the respiration rate, which can be detected 2-4 hours after soaking in the case of peas. After this initial rise the respiration rate in peas reaches a steady value which is maintained for several hours. At about the time that the testa is broken by the radicle there is a further rapid rise in respiration rate, suggesting that in the initial phases of germination gaseous exchange is limited by the testa. By contrast, in barley and wheat there is a fairly uniform rise of respiration rate during germination.

The changes in carbon dioxide output (Q_{CO_2}), of oxygen uptake (Q_{O_2}) and of the respiratory quotient (RQ) during germination provide an indication of the type of respiration occurring (i.e.

whether aerobic or anaerobic) and of the nature of the respiratory substrate, i.e. whether carbohydrate, fat or protein. During the early stages of germination of peas respiration appears to be predominantly anaerobic, owing to the restriction of oxygen uptake by the testa, and ethanol may accumulate in the tissues.

Glycolytic enzymes appear to be present in dry seeds and become active during imbibition. On the other hand, although preformed mitochondria are present in dry seeds, they are non-functional. Consequently the tri-carboxylic acid cycle and terminal oxidation are not active, and hence oxidative phosphorylation cannot occur immediately upon imbibition. However, as seeds proceed to imbibe water the structure of the mitochondria develops, with the formation of cristae, and at the same time oxygen uptake and the enzymes of the tri-carboxylic acid cycle become active, and the electron transport chain is completed.

Among the earliest and most striking events occurring when a seed is allowed to imbibe water is a very rapid increase in ATP levels. However, this rise in ATP cannot be due to oxidative phosphorylation and it is unlikely that glycolysis becomes active quickly enough or at a high enough rate to account for the rise in ATP. It has been suggested that the ATP may come from the dephosphorylation of phosphoproteins, but there is no direct evidence in support of this hypothesis at present.

There is evidence that germination depends upon the amount of metabolically available energy in the cells, and this is related to the "energy-charge" calculated as

$$\frac{[ATP] + \frac{1}{2}[ADP]}{[ATP] + \ \ [ADP] + [AMP]}$$

The energy charge of germinating wheat embryos increases during the early stages of germination, and the very early, rapid formation of ATP may constitute a regulatory mechanism.

Although the tricarboxylic acid cycle and terminal oxidation pathways are not active during the initial stages of germination, reference has already been made (p. 271) to evidence that the pentose phosphate pathway may be important in the initiation of germination of dormant cereal grains and other seeds. The possible role of the pentose phosphate pathway in germination is obscure, but it may be to produce intermediates which are essential as building blocks for various synthetic processes.

Germination is accompanied by an increase in activity of a wide diversity of enzymes present initially in the seed and by the appearance of new enzymes. Some enzymes, which were produced during seed development and which are present in the dry seed, become active instantaneously on imbibition. A second group of enzymes is apparently present in dry seeds in an inactive form and activity only appears as germination progresses; apparently such enzymes become activated by various mechanisms. A good example of the synthesis of a new enzyme during germination is provided by the *de novo* synthesis of α-amylase in barley aleurone described earlier (p. 96).

Protein synthesis, as detected by the incorporation of amino acids, such as ^{14}C-leucine, into polypeptides can be shown to occur within a short time (varying from a few minutes to several hours) after imbibition. Moreover, in wheat embryos there is a rapid formation of functional polyribosomes within 30 minutes of the start of imbibition. RNA synthesis can also be detected within 60 minutes of imbibition of rye embryos. However, several lines of evidence indicate that early protein synthesis is not dependent upon newly synthesized RNA, and that all the major RNA fractions required for protein synthesis are present in the dry seed. Thus, ribosomes from dry seeds can support *in vitro* protein synthesis, provided that they are supplied with messenger RNA, together with ATP and certain other cytoplasmic fractions, which include amino acids, tRNA and various enzymes and other factors involved in protein synthesis. There is also strong evidence suggesting that dry seeds contain 'stored' RNA, including the following observations:

(1) Protein synthesis can be shown to take place *in vivo* in certain seeds, even when RNA synthesis is inhibited.

(2) Polysomes form rapidly after imbibition in the absence of RNA synthesis or when the latter is inhibited.

(3) A fraction can be extracted from dry wheat embryos which can support *in vitro* protein synthesis.

(4) RNA fractions rich in poly-adenine (a characteristic of mRNA—see p. 309) have been isolated from dry seeds.

(5) *De novo* synthesis of specific enzymes when RNA synthesis is inhibited has been demonstrated for cotton embryos.

This evidence has been taken to indicate that early protein synthesis in various seeds is supported by preformed, "long-lived" messenger RNA, which was formed during seed development and stored in the dry seed. Studies on developing cotton embryos suggest that mRNA coding for the enzyme protease is present during the last 20 days of seed development, and yet synthesis of the enzyme does not occur at that time. It would appear that the translation of some mRNAs is inhibited in the developing embryo, possibly by the presence of endogenous abscisic acid.

Although in the initial stages of germination protein synthesis apparently depends upon preformed m-RNA, there is no doubt that before long the embryo becomes dependent upon newly-formed mRNA. Initial expansion of the radicle within the seed generally occurs by cell elongation. The stage at which DNA synthesis and mitosis occur in the radicle of germinating seeds varies considerably between species but, in general, these processes commence later than RNA and protein synthesis.

FURTHER READING

General

Bewley, J. D. and M. Black. *Physiology and Biochemistry of Seeds,* Vol. 1, Springer-Verlag, Berlin, 1978.
Clutter, M. E. (Ed.). *Dormancy and Developmental Arrest,* Academic Press, New York, 1978.
Khan, A. (Ed.). *The Physiology and Biochemistry of Seed Dormancy and Germination,* North-Holland Publ. Co., Amsterdam, 1977.
Mayer, A. M. and A. Poljakoff Mayber. *The Germination of Seeds,* 2nd ed., Pergamon Press Ltd., Oxford, 1976.
Villiers, T. A. *Dormancy and the Survival of Plants,* Edward Arnold, London, 1975.

More Advanced Reading

Ching, T. H. Metabolism of germinating seeds. In: *Seed Biology Vol. II,* T. T. Kozlowski (ed.), Academic Press, New York and London, 1972.
Heydecker, E. (Ed.). *Seed Ecology,* Butterworths, London, 1973.
Kozlowski, T. *Seed Biology,* Vols. I and II, Academic Press, New York and London, 1972.
Mayer, A. M. and Y. Shain. Control of seed germination, *Ann. Rev. Plant Physiol.* **25**, 167, 1974.
Roberts, E. H. (Ed.). *Viability of Seeds,* Chapman & Hall, London, 1972.
Taylorson, R. B. and S. B. Hendricks. Dormancy in seeds, *Ann. Rev. Plant Physiol.* **28**, 331, 1977.
Wareing, P. F. and P. F. Saunders, Hormones and dormancy, *Ann. Rev. Plant Physiol.* **22**, 261, 1971.

12

Senescence and Abscission

12.1 INTRODUCTION

In common with all multicellular organisms, higher plants are mortal and the life of the individual plant is ultimately terminated by death. Before the death of the whole plant has occurred, however, it is likely that there will have been earlier death of a number of its organs and tissues. As a result of the activity of the apical meristem, the upper part of the shoot shows a prolonged embryonic condition, while at the same time senescence and death is occurring in the older lateral organs, notably the leaves, flower parts and fruits. In many plants the death of organs has a pronounced seasonal character and there is a regular annual loss of part of the shoot system. In trees, this annual loss is mainly confined to the lateral organs mentioned, but in some herbaceous

perennials, e.g. dock (*Rumex*), nettle (*Urtica*), bracken (*Pteridium aquilinum*), the whole above-ground part of the shoot may die each year. In annual species, the whole plant, except the seeds, dies after flowering and fruiting. Thus, the death of plant parts is a regular feature of the annual cycle of growth and is much more common in plants than in animals.

Death of an organ or of the whole plant is always preceded by the process of *senescence,* which may be regarded as the final phase in development that leads to cellular breakdown and death. We may conveniently distinguish between *organ senescence* and *whole plant senescence.* In most plants each leaf has only a limited life span so that as the shoot continues to grow in height, the older leaves at the base tend to senesce and die progressively (Fig. 12.1). This pattern of senescence has been described as *sequential senescence,* and it must be distinguished from the *simultaneous* or *synchronous senescence* of leaves of temperate deciduous trees, which is so conspicuous in the "fall". *Fruit senescence* is seen during ripening of both succulent and non-succulent fruits. The ripening of succulent fruits is a complex process which ultimately terminates in the senescence and decay of the tissues.

Before considering plant senescence it is necessary to distinguish between *monocarpic* species, which flower and fruit only once and then die, and *polycarpic* plants, which flower and fruit repeatedly. Monocarpic species include all annual and biennial plants and also a certain number of

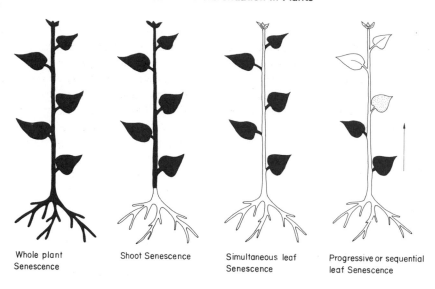

Fig. 12.1. Types of plant- and leaf-senescence.

perennial plants which grow vegetatively for a number of years and then suddenly flower, fruit and die. Examples of this latter type of plant are seen in the "Century Plant" (*Agave*) and bamboo, both of which may grow vegetatively for many years, before the single reproductive phase occurs. Thus, in monocarpic species, death of the whole plant is closely connected with reproduction and is evidently gentically determined to occur at this stage in the life cycle. By contrast, in polycarpic species, which include both herbaceous perennials and woody plants, death of the whole plant is not normally associated with reproduction and there is great variation among the different individuals of a given species with respect to the length of the life span. Thus, in dicotyledonous woody plants, which are all polycarpic, the individual tree normally lives for many years, and there is no universal life span characteristic of the species. Indeed, apart from accident and disease, there would seem no reason why a tree should not live indefinitely, although there would no doubt be mechanical problems when the branches become too heavy to support themselves. In monocarpic plants, on the other hand, death is destined to occur at a given point in the life cycle, and we may speak of *programmed death* in such plants.

What determines the life span of an individual cell, an organ or a whole plant? It might be suggested that a given cell has only a limited life span, which is determined by factors inherent in the cell itself. The loss of viability of seeds appears to be due to chromosome breakage within individual cells (p. 277), but it cannot be assumed that changes occurring in the dehydrated tissues of seeds in dry storage occur also in actively metabolizing cells and are a cause of senescence in the later stages of the plant cycle, although it has been suggested that chromosomal changes are important in the ageing and senescence of animal cells.

In some types of tissue, differentiation involves the early death of certain cells, such as those of vessel elements in the xylem, whereas neighbouring parenchymatous cells may remain living for many years. The changes occurring in the protoplast during the differentiation of a vessel element may correspond closely to those occurring later in the constituent cells of a senescent organ, such as a leaf. However, the process of vacuolation and enlargement does not necessarily involve degenerative changes, since parenchymatous cells may live for many years, as in those of the pith and rays of some woody plants. Thus, it would seem that the maximum potential life of many types of differentiated plant cell is seldom reached in herbaceous plants, and that senescence and death do not occur on account of factors intrinsic in the individual cells, but because of conditions prevailing within the organ or organism as a whole. For example, sequential leaf senescence seems to be caused by competition between mature leaves and the growing regions of the shoot, and if a leaf is removed and allowed to form roots in the petiole it may live very much longer than if it had remained attached to the parent plant (p. 287). Thus the rate of senescence of plant organs is often under control of the whole plant and is not simply determined by intrinsic characteristics of the cells of that organ. However, it would appear that certain organs show inherent senescence processes, which are not under the control of the whole plant; thus, flowers and fruits undergo senescence whether they are allowed to remain on the parent plant or not.

In addition to the various "internal" factors of the plant which are involved in the regulation of senescence, a number of external factors may affect the rate of senescence, including drought, mineral nutrition, light intensity and daylength, and disease. We shall mainly be concerned with a consideration of the internal factors, but the importance of environmental factors must also be borne in mind.

12.2 THE BIOLOGICAL SIGNIFICANCE OF SENESCENCE

If we are correct in regarding senescence in monocarpic plants as "gentically programmed", then this implies that the process of senescence has arisen as a result of natural selection and that it has certain biological advantages to the species. What are these advantages? Why, for example, should it be any advantage to an annual species for the whole shoot to senesce during the development and ripening of the fruit—why should the leaves and other parts of the shoot not remain green, as they do during the ripening of the fruit in

many polycarpic plants? The answer to this question probably lies in the fact that during the ripening of annual plants there is considerable breakdown of proteins to amino acids, which are exported from the leaves to the developing seeds and reutilized there as reserve material. For example, in the oat plant a large proportion of the nitrogen in the senescing leaves is transported to the fruits and accumulated there. Thus, recovery of nutrients from senescing organs constitutes a valuable saving to the rest of the plant. Similar export of reserve material occurs during the simultaneous senescence of the leaves of trees in the fall, the exported material being stored in the stem in this case. However, another advantage of leaf fall in deciduous trees probably lies in the resulting reduced rate of transpiration, which is probably essential for survival in climates in which the soil is frozen during the winter (p. 260), while at the same time the return of leaf material and its breakdown in the surface litter releases mineral nutrients to the soil which are available for reutilization.

Sequential senescence of the basal leaves of a shoot not only releases reserves of nitrogenous and other substances for the young growing leaves, but may also bring about a saving of carbohydrates and other photosynthates where the basal leaves are heavily shaded and hence might actually become "parasitic" upon the plant by importing photosynthates from other parts which they would only consume in useful respiration. Thus, it is not difficult to see how senescence, as an active "programmed" process, may have several advantages.

12.3 THE MECHANISM OF SENESCENCE

Our present knowledge of the physiology and biochemistry of senescence is derived mainly from studies on leaves (both attached leaves undergoing sequential senescence and detached leaves or leaf discs) and on ephemeral flowers such as those of *Ipomoea tricolor,* the corolla of which opens and undergoes senescence within the space of only 24 hours (Fig. 12.10).

12.3.1 Sequential leaf senescence

The first visible sign of senescence is yellowing of the leaf, due to the breakdown of chlorophyll which renders visible the other leaf pigments, particularly the xanthophylls and carotenoids. A study of the fine structure of senescing leaves shows that there is progressive degeneration of the membrane structure of the grana of the chloroplasts, accompanied by the appearance of dense globules of lipid material (probably formed from the broken-down membranes) in which the carotenoid pigments dissolve. Other early changes involve the degeneration of the endoplasmic reticulum and the gradual disappearance of the ribosomes. The mitochondria retain their structure during the early stages of senescence, but later they undergo degeneration. The cells of the fully senescent bean leaf still retain an intact plasmalemma but the tonoplast disappears, and the structure of the cytoplasm and nucleus is almost completely lost, leaving the chloroplasts represented by vesicles containing lipid globules.

These structural changes in the cells of the senescing leaf are accompanied by changes in composition and metabolic activity. The protein content of the leaf declines progressively, as a result of the breakdown of proteins to amino acids and amides (Fig. 12.2). There is also a progressive

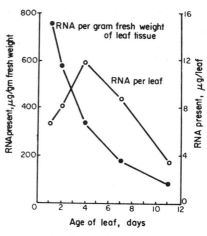

Fig. 12.3. Changes in ribonucleic acid (RNA) content of attached leves of pea (*Pisum sativum*) from completion of leaf expansion. (From R. M. Smillie and G. Krotov, *Can. J. Bot.* **39**, 891, 1961.)

decline in the RNA content of the leaf, with a particularly marked fall in ribosomal RNA (Fig. 12.3).

These degenerative changes are reflected in the rate of photosynthesis and respiration of the leaf. In *Perilla,* the rate of photosynthesis declines gradually from the time of full leaf expansion, accelerating during the later stages of senescence (Fig. 12.4). In this species the trend in photosynthetic rate follows closely the level of the soluble

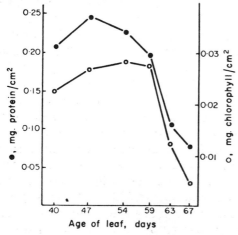

Fig. 12.2. Changes in protein and chlorophyll content of attached leaves of *Perilla frutescens* from expansion to abscission. (From H. W. Woolhouse, *Symp. Soc. Exp. Biol.* **21**, 179, 1967.)

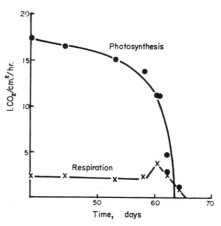

Fig. 12.4. Changes in rates of photosynthesis and respiration of attached leaves of *Perilla frutescens* from completion of expansion to abscission. (From H. W. Woolhouse, as for Fig. 12.2.)

protein fraction of the chloroplasts known as "Fraction 1" containing the enzyme, ribulose-1,5-diphosphate carboxylase, which catalyses the process of carbon dioxide fixation. The trend in respiration rate seems to vary according to the species, but in some the rate appears to remain fairly constant until the final stages of senescence, when there is a sharp rise in respiration rate, corresponding to the "climacteric" observed during fruit ripening and senescence.

The question arises as to what initiates and controls the degradative changes occurring during leaf senescence. The observation that, in some species at least, the respiration rate remains constant during the early stages of senescence suggests that it is not changes in respiratory metabolism which cause senescence. On the other hand, we have seen that a constant concomitant of senescence is marked decline in the protein and RNA contents of leaves, and close attention has been paid to these changes as possibly indicating the "key" processes in senescence. Now, it has been shown that a certain proportion of the leaf protein undergoes continuous "turnover", i.e. the protein is being continuously synthesized and broken down, so that the overall rate of change in protein content represents the net differences in the rates of these two processes. Where there is such a continuous turnover, a decline in protein content may reflect a fall in the rate of synthesis or a rise in the rate of breakdown, or both.

The breakdown of protein is brought about by proteolytic enzymes. Studies on changes in proteolytic enzyme activity have given no indication of increased activity during leaf senescence, and hence it would appear that the decline in protein content is primarily due to a reduced rate of synthesis. One possibility that has been suggested is that the senescent leaf retains its full capacity for protein synthesis and that the rate of synthesis is limited by lack of amino acids in the leaf. In a healthy, green leaf the amino acids released by protein breakdown are reutilized in further protein synthesis. But it has been suggested that in a senescent leaf amino acids are exported to other parts of the plant so rapidly that there is no "pool" of free amino acids available for protein

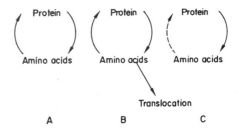

Fig. 12.5. Diagram to illustrate (A) turnover of leaf protein, (B) the translocation hypothesis, and (C) the hypothesis of a defect in protein synthesis. (From E. W. Simon, *Symp. Soc. Exp. Biol.* **21**, 215, 1967.)

synthesis, so that there is a decline in protein content (Fig. 12.5). It is, in fact, found that there is no appreciable accumulation of free amino acids in senescent attached leaves, no doubt due to continuous export. However, it is then necessary to explain why the supply of amino acids should be limiting in a senscent leaf and not in a normal one.

Now, there is evidence that the sequential senescence of leaves arises from competition for metabolites and nutrients between old leaves at the base of the stem and young growing leaves in the apical regions. This conclusion is indicated by the fact that if a plant is decapitated and the axillary shoots are removed then the senescence of the older remaining leaves is greatly retarded and indeed leaves which are already showing signs of senescence may undergo a recovery and become green again, even if they were previously showing yellowing. The uppermost leaf of a decapitated plant may also undergo considerable growth and become abnormally large. Thus, it would appear the sequential senescence is a "correlative phenomenon", and shows resemblances to apical dominance (p. 133). Sequential senescence is more pronounced under conditions of mineral nutrient deficiency—for example, plants grown in too small pots frequently show marked senescence of the basal leaves, under which conditions there is presumably severe competition between young and old leaves for available nutrients, and in this competition the young leaves are evidently at an advantage. However, the senescence of the older leaves can be retarded or reversed by the application of nitrogenous nutrients, such as ammonium nitrate.

Thus, it is possible that the competition between young and old leaves results in a rapid rate of transport of amino acids from old to growing leaves and hence to a lower pool of these metabolites being available for protein synthesis within the old leaves. It is an essential part of this hypothesis that there should be active protein turnover, but in fully expanded *Perilla* leaves the "Fraction 1" protein is found to have zero turnover rate, i.e. there is no continuous synthesis and breakdown, and yet during senescence Fraction 1 declines more rapidly than a second fraction "Fraction 2", which does show active turnover.

Another postulate of the hypothesis we are considering is that the capacity for protein synthesis remains relatively unimpaired during leaf senescence. A convenient method for measuring the rate of protein synthesis is to determine the rate of incorporation of radioactive amino acids, such as ^{14}C-leucine, into protein. Similarly, the rate of RNA synthesis can be followed by the rate of incorporation of an RNA precursor, such as ^{14}C-adenine. Studies of this type have shown that the capacity of tobacco leaves to incorporate ^{14}C-leucine and ^{14}C-adenine declines during senescence, although quite yellow leaves retain some capacity to synthesize certain enzymes, such as peroxidase and ribonuclease (which brings about the breakdown of RNA). It might be argued, however, that the decline in capacity for protein synthesis is the *result,* rather than the cause of senescence. Overall, nevertheless, it seems that protein metabolism in senescing attached leaves may be viewed as an unbalanced turnover reaction, with catabolism exceeding anabolism.

Thus we see that some evidence supports the hypothesis that the decline in protein content of attached senescent leaves is due to limiting levels of amino acids, rather than to a reduced capacity for protein synthesis, but that the evidence is not conclusive. Furthermore, it is known that when the young leaves are removed from a plant, the remaining older leaves "regreen" and show a marked and rapid rise in RNA synthesis which precedes a rise in leaf protein level. Within 12 hours of removal of the young leaves from *Perilla*

plants the incorporation of $^{32}PO_4$ into chloroplastic ribosomal RNA is stimulated in older leaves, and this change occurs before an effect is seen in the cytoplasmic ribosomes of the leaf. Results such as these suggest that very early events in leaf senescence involve changes in nucleic acid metabolism within the chloroplasts.

We shall now consider an alternative approach to the problem of leaf senescence, in which use has been made of leaves, or portions of leaves, which have been isolated from the parent plant. Excision of a leaf usually results in the immediate onset of senescence processes in its tissues, and detached leaves or leaf discs therefore provide convenient material for experiments under controlled conditions, without the additional complication of correlative influences from other parts of the plant.

12.3.2 Senescence of detached leaves

So far, we have considered the natural senescence of leaves still attached to the plant, but it has been known for many years that when a green leaf is detached from its parent plant it rapidly deteriorates and shows signs of accelerated senescence.

As with attached leaves, the visible signs of senescence are accompanied by a decrease in the protein and RNA contents of the leaf. Protein breakdown commences remarkably soon after the leaf has been detached; for example, protein breakdown is detectable 6 hours after excision of barley leaves. The breakdown of protein commences at the same rate whether the leaves are kept in the light or the dark, but the rate of breakdown later decreases in the light, whereas it continues at a high level in the dark (Fig. 12.6). The initial equal rates of breakdown in light and dark suggest that senescence is not triggered off by carbohydrate deficiency, although this factor may be important during the later stages of senescence in the dark.

In detached leaves, protein degradation leads to the accumulation of amino acids and amides in the leaf, since they cannot be exported, as is the case with attached leaves, although there may be

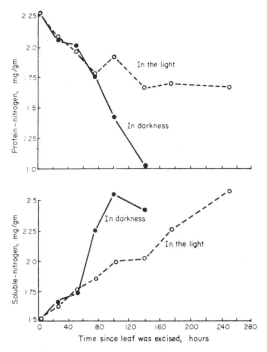

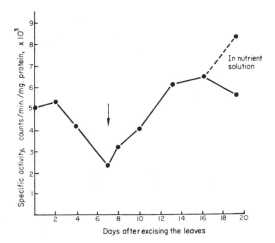

Fig. 12.6. Changes in protein and soluble nitrogen in *detached* tobacco leaves. Following excision, protein-N declines and soluble-N increases. Leaves kept in darkness show these changes to a greater extent and are yellow and senescent after approximately 100 hours. (From H. B. Vickery *et al., Connecticut Agric. Exp. Sta.* (*New Haven*) *Bull.* **374**, 557, 1935.)

Fig. 12.7. Changes in capacity for protein synthesis (as measured by incorporation of ^{35}S-methionine into protein) in leaves of *Nicotiana rustica*. Note initial decline in capacity for protein synthesis, followed by recovery when roots appeared on petiole (indicated by arrow). (After von. B. Parthier, *Flora, Jena,* **154**, 230, 1964.)

accumulation at the base of the petiole where this is present. Thus, senescence can occur in detached leaves, or in leaf discs, even though there is a high level of amino acids within the tissues. Hence, under these conditions the reduced protein synthesis cannot be attributed to lack of amino acids. On the other hand, detached leaves show reduced capacity for protein synthesis, as indicated by the reduced capacity to incorporate ^{14}C-leucine into protein. Indeed, leaf discs which have remained in the dark for several days lose this capacity completely. Thus, there seems no doubt that the decline in protein content observed in detached leaves arises from a reduced capacity for protein synthesis.

It was observed that the rapid rate of protein breakdown in detached leaves is arrested if they are allowed to form roots (Fig. 12.7). Indeed, leaves which have been planted in soil and induced

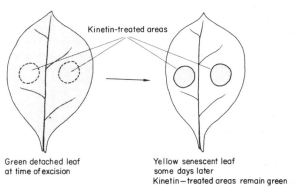

Fig. 12.8. Effects of kinetin on leaf senescence.

to form roots on the petiole will live for considerable periods. On these grounds, Chibnall suggested that roots must supply some "factor" which is necessary for the maintenance of protein synthesis in the leaves, but the nature of this "root factor" remained unknown. However, it was later found that kinetin will prevent the senescence of detached leaves if applied to them in aqueous solution. Thus, if a drop of kinetin solution is applied to a senescing tobacco leaf, then the area of leaf which received the kinetin will remain green, although the surrounding leaf tissue continues to yellow (Fig. 12.8). Similarly, if discs of radish or *Xanthium* leaves are placed on a solution of kinetin they still retain their green colour after the control discs (on water) have become fully senescent. Moreover, electron microscopic studies show that the kinetin-treated discs still have the normal structure of a green leaf.

Thus, senescence in detached leaves can be prevented either (1) by the presence of roots, or (2) by application of kinetin and other synthetic cytokinins. It was, therefore, natural to consider whether the natural "root factor" which delays senescence is an endogenous cytokinin. It is of great interest, therefore, to discover that cytokinins are present in the root exudate from sunflower, grape and other plants, and this suggests that leaves may depend on the supply of endogenous cytokinins from the roots, for the maintenance of a normal, green condition. We shall return to this question later (p. 292).

Not only do cytokinins delay senescence of detached leaves, but they also cause a rapid increase in the rate of RNA and protein synthesis only a few hours after their application (Fig. 12.9); incorporation of precursors, such as

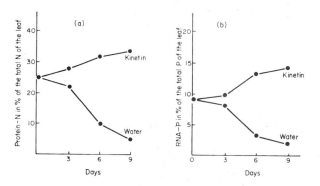

Fig. 12.9. Changes in (a) protein and (b) RNA content in kinetin- and water-treated halves of undivided leaves of *Nicotiana rustica*. (After R. Wollgiehn, *Flora, Jena,* **151,** 411, 1961.)

^{14}C-adenine or ^{14}C-cytidine, into all fractions of RNA appears to be increased by kinetin. Since protein synthesis involves the RNA "machinery" of the cell, the effect of cytokinins on protein synthesis is readily understood if their primary effect is upon RNA synthesis, but they may also delay senescence through their effect on the mobilization of metabolites (p. 291). The possible mode of

action of cytokinins in stimulating RNA synthesis was discussed in an earlier chapter (p. 94).

With most species, gibberellins do not affect the rate of senescence of leaf discs, but in *Taraxacum, Rumex, Tropaelum* and a few other herbaceous species, gibberellins are more effective in delaying senescence than cytokinins. On the other hand, auxins are found to be relatively ineffective in delaying senescence in the leaves of herbaceous plants, but they are active in this respect in the leaves of some deciduous woody perennials (e.g. cherry) and in pericarp tissues of dwarf beans (*Phaseolus vulgaris*). Applications of gibberellins may also delay the onset of senescence in leaves of some deciduous trees.

Contrasting with the senescence-delaying effects of cytokinins, gibberellins and auxins, both abscisic acid (ABA) and ethylene appear to accelerate the processes of senescence in leaves. Exposing leaves on intact plants, or detached leaves or leaf discs, to ethylene gas or to ABA usually results in accelerated yellowing. Fruit senescence (or ripening) is similarly speeded up by ethylene, and it is well established that ethylene plays an important role in the natural ripening of succulent fruits. Both ABA and ethylene inhibit RNA and protein synthesis in isolated leaf discs of many species, which emphasizes that their effects in senescence are opposite to those of cytokinins and gibberellins.

12.3.3 Senescence in flowers

The evanescence of flowers is proverbial, and is normally a consequence of senescence in the corolla. In recent years, studies of the biochemical and physiological bases of flower senescence have tended to concentrate upon certain shorlived, or ephemeral, flowers such as those of *Ipomoea tricolor* (Fig. 12.10). As also seen in leaves, it has been found that senescence in these flowers is accompanied by rapid falls in the levels of protein, RNA and DNA (Fig. 12.11). Exposure of newly opened flowers to ethylene causes petal fading and other senescent changes within 90 minutes, and it

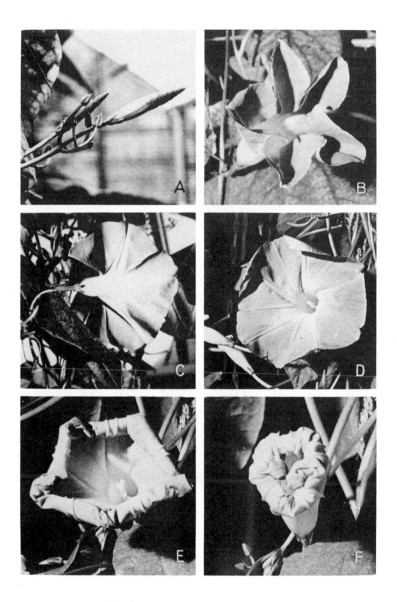

Fig. 12.10. Development of the ephemeral corolla of *Ipomoea tricolor*. A. Mature flower bud in the early morning of the day of flowering. B. Anthesis at 6 a.m. C. Rigid ribs responsible for the shape of the funnel. D. First signs of fading and wilting in early afternoon. E. Partially rolled-up funnel at 5 p.m. F. Corolla in the morning of the day after flowering. (From Ph. Matile, *The Lytic Compartment of Plant Cells,* Springer-Verlag, Wien and New York, 1975. Original prints kindly provided by Professor Dr. Ph. Matile.)

has been found that endogenous ethylene production rises at the same time that natural fading occurs and RNAs levels rise.

The time-course of decreases in protein, RNA and DNA levels can be compared with changes in the levels of activity of hydrolytic enzymes such as proteinase, RNase and DNase (Fig. 12.11), and it can be seen that the breakdown of nucleic acids occurs at the same time that nuclease activity increases dramatically whereas the level of proteinase is practically unchanged throughout

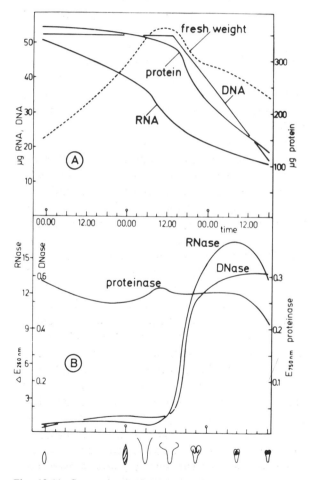

Fig. 12.11. Senescence in the ephemeral flowers of *Ipomoea tricolor*. *Top:* contents of protein and nucleic acids. *Bottom:* relative activities of proteinase and nucleases per corolla. The sketches below the time axis indicate the changing shape of the corolla—see also Fig. 12.10. (From Ph. Matile, *The Lytic Compartment of Plant Cells,* Springer-Verlag, Wien and New York, 1975. Original print kindly supplied by Professor Dr. Ph. Matile.)

senescence. It seems clear that the decrease in protein levels which occur during senescence is not brought about by changes in total proteinase activity present in the cells, but rather by some process which allows proteinases to act on the cell proteins. The senescing cells of the corolla, like those of senescing leaves, retain the capacity for synthesis of certain proteins (e.g. the RNase that rises in activity during senescence is known to be synthesized *de novo*) and overall loss of protein therefore reflects an imbalance in protein turnover, as referred to previously (p. 286), with catabolism exceeding anabolism.

There is accumulating evidence that many of the hydrolytic enzymes present in plant and animal cells are normally compartmentalized, in that they are located in particular regions and kept separate from other cellular constituents. This prevents the occurrence of *autolysis* (uncontrolled self digestion) in the cell. The regions, or compartments which contain these hydrolyses are grouped together under the general term, the *lytic compartment,* which includes organelles such as *lysosomes.* As long as the lytic compartment remains intact, breakdown of cellular components is under control and is known as *autophagy.* The phenomenon of autophagy probably involves transport of substrate molecules across lytic membranes to come into contact with the hydrolases, followed by transport of hydrolysis products out of the lytic compartment. Because metabolism continues in senescent tissues, including protein synthesis, the lytic compartment is probably intact during most of the senescence process. The final step in senescence does nevertheless, appear to be that of autolysis, involving rupture of the membranes of the lytic compartment.

12.3.4 Whole plant senescence

As we have seen, in monocarpic species, such as wheat (*Triticum*), soybeans (*Glycine max*) and French beans (*Phaseolus vulgaris*), senescence of the whole plant is usually associated with the development of fruits. Thus, in the French bean plant, as the pods and seeds grow and approach their full size, there is perceptible yellowing of the leaves. When fully developed, the fruits yellow and ultimately lose water as they ripen, and at the same time the leaves and stems become progressively more senescent and in due course they die. Not only are plant senescence and fruit development closely correlated in time, but there does indeed appear to be a causal relationship between the two processes, as shown by the fact that removal of all flowers and young fruits from bean plants greatly delays senescence of the remainder of the plant. Moreover, the same effect is observed if, instead of removing the whole fruits, only the seeds are removed surgically from the pods. Thus, the presence of developing seeds appears to control plant senescence, suggesting that some "signal" is sent out from the developing seeds, which regulates senescence in other parts of plant.

Now, the developing seeds of bean increase steadily in dry weight, and food reserves, including protein and starch, accumulate in the developing cotyledons. It is known that, at the same time, the breakdown of proteins and carbohydrates is occurring in the leaves and that amino acids, sugars and other metabolites are transported to the developing fruits. Similar effects are seen clearly in other species. It was, therefore, suggested by Molisch that developing seeds cause senescence in other parts of the plant by mobilizing and accumulating not only carbohydrates but also amino acids and other substances derived from the leaves.

The accumulation of reserve materials into fruits, tubers and other storage organs is well known. Similarly, actively growing meristematic regions, such as young leaves, are known to be able to mobilize nutrients from other parts of the plant. Such centres of mobilization are often referred to as "sinks" for nutrients. The manner in which mobilization by such sinks is achieved is not understood, since the mechanism of phloem transport is itself still not understood. However, in recent years increasing evidence has appeared indicating that growth hormones may play an important role in mobilization effects. Thus, when kinetin is applied to a small area on a senescing tobacco leaf, it is found that radioactive amino acids applied to the other parts of leaf are accumulated at the point of application of kinetin. Thus, metabolites appear to be "attracted" towards regions which have been treated with kinetin. It is possible that this effect of kinetin on the mobilization of metabolites is a secondary one, arising from the stimulation of protein synthesis at the point of kinetin application, so that a metabolic "sink" is created, which in turn causes movement of metabolites towards this point. However, Möthes was able to show that if aminoisobutyric acid (which is not a naturally occurring substance and is not incorporated into proteins) is applied to a part of the leaf away from the point of kinetin application, it accumulates in the latter region, just as the natural amino acids do. Thus, on this evidence it would seem that kinetin may affect the movement of metabolites independently of any effect it has on protein synthesis.

There is evidence that auxin-directed transport may also play an important role in the movement of metabolites in ripening bean plants. We have seen (p. 135) that metabolites tend to accumulate in regions of high auxin content and it is known that developing bean seeds are rich sources of auxin. Thus, it is possible that the movement of reserve materials from the leaves into the developing seeds may be an example of auxin-directed transport. Evidence in support of this hypothesis has been obtained in the following experiment. The fruits were removed from young bean plants, and a decapitated peduncle (fruit stalk) was allowed to remain on each plant. In some of the plants indole-acetic acid (IAA) in lanolin was applied to the peduncle stump and in other plants (controls) plain lanolin was applied. It was shown that when radioactive phosphorus, ^{32}P (in the form of orthophosphate), was applied to the stem bases, it very rapidly moved up into the peduncles to which IAA had been applied, but very little moved into the peduncles to which only plain lanolin was applied.

The hypothesis that plant senescence is brought about by the mobilization of nutrients by developing seeds would seem to be consistent with many observations, and if we postulate that the leaves become depleted of amino acids as a result of their export to the seeds, we have essentially the same hypothesis as was suggested above for the sequential senescence of leaves. However, there are certain observations which are difficult to reconcile with this hypothesis. Thus, it is found that in spinach (*Spinacia oleracea*), which is a dioecious species (with separate male and female plants), senescence of the *male* plants follows flowering, just as in female plants, although such male plants do not, of course, carry any developing fruits; moreover, removal of the male flowers delays senescence of the leaves. Further, in an experiment with *Xanthium pensylvanicum,* it was found that if all buds were removed from plants before exposure to short days, so that no flowers could be formed, the leaves of these plants later senesced at the same time as did those of plants which had been allowed to form flowers and fruit. On these and other grounds, the hypothesis that plant senecence can be explained simply in terms of the mobilization of nutrients by developing seeds is rejected by many workers.

Since protein synthesis is controlled by the RNA apparatus of the cell, it is entirely possible that the presence of developing seeds brings about protein breakdown in the leaves through an effect upon their RNA metabolism. For example, some fractions of RNA undergo rapid turnover, and it may be that certain RNA precursors are exported from the leaf to the developing seeds, and so are not available for reincorporation into RNA in the leaf. A different explanation has been suggested, as follows. We have seen that leaves undergo rapid protein and RNA breakdown when they are separated from the plant and it has been suggested that they are dependent on a continuous supply of cytokinin from the roots for normal RNA metabolism. Now, seeds are found to be rich in cytokinins. It is not known whether the total cytokinin content of seeds is synthesized there, or whether cytokinins are mobilized there from other parts of the plant. If the latter were the case, then it is possible that cytokinins produced in the roots are directed away from the leaves in the presence of developing seeds, with the result that the normal maintenance of the RNA apparatus of the leaf

is not possible and hence protein synthesis is prevented, as in a detached leaf. One observation which is against this latter hypothesis is the fact that the senescence of attached leaves cannot normally be arrested by application of kinetin, suggesting that they are not deficient in cytokinin and hence their senescence must be due to some other cause. Thus, it would seem possible that the causes of senescence in a detached leaf are not the same as those of an attached leaf which undergoes sequential senescence or in response to fruiting.

It will be clear that although a number of interesting approaches to the problem of senescence in plants are being made, it is too early to be able to present a single overall hypothesis which will account for all the facts. However in general one can recognize that the initiation of senescence involves an imbalance in the relative levels of growth hormones, and that this change in hormonal status may be caused by either an environmental stimulus (such as short days) or by internal factors such as competiton between spatially separated organs of the plant. Once set in motion, senescence involves autophagy and eventually autolysis.

12.4 SYNCHRONOUS LEAF SENESCENCE

The synchronous senescence of leaves seen in deciduous woody plants in the autumn or "fall" is so striking that is has given its name to this season of the year, in America at least. This type of leaf senescence differs rather remarkably from sequential senescence in two respects. Firstly, it is primarily controlled by environmental rather than "internal" factors, such as competition between young and old leaves. Secondly, it appears to involve rather different hormonal factors.

Two environmental factors appear to be involved in determining the onset of senescence in deciduous woody plants, namely, daylength and temperature. It has long been known that short days tend to promote leaf senescence in woody plants such as *Liridodendron tulipifera* and *Ailanthus altissima.* Moreover, there have frequently been reports of delayed leaf fall in trees growing near street lights, so that they were exposed to long days as the natural daylength shortened. However, if seedlings of woody plants are rendered dormant by placing them under short-day conditions in a warm greenhouse, many species are found to retain their leaves in a green, healthy condition for several weeks at least. Thus, it seems likely that normal leaf fall is determined by shortdays in association with the low temperatures occurring in the fall, but precise experimental data on this matter are still lacking.

Although environmental factors are very important in controlling synchronous leaf senescence, influences from the rest of the plant are evidently also operating, since if a disc is nearly cut out of a cherry leaf by means of a cork borer, leaving only a small connection with the rest of the lamina, this disc remains green long after the remainder of the leaf has become senescent. Thus, it would seem that synchronous, as well as sequential senescence, depends upon the export of materials from the leaf.

We have seen that the senescence of excised leaves or leaf discs of herbaceous plants can be delayed by kinetin, and sometimes by gibberellin, but not by auxins. In woody plants, however, the reverse appears to be true. Thus, if a drop of the auxin, 2,4-D, is applied to cherry leaves in the fall, the areas receiving the auxin remain green long after the remainder of the leaf has become yellow. Similar results may be obtained with detached leaves of woody plants at all times of the year. Gibberellin will delay senescence of leaves of ash (*Fraxinus excelsior*). Just as application of kinetin and gibberellin enables the leaves of herbaceous plants to maintain their capacity for RNA and protein synthesis, so does 2,4-D for leaves of woody plants. Thus, it would seem that endogenous auxins probably play an important role in the natural senescence of leaves of trees. It is very likely that the influence of daylength on leaf fall is exerted through its effect on endogenous hormone levels, since there are lower levels of endogenous auxins and gibberellins in the leaves of woody plants under short days than under long days (Chapter 11).

We know little about factors controlling leaf senescence in evergreen broad-leaved trees and conifers. Some evergreen species may retain their leaves for several years. In some cases senescence of the older leaves occurs when a new suite of leaves is put out, suggesting that competition between old and new leaves may be important in such species.

12.5 HORMONES AND ABSCISSION

Shedding of both vegetative and reproductive organs and tissues is a naturally occurring process in all plants. Parts of plants are lost either by a process of death and withering, as in most monocotyledons and in plants such as potato with subterranean organs of perennation, or by the formation of abscission layers. The organ or tissue lying distal to the abscission layer falls to the ground by its own weight or by the action of an external force such as wind.

Abscission, or separation, processes are responsible for the shedding of tissues such as those of the bark of trees and the periderm in roots. However, because the loss of such tissues is usually continuous but slow, the phenomenon does not lend itself readily to experimental studies. In contrast, the abscission of leaves, flowers and fruits tends to be much more rapid and dramatic, and is often of importance in agriculture and horticulture. These facts have resulted in very much more attention being devoted to gaining an understanding of how abscission is achieved and regulated in such organs.

12.5.1 Leaf abscission

In most plant species there comes a time during the life of each leaf when it is shed from the stem. This occurs most obviously at the end of the growing season in temperate regions of the world, but leaf fall is not confined to autumn. All through the summer in temperate areas, and all through the year in the tropics, there is a less conspicuous but continuous dropping of older leaves from

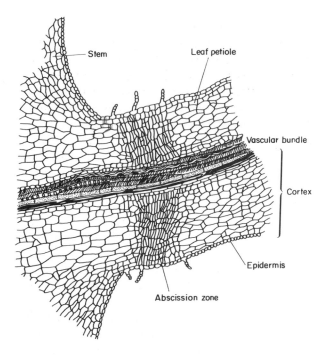

Fig. 12.12. Longitudinal section through the abscission region of the base of the petiole of a leaf of a typical dicotyledonous plant. (From J. Torrey, *Development in Flowering Plants,* Macmillan, New York, 1967.)

plants. The process by which leaves (and also other organs such as fruits and flowers) are removed from the plant is known as *abscission.* The act of abscission of an organ, such as a leaf, is usually achieved by the formation of a *separation* or *abscission layer* at the base of the petiole. This is a thin plate of cells orientated at right angles to the axis of the petiole (Fig. 12.12). The walls of the separation layer cells become softened and gelatinous, so forming a weak region, which readily breaks under the strain of wind-induced movement of the leaf.

It is a general feature of abscission that, before the shedding process occurs, the organ that is to be discarded undergoes senescence. Senescence in a plant organ, such as a leaf, normally commences in its distal parts and spreads to the proximal regions. The last cells of the organ to undergo senescence are those situated immediately adjacent to the abscission zone, and at this stage the separation point is often clearly visible as a yellow/green junction. This junction therefore represents an interface between two physiologically dissimilar tissues, and it is the development of this interface at the abscission zone which signals the start of the abscission process.

To begin to understand the process of abscission it is necessary to identify the particular characteristic of senescent tissue which initiates the biochemical events of abscission. Over the past 10 years it has become increasingly clear from research that it is the relatively high levels of ethylene produced by the senescent cells distal to the abscission zone which is the important regulatory factor. The ethylene generated in the senescing tissues initiates the separation process in the abscission zone cells. In many plant species the whole of a senescent leaf produces large amounts of ethylene, whereas in others only senescing tissues of the petiole show a rise in ethylene production. In either case, however, the effect is to increase the ethylene concentration at the site of the abscission zone in the petiole.

Abscission of leaves can be artifically induced by exposing plants to ethylene at a concentration as low as $0.1 \ \mu l \ l^{-1}$, but the leaves of different ages vary considerably in their sensitivity to the gas. The oldest leaves are the first to abscind, with the remainder being shed sequentially so that the youngest are the last to fall. This sequential pattern of abscission response to exogenous ethylene appears to be related to variations in auxin levels in leaves, but the mechanisms by which auxin affects senescence and ethylene-sensitivity are not yet understood.

Many of the experiments conducted on leaf abscission have involved the use of isolated excised abscission zones (usually a nodal explant bearing petioles of each of which the lamina has been removed). With such an experimental system it has been found that application of an auxin, such as IAA, to the cut end of a petiole immediately after removal of the lamina will delay abscission, but that if the auxin application is delayed for some hours after lamina excision, then abscission may be accelerated rather than retarded. Because of this, it has been suggested that one can distinguish between phases in the abscission process —an initial "Stage 1" which is inhibited by auxin, and a "Stage 2" which may be promoted by auxin. It is likely that the explanation for the occurrence of these two stages is that during Stage 1 little or no senescence has occurred and applied auxin inhibits the senescence processes, but that by the time Stage 2 has been reached senescence is already occurring in the petiole of the explant and auxin is not able to bring it to a halt. As we have already considered, senescing tissues produce relatively large amounts of ethylene, and it is also known that auxin can promote ethylene synthesis. Thus, nodal explants in the senescent Stage 2 condition respond to added auxin by producing even more ethylene, and the abscission zone cells are already in an ethylene-sensitive condition, so that the result is an acceleration of the abscission process.

In summary, our present understanding of the regulation of leaf abscission envisages the following sequence of events: (a) the basipetal development of senescence of the lamina in response to an environmental stimulus or some internal "signal" (b) the development of sensitivity to ethylene in the cells of the abscission zone, (c) a rise in ethylene levels in senescent cells, particularly those immediately distal to the abscission zone, to a critical level of approximately 1 nl g^{-1} hr^{-1}, and (d) a series of biochemical and physiological responses to this ethylene in the cells of the abscission zone and in the nonsenescent cells lying immediately below the zone, which culminate in cell separation in the abscission layer. We consider below the cellular events which lead to cell separation, but before doing so it is appropriate to raise the question of what it is which controls the rate of production of ethylene in senescing cells distal to the abscission zone.

Evidence exists that during senescence of leaf cells there is release of a *non-volatile* substance which accelerates abscission. This has been termed the "senescence factor", or "SF", by Osborne. It has been suggested that abscisic acid (ABA) may be the SF, since applications of ABA can cause leaf senescence and abscission in a number of plant species. However, several lines of evidence argue against identifying SF as ABA. Thus, changes in endogenous ABA levels in leaves do not correlate well with time of leaf abscission, and it has been found that although exogenous ABA appears to affect directly the rate of ethylene production in some species, in others it appears to act by accelerating senescence directly but only indirectly inducing an increase in ethylene synthesis and rate of abscission. The SF, in contrast, is known to both accelerate abscission and have an *immediate* stimulatory effect on ethylene production. It would appear, therefore, that SF is a regulator of ethylene biosynthesis. The SF is present in both young healthy leaves as well as in old senescing leaves. However, SF is in a non-water soluble form in young leaves (but is extractable with organic solvents) but is water-soluble in senescing leaves. The significance of these observations on SF is not at all clear at the present time, but it is tempting to speculate that the release of SF in leaf cells detected by Osborne may be the same phenomenon as the release of homocysteine from a cellular compartment in senescing flowers of *Ipomoea tricolor* that results in increased ethylene synthesis (p. 69).

The act of leaf separation itself, which is the response induced by ethylene at an appropriate concentration, consists of two distinct processes: (a) enhanced cell growth in the region immediately proximal to (below) the separation layer, and (b) the synthesis and secretion of cell-wall-digesting enzymes by the same cells, which degrade the walls of the senescent cells above and lead to the separation of the two tissues. Growth of the cells proximal to the separation point is marked by rises in their rates of RNA and protein synthesis. A significant effect of this growth is to increase the physical pressure on the cells of the abscission layer, and also to provide the basis for the formation of a wound, or scar, tissue to seal the rupture surface after the leaf has been shed. The wall-degrading enzymes synthesized and secreted by the growing proximal cells consist of a range of appropriate polysaccharide hydrolases, including endopolygalacturonase and cellulase (ß-1,4-glucanase), whose effects would be expected to be the weakening of cell to cell cohesion. The effect of ethylene on these enzymes is to enhance both their rates of synthesis and secretion through the

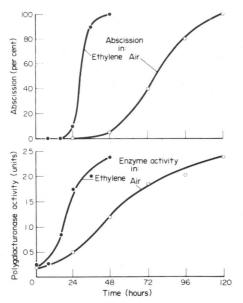

Fig. 12.13. A comparison of the time-courses of development of polygalacturonase activity and occurrence of abscission in *Citrus sinensis* leaf explants incubated in either air or in air containing 10 ppm ethylene. (Adapted from J. Riov, *Plant Physiol.* **53**, 312-16, 1974.)

plasmalemma of the proximal cells (Fig. 12.13). It is intriguing to note that the enzymes appear to be secreted only, or mainly, from one side of the layer of proximal cells; that which abuts the thin layer of abscission cells.

Once the walls of the abscission zone cells have been sufficiently weakened by the action of polysaccharide hydrolases, the final act of breakage is probably purely mechanical, aided not only by external forces such as wind action, but also by the cell enlargement on the proximal side of the separation layer and dehydration of the senescent tissues on the other, which together provide shearing forces in the abscission layer.

12.5.2 Fruit abscission

Abscission phenomena are also shown by flowers and fruits. Thus, an abscission zone is commonly found at the base of the pedicel of the flowers in many species, if pollination and fertilization fail to take place. Similarly, even when successful fertilization has occurred, an abscission zone, leading to fruit-drop, may be formed at various stages during the development of the fruit.

This is well seen in certain varieties of apple, in which there may be three peak periods of fruit drop: (1) immediately after pollination ("post-blossom" drop), (2) soon after growth of the young fruits ("June drop"), and (3) during ripening ("pre-harvest drop").

For some species it has been shown that the periods of fruit-drop coincide with periods of low auxin content in the fruit and conversely the times of low drop rate occur when the auxin content is high. This situation is found in the blackcurrant, in which we have already seen that it is possible to correlate growth rate with the levels of two auxins, one acidic and one neutral (Fig. 5.19). It has been shown that there is a second acidic auxin the levels of which show a different pattern of variation which correlates inversely with the rate of fruit abscission (Fig. 12.14). This situation is analogous to that in leaves, therefore, where the formation of an abscission layer is associated with diminished auxin levels in the lamina.

On the other hand, in other species it has not been possible to correlate the variations in rate of fruit-drop with auxin levels and it seems likely that in such cases other factors play a role in determining fruit-drop. Indeed, there is little doubt that ethylene is involved in the control of fruit, as well as leaf, abscission, for this substance promotes fruit senescence and abscission.

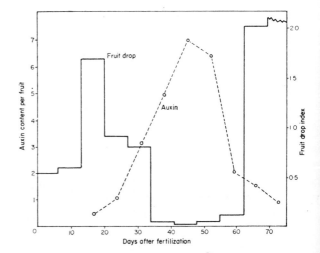

Fig. 12.14. Changes in the content of an unknown acidic auxin in the blackcurrant berry, in relation to the rate of fruit drop. (From S. T. C. Wright, *J. Hort. Sci.* **31**, 196, 1956.)

FURTHER READING

General

Abeles, F. B. *Ethylene in Plant Biology,* Academic Press, New York and London, 1973.

Leopold, A. C. and P. E. Kriedemann. *Plant Growth and Development,* 2nd ed., McGraw-Hill, New Yor, 1975.

Pitt, D. *Lysosomes and Cell Function,* Longman, London, 1975.

Woolhouse, H. W. *Ageing Processes in Higher Plants,* Oxford Biology Readers, Oxford University Press, Oxford, 1972.

Thimann, K. V. (Ed.). *Senescence in Plants,* Vols. I & II, C.R.C. Press, U.S.A., 1980.

More Advanced Reading

Articles by H. W. Woolhouse, E. W. Simon, R. Woolgiehn, D. J. Osborne, P. F. Wareing and A. K. Seth, and D. J. Carr and J. S. Pate. In: *Aspects of the Biology of Ageing, Symp. Soc. Exp. Biol.* **21,** 1967.

Beevers, L. Senescence. In: *Plant Biochemistry,* 3rd ed. (ed. J. Bonner and J. E. Varner), pp. 771-794, Academic Press, New York, 1976.

Carns, H. R. Abscission and its control, *Ann. Rev. Plant Physiol.* **17,** 295, 1966.

Hanson, A. D. and H. Kende. Methionine metabolism and ethylene biosynthesis in senescent flower tissue of Morning-glory, *Plant Physiol.* **57,** 528-37, 1976.

Kende, H. and A. D. Hanson. Relationship between ethylene evolution and senescence in Morning-glory flower tissue, *Plant Physiol.* **57,** 523-7, 1976.

Matile, Ph. *The Lytic Compartment of Plant Cells,* Springer-Verlag, Wien and New York, 1975.

Sacher, J. A. Senescence and postharvest physiology, *Ann. Rev. Plant Physiol.* **24,** 197-224, 1973.

Thomas, H. and J. L. Stoddart. Leaf senescence, *Annual Rev. Plant Physiol.* **31,** 83-111, 1980.

Various articles by: T. T. Kozlowski; B. D. Webster, F. T. Addicott and J. L. Lyon; D. J. Osborne; W. F. Millington and W. R. Chaney; G. A. Borger; and G. B. Sweet, In: T. T. Kozlowski (ed.), *Shedding of Plant Parts,* Academic Press, New York and London, 1973.

SECTION IV

Molecular and General Aspects of Differentiation

Introduction

We have now considered plant development from several rather different standpoints. In the first two chapters we considered development from the viewpoint of the experimental anatomist and morphologist. Then we examined the action of growth hormones and discussed their roles in various aspects of growth and development. Finally we considered the physiology of flowering, senescence and dormancy, bringing in the effects of external factors, such as daylength, and internal factors, especially hormones. In the present chapter we shall examine the molecular and more general aspects of development.

A brief summary of the present state of knowledge regarding the structure of the eukaryote genome and of the procesess involved in the control of gene expression is presented for the benefit of readers who are not familiar with this subject, and this information forms the basis for further discussion of other general aspects of differentiation, especially the importance of cell determination in plant development.

13

Gene Expression and Cell Determination in Development

13.1 GENES AND DEVELOPMENT

So far we have paid little attention to the genetical aspects of development, and yet every organism is, by definition, the product of the interaction between its genetic potentialities and the environment, and in the final analysis development has to be described in terms of activities of genes.

The fact that developmental processes are basically gene-controlled is self-evident, since genetic variations are known which affect almost every aspect of development, ranging from external morphology (leaf and fruit shape, flower colour, etc.) and internal anatomy to physiological characters, such as growth rate, flowering time and length of dormancy period. Thus, there seems no doubt that the pattern of development followed by any individual is determined primarily by the "programme" laid down in its genetic code.

During development there is progressive differentiation of organs and tissues, giving rise to a wide range of different types of cell. Not all the genes of the total gene complement are expressed all the time and in all parts of the plant, however. Thus, the genes controlling flower development are not normally expressed in the embryo, nor during the purely vegetative phase of development. However, we know that the cells of a vegetative organ such as a leaf contain the genes required for flower development, since a complete new flower-bearing plant may be regenerated from leaf cells in certain species. Since, therefore, differentiation in plants apparently does not entail any genetic (i.e. inheritable) differences between the nuclei of various types of cell and tissue, it must involve differences in *gene expression* in different parts and at different stages of the life cycle.

Since development is such an orderly process it must require that the right genes are expressed in the right cells at the right time. That is to say, development is essentially a process involving selective gene expression, and the concept that development involves the activity of specific groups of genes, which in turn control the synthesis of the enzymes and other proteins characteristic of specialized cells, is called the *variable gene expression theory* of differentiation. There must be some means of regulating the expression of specific genes during cell differentiation and, since development is a very orderly process, there must be some mechanism for determining the sequence of gene expression. Therefore we have to consider how selective gene expression may be achieved.

Gene expression involves (1) transcription of the DNA sequences to form nuclear RNA, (2) processing of nuclear RNA to form messenger RNA, (3) translation of the information in the m-RNA into amino acid sequences in the polypeptide chain of enzyme and structural proteins, (4) activity of the enzymes in various reactions to give the final products of gene expression. The essential features of these processes are too well known to require detailed elaboration here. Before we can discuss the control of selective gene expression in develop-

ment, however, we need to describe the structure and organization of the genome in Eukaryotes.

13.2 THE COMPOSITION OF THE EUKARYOTE GENOME

The genome of a prokaryote, such as the bacterium *Escherischia coli,* consists of a single chromosome, which is a circular DNA double helix lying free in the cytoplasm; during cell division the two double stranded DNA chromosomes, formed as a result of replication, are distributed to the two daughter cells without mitosis.

By contrast, eukaryote genomes are very much more complex and involve a vastly greater amount of DNA. Moreover, the plastids and mitochondria in plant cells also contain their own complements of DNA (p. 310). It is estimated that the total length of the nuclear DNA in a single cell of *Vicia faba* amounts to about 9 m, so that if it replicated as a continuous chain the distribution of the two replicated genomes during cell division would present insuperable mechanical difficulties. This problem has been met by packaging the genome into a number of separate chromosomes in which the DNA chains are greatly compacted by coiling and "supercoiling". Moreover, in the resting cell the genome does not lie free in the cytoplasm, as in prokaryotes, but is enclosed in a nucleus which is separated from the cytoplasm by a nuclear membrane. Communication between the nucleus and the cytoplasm is maintained through pores in the nuclear membrane. During the prophase of mitosis or meiosis the DNA becomes greatly condensed into tight coils—indeed, coiling can be detected at various levels of organization, from the molecular level, to the supercoils visible in the light microscope.

Relatively pure preparations of interphase chromosomal material can be made and is referred to as "chromatin". Analysis of chromosomes or chromatin (p. 312) shows that they consist not only of DNA, but also of a large number of proteins, which constitute 70 per cent of the total material present, together with varying amounts of RNA.

The proteins present in chromatin fall into two main classes: (1) basic proteins, known as histones and (2) acidic proteins.

The histones are rich in the basic amino acids lysine and arginine and show little heterogeneity, there being only five main classes, which vary in their contents of lysine and arginine. There is very little variation between different species, in the structure of these histones, suggesting that they play a very central role, and have been conserved during evolution.

The amount of histone in chromosomes is approximately equal to the amount of DNA, to which the histones are bound by electrostatic attraction between the negatively-charged phosphate groups of the DNA and the positively charged basic amino acids of the histones. It appears that the presence of histones bound to the DNA causes supercoiling brought about by interactions, including hydrogen bonding and electrostatic repulsion and attraction, which takes place between the amino acids in histone molecules.

The non-histone, acidic proteins present in chromatin vary greatly in size and composition, so that hundreds of different proteins can be isolated. By contrast with the histones, the acidic proteins from different animal tissues are variable and show tissue-specific binding to DNA. Many of the acidic proteins are phosphorylated, and the degree of phosphorylation varies during the cell cycle and also during the development of animals. Phosphorylation may result in conformational changes which in turn may affect the binding of the acidic proteins to DNA or histones. Phosphorylation is brought about by enzymes known as protein kinases, of which many different kinds occur in nuclei.

The acidic proteins also include a number of DNA and RNA polymerases. In general, tissues or organs which exhibit high metabolic activity are characterized by a high content of acidic chromatin proteins, whereas metabolically inactive tissues contain lower amounts. As we shall see later, the acidic proteins of chromatin may play a role in the regulation of gene expression.

It has been shown that the DNA-protein complexes are organized in a series of discrete spherical particles, the "nucleosomes", like beads on a necklace (Fig. 13.1). Nucleosomes contain a histone core consisting of two molecules of each of 4 different type of histone, and a length of DNA containing 150-250 nucleotide pairs. The DNA is located on the surface of the nucleosomes. Small lengths of "linker" DNA associated with the fifth histone connect one nucleosome with the next.

Fig. 13.1. A model for the structure of the nucleosome, in which DNA is wrapped around the surface of a flat protein cylinder consisting of histones. (From R. A. Laskey and W. C. Earnshaw, *Nature*, 286, 763, 1980.)

13.3 THE ORGANIZATION OF THE EUKARYOTE GENOME

In bacteria, such as *E. coli*, the chromosome contains 10^6 nucleotide pairs, corresponding to about 1000 genes, whereas the nuclei of plants and animals contain vastly greater amounts of DNA. For example, the diploid nucleus of *Vicia faba* contains 29×10^{-12}g of DNA. One $\times 10^{-12}$g of DNA corresponds to 1×10^9 nucleotide pairs, and is sufficient to code for $8\cdot5 \times 10^5$ proteins each of molecular weight 50,000. Therefore the nuclear DNA content of *Vicia faba* is sufficient to code for about 25×10^6 genes!

On the other hand, genetical studies on both plants and animals, and studies on the salivary gland chromosomes of *Drosophila*, in which each band probably corresponds to a single gene, indicate that the number of functional genes coding for proteins does not exceed more than a few thousand. Another strange phenomenon is that various closely related plant species differ enormously in their DNA content. Thus, *Vicia faba*

306 Growth and Differentiation in Plants

contains 7 times more DNA per cell than the closely related species, *V. sativa,* although there is no
evident reason why the former species should require many more genes.

The clue to these paradoxes has come from
studies on the renaturation behaviour of prokaryote and eukaryote DNAs. The two strands of
the DNA double helix can be separated by heating
or by treatment with alkali and can be re-natured,
i.e. allowed to reform the double-stranded condition under suitable conditions of pH and
temperature. Only complementary single strands
will pair and renature, and the rate of renaturation
is dependent upon the concentration of the complementary sequences, the greater the concentration of a sequence, the more frequent being the
collisions and hence the faster the rate of renaturation. Since single-stranded DNA has a higher
absorption coefficient for ultra violet light than
double-stranded DNA, the rate of renaturation
can be followed by recording the decrease in absorbance at 260 nm in a spectrophotometer. The
rate at which the DNA renatures is plotted as a
percentage against concentration × time expressed in moles of nucleotides × seconds per litre
(COT).

When this procedure is carried out with
bacterial DNA, the COT plot gives a smooth curve
(Fig. 13.2). DNA from eukaryotes, however,
shows a different pattern. At low DNA concentrations and short incubation times, a large proportion of single stranded DNA renatures, and as the
COT value increases, additional double-stranded
molecules are formed to produce a biphasic curve.
The rapid renaturation at low COT values shows
that some of the base sequences in eukaryotes are
highly repetitive, i.e. with sequences repeated
10,000 times or more.

Studies of this sort have shown that eukaryote
genomes contain various types of DNA sequences,
with different degrees of repetition of particular
nucleotide sequences. One type ("highly
repetitive") is made up of rapidly renaturing DNA
sequences, which are present as 10^5-10^6 copies;
another is an *intermediate* repetitive class with
1,00 to 10^5 copies; there is a *moderately repetitive*
class of 20-50 copies, and a single copy (*non-*

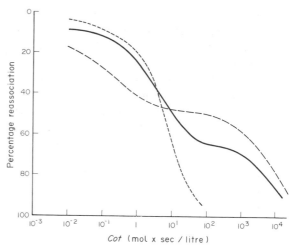

Fig. 13.2. COT reassociation curves for the DNA of two
eukaryotes, *Lathyrus sativus* (—) and the calf (– –), and of the
prokaryote *E. coli* (----). (From H. Rees and R. N. Jones,
Chromosome Genetics, Edward Arnold, London, 1977.)

repetitive) class where there are only 1-2 copies.
Thus, a high proportion of the DNA of eukaryote
genomes is represented by multiple copies of
certain sequences, and much of the variation in
amounts of DNA between species can be
attributed to varying amounts of repetitive DNA.

Some types of repetitive sequence are found to
be "clustered" within specific parts of the
genome. For example, the most highly repetitive
DNA sequences are the so-called "satellite"
DNAs that are obtained as a distinct band by density gradient centrifugation of total DNA. This
satellite DNA consists of more than 10^6 copies of a
short, simple nucleotide sequence. The satellite
DNA can be used as template to make radioactive
complementary RNA and when such a copy of
mouse satellite DNA is applied to preparations of
mouse chromosomes it is found that the centromere region of each chromosome has become
labelled, suggesting that satellite DNA may be
related to the functioning of the centromeres.
However, similar experiments carried out with
plant satellite DNAs have led to labelling at different positions on the chromosomes, so it would
appear that the function of this class of repetitive
DNA is not necessarily related specifically to the
centromere. Thus the function of satellite DNA is

still not known but it does not code for any proteins in the cell. It has been suggested that it may have a structural role or may be involved in chromosome pairing.

Another class of repetitive DNA which is found in a clustered arrangement is constituted by the genes coding for ribosomal RNA, which comprise up to 90 per cent of the total RNA of the cell. The nucleolus is the principal site of ribosome synthesis and using a similar technique to that described for satellite DNA it has been shown that most of the genes coding for ribosomal RNA are localized in the nucleolus in clusters of several thousand copies in plant cells.

There is also evidence for repetition of genes coding for histones, which are present in 400-1200 copies in the sea-urchin egg, again in a clustered arrangement. Thus, the sections of the genome concerned with ribosome and chromosome structure appear to occur in highly reiterated sequences.

There is another group of repeated sequences which are apparently dispersed, and not clustered,throughout the genome. This has been shown by the results of experiments in which the extent of renaturation has been measured as a function of the size of the DNA fragments used; such studies show that most of these repetitive sequences are distributed throughout the genome and are apparently interspersed between non-repetitive (single-copy) DNA. It is generally held that the single copy DNA sequences include both structural genes, coding for proteins, and other genes of unknown function. Studies of this type with the tobacco plant indicate that: (1) the majority of repetitive sequences are short, averaging 300 nucleotide pairs; (2) each of the repeated sequences is separated by a unique DNA sequence 1400 nucleotides in length, but with some 4000 nucleotides long. Variations in the sizes of the repeats and unique DNA sequences exist in different animals and plants, some examples of which are shown in Fig. 13.3.

The development of elegant techniques for "cloning" DNA, to yield large amounts of reproducible, specific DNA fragments (Fig. 13.4) has recently opened up the whole field of genome

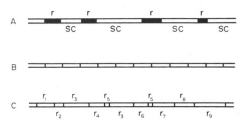

Fig. 13.3. Three different arrangements of repeated DNA common in higher plant chromosomes. A, short pieces of repeated DNA (r) 50-200 base pairs long are interspersed with single copy DNA (Sc) 200-4000 base pairs long. B, essentially identical repeating units one in tandem arrays. C, unrelated short repeats (r_1 to r_9) are interspersed with each other in different permutations. (From R. Flavell, 1980.)

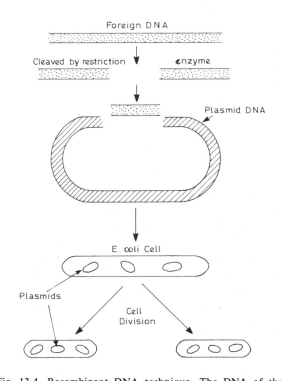

Fig. 13.4. Recombinant DNA technique. The DNA of the "foreign" organism is first treated with restriction enzymes, which cleave the double-stranded molecule at particular nucleotide sequences on a random basis. The same enzyme is then used to cleave the DNA of a plasmid isolated from *E. coli* bacteria. Under suitable conditions the ends of a sequence from the foreign DNA can be coupled with a sequence from the plasmid to restore the circular form of the plasmid; this can then be inserted into a cell of *E. coli,* where it is replicated. In this way a large number of "cloned" copies of the inserted nucleotide sequence from the foreign DNA can be obtained. (Adapted from C. Grobstein, *Sci. Amer.* 237 (1), 22, 1977.)

structure, organization and function. Cleavage of double-stranded DNA with one of a large number of "restriction" enzymes, nucleases which recognize and cut at very specific short nucleotide sequences (4 to 6 nucleotide pairs), produces highly reproducible fragments of DNA, the ends of the two DNA strands usually being offset due to the specificity of the cutting site in a double-stranded molecule whose strands are complementary in base composition. If plasmid DNA (self-replicating, extrachromosomal double-stranded units commonly present in bacterial cells) are cleaved by the same restriction enzyme, then the ends generated on the DNA and the plasmid are identical such that when the DNA and plasmid fragments are allowed to rejoin, DNA fragments will be inserted into the plasmid sequence at a low frequency. The DNA is usually inserted into a plasmid gene conferring some selection property, such as an antibiotic resistance gene which allows bacteria containing the plasmid to grow in the presence of the antibiotic.

The ligated (rejoined) plasmids are returned to bacterial cells, and bacteria containing those plasmids in which the inserted DNA has inactivated the "selection" gene are selected from those containing normal plasmids, or those containing no plasmid, by their growth properties on a range of antibiotic-containing media. Many copies of the plasmid are replicated in the bacteria, so that relatively large amounts of the inserted DNA fragment may be "grown" and reisolated simply by cleavage with the initial restriction enzyme, and separation of the cleavage products by gel electrophoresis. The use of this recombinant DNA technology has revolutionized study on the genes.

13.4 THE PROCESSING OF RNA TRANSCRIPTS

As we have already seen (p. 307) in most cells the nucleolus is the principal organelle for the synthesis of ribosomes and the genes coding for ribosomal RNA (rRNA) occur in the nucleolus organizer.

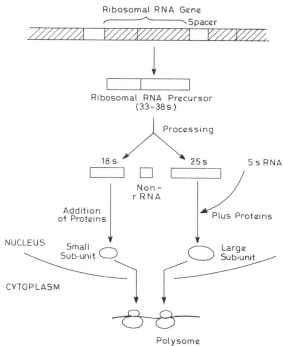

Fig. 13.5. Scheme to illustrate the synthesis and processing of ribosomal RNA precursor to form ribosomes.

Ribosomal RNA includes two major molecules differing in size, known as 25S and 18S in plants (28s and 18s in animals), together with a smaller 5S unit. (The size of a macromolecule is expressed in sedimentation units (S), and is determined by measuring its rate of sedimentation during centrifugation at high speed.) Although the 25S and 18S RNA are coded by separate genes, it appears that these are closely linked, since the first step in ribosomal RNA synthesis is the formation of large (33 to 38S) RNA precursors which contain the 18S and 25S sequences (Fig. 13.5). The large precursor molecules never leave the nucleus but are converted to smaller units by special processing enzymes. The 18S and 25S sections of the DNA are arranged in a sequence separated by a small section of DNA that is transcribed into the precursor, but then removed during processing, and there are also longer sequences of DNA associated with the genes which are never transcribed as an RNA copy, known as spacer-DNA.

The 5S RNA which forms part of the larger subunit of the ribosome appears to be transcribed separately: there are several genes coding for 5S RNA but these do not appear to be located in the nucleolus.

Messenger RNA (mRNA) is also formed as a result of the transcription of DNA in the nucleus. However, the RNA units which are initially formed in the nucleus are much larger than the forms of mRNA which pass from the nucleus into the cytoplasm, and hence it is apparent that, as with ribosomal RNA precursor, the large initial transcripts must be "processed" before they can form the messengers. The initial large transcripts vary greatly in size (i.e. in the number of nucleotides they contain) and hence they are referred to as *heterogeneous* nuclear RNA (hnRNA).

Recent discoveries have revealed surprising facts about the processing of hnRNA in animal tissues. Analyses of eukaryotic genes have shown that the coding sequences in DNA are not continuous but are interrupted by one or more lengths of "silent" DNA, which do not code for proteins; further studies have shown that these silent sequences are excised during the processing, and the remaining sections are "spliced" together to form mRNA. Thus, heterogeneous nuclear RNA consists of the long transcriptional products out of which the much smaller ultimate messengers are spliced.

It is too early to apply these new concepts to the problem of the regulation of gene expression in eukaryotes, but it is clear that there must be a very accurate enzymatic mechanism for cutting out the excised segments of hnRNA; the processing of ribosomal RNA must involve a similar mechanism. The splicing of the remaining fragments to form mRNA must involve "ligase" enzymes.

13.5 MESSENGER RNA

In the past, the synthesis of mRNA has proved very difficult to investigate due to problems of identification and purification of mRNA. Moreover, specific mRNAs are frequently present in very low amounts in the cell. However, a number of advances in technique have now over-come many of these difficulties and our knowledge of mRNA is now advancing rapidly. Firstly, the development of an *in vitro* protein synthesizing system has made it possible to test the capacity of an RNA fraction to direct the synthesis of specific proteins. Secondly, the discovery that many RNAs contain sequences of polyadenylic acid (poly A) has provided a means of purifying that fraction of mRNA.

Plant cells normally produce a large number of different proteins, so that there is a corresponding large number of mRNAs, each of which is present in relatively small amounts. However, where a cell produces a large amount of a single protein, as in the case of the leghaemoglobin produced in soybean root nodules, large amounts of the corresponding mRNA are present and can be isolated. This has been done for leghaemoglobin and the purified mRNA has been used in a cell-free system from wheat germ to produce *in vitro* synthesis of a protein fraction, which has been identified as leghaemoglobin. Similar results have been achieved with the mRNA of the small subunit of the enzyme ribulose-1,5-biphosphate carboxylase and phenylalanine ammonia lyase.

Recombinant DNA technology is also applicable to the study of RNA molecules, since it is possible to synthesize a single-stranded DNA copy (cDNA) of the mRNA with the enzyme reverse transcriptase and then to use *E. coli* DNA polymerase to form the second complementary DNA strand, so yielding double stranded DNA which can be inserted into plasmids and cloned as previously described (p. 308). In this way it has been possible to establish the nucleotide sequence of a number of specific eukaryote mRNAs, and such "probes" of DNA complementary to specific mRNAs greatly facilitate identification and subsequent isolation of the particular gene from the genome.

In the tobacco plant, approximately 60 per cent of mRNA mass is accounted for by a relatively few specific types, represented by several thousand copies per cell. However, the remaining 40 per cent of the mRNA mass is constituted by species which are represented by only a few copies per cell ("rare class mRNA") and the number of different species is very large—for example, in the tobacco leaf the number of specific mRNAs is estimated to be about 27,000 and at least 60,000 diverse structural genes are estimated to be expressed in the whole tobacco plant during its life cycle!

13.6 ORGANELLE DNA

Mitochondria and chloroplasts are apparently not formed *de novo,* but are formed from pre-existing organelles and are transmitted from one cell generation to the next by partition to the two daughter cells following mitosis.

Carefully purified preparations of chloroplasts and mitochondria have been shown to contain DNA and, indeed, strands of DNA can be seen in sections of these organelles under the electron microscope. Organelle DNA has been shown to differ from nuclear DNA in base composition, and in a number of cases it has been shown to be circular. It would appear, therefore, that chloroplasts and mitochondria have their own genomes which are independent of the nuclear genome; indeed, isolated chloroplasts and mitochondria can incorporate radioactive amino acids into proteins, indicating that they contain the necessary apparatus for protein synthesis. However, the organelle genomes are quite limited in size and a considerable part of each genome consists of templates for the production of ribosomal and transfer RNA.

Organelle ribosomes are smaller than cytoplasmic ribosomes and are different from the latter in their sensitivity to drugs such as puromycin, cycloheximide and chloramophenicol. Organelles code for their own ribosomal RNA, but the proteins of organelle ribosomes are apparently made on cytoplasmic ribosomes and are therefore probably not coded for by the organelle genome. DNA/RNA hybridigation studies indicate that at least some of the organelle RNAs are coded for by the chloroplast and mitochondrial DNA.

The chloroplast DNA codes for certain proteins, such as the large sub-unit of the photosyn-thetic enzyme, ribulose bis-phosphate carboxylase. However, although some organelle components are produced internally, the majority of proteins, including DNA and RNA polymerases, present in them must come from the cytoplasm.

13.7 THE CONTROL OF GENE EXPRESSION IN BACTERIA

Some enzymes of bacteria are present all the time and are said to be constitutive. On the other hand, other enzymes are only formed when their substrate is present in the external medium. For example, when the bacterium, *Escherichia coli,* is grown in the absence of a galactoside (i.e. a compound containing the sugar, galactose, linked to another, non-sugar molecule), only traces of the enzyme ß-galactosidase are formed, but as soon as a galactoside is added, the rate of synthesis of this enzyme increases enormously. This type of enzyme is said to be *inducible.* Removal of the substrate results in the almost immediate cessation of enzyme synthesis.

Contrasting with this process of enzyme induction is the *repression* of enzyme synthesis seen in other enzyme systems. For example, when *E. coli* is grown in the absence of the amino acid histidine, the enzymes involved in the production of histidine are actively synthesized. As soon as histidine is added to the medium the enzymes cease to be synthesized. In this instance, therefore, there is *repression* of enzyme synthesis. This phenomenon is called "end-product repression", since the product of the sequence of reactions, in this case histidine, inhibits the formation of the enzymes concerned with its own biosynthesis. The result is that if the end-product is supplied from outside, it inhibits its own synthesis, and hence the cell ceases to make any more of the compound. Such regulation of the synthesis of an enzyme will control the cellular level of enzyme provided the enzymes are normally turning over, i.e. being continuously metabolically degraded, but will not operate for stable enzymes.

The phenomenon of enzyme induction and repression led Jacob and Monod to propose a

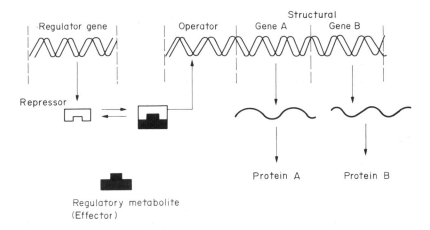

Fig. 13.6. Scheme illustrating the control of protein synthesis according to the theory of Jacob and Monod (see text).

general theory for the control of gene activity in bacteria, which we shall now describe briefly. The gene which determines the structure of a specific enzyme protein is referred to as a *structural gene*. The synthesis of m-RNA is assumed to be initiated only at certain regions or *operators* of the DNA strands (Fig. 13.6). In some instances a single operator may control the transcription of several adjacent structural genes into m-RNA. The segment of DNA thus controlled by a single operator is referred to as an "operon", which may contain one or several structural genes.

The rate of transcription of structural genes is controlled by other genes referred to as *regulator genes*. A regulator gene forms a protein known as a *repressor*. The repressor formed by a given regulator gene is assumed to have an affinity for certain specific operator genes, with which it binds. This combination blocks the production of m-RNA by the whole operon controlled by the operator, and therefore prevents the synthesis of the specific proteins controlled by the structural genes of the operon.

A repressor (R) has the property of reacting with certain small molecules, called "effectors" (F) expressed as follows:

$$R + F \rightleftharpoons RF$$

In inducible systems only the R form of the repressor is active and blocks the transcription of

the operon. The presence of an effector (or inducer) inactivates the repressor and therefore allows messenger-RNA synthesis to proceed. By contrast, in repressible systems only the combined RF form of the repressor is active. Synthesis of m-RNA by the operon, allowed in the absence of the effector (or repressing metabolite), is, therefore, prevented in its presence. These ideas are summarized in Fig. 13.6.

The repressor may be an "allosteric" protein, with two active sites, one able to react with the operator and one to react with the inducing or repressing molecule. It is supposed that when the repressor combines with an inducer its shape is deformed somewhat, so that it can no longer react with the operator, which is thereby "derepressed".

13.8 CONTROL OF GENE EXPRESSION IN EUKARYOTES

It is tempting to assume that the type of control of gene expression established in bacteria is also applicable to eukaryotes. However, there is very little evidence that operons occur in eukaryotes, in which genes affecting the same phenotypic character do not necessarily occur in close proximity and may be located in several different chromosomes.

However, genes are known in maize which appear to show some of the characters of regulator genes. Thus, a locus in maize known as "Activator" (Ac), appears to be a master locus for a second locus, "Dissociation" (Ds), which is unable to function in the absence of Ac. Ds, in turn, affects the expression of a number of other genes, and hence is analogous to an "operator" gene in bacteria, while Ac may be regarded as the "regulator". For example, under certain conditions Ds causes C, which gives coloured aleurone in the grains, to behave as if it were the colourless recessive allele, c. Thus Ds appears to "repress" the action of C. On the basis of these and other observations, McClintock has postulated that the chromosomes of higher plants contain both genes and "controllers", which regulate the action of the genes.

The study of the control of gene expression in eukaryotes is proving difficult, because of the size and complexity of the genome involving both single copy and reiterated sequences, coupled with the presence of other substances, including histones, acidic proteins and RNA, in addition to DNA, in chromatin. Consequently, the precise mechanism of control of gene expression in eukaryotes can only be a matter of speculation at present, and the widely differing theories put forward are an indication of our lack of firm knowledge of the processes involved.

However, there is good evidence from studies involving DNA-RNA hybridization that different tissues contain qualitatively different repetitive and single copy mRNA transcripts. Thus, studies with the tobacco plant have shown that there are about 25,000-30,000 different species of mRNA in each of the main types of tissue studied (leaf, stem, root, petal, ovary, anther), and of these about 8,000 species of mRNA are common to all types of tissue. The remaining diverse species of mRNA are apparently characteristic of each type of tissue, so that each tissue possesses a unique set of mRNAs which correspond to thousands of diverse structural genes.

This conclusion implies that, in any given type of tissue, only a small proportion of the total nuclear DNA is represented by mRNA in the cytoplasm, and this is generally assumed to indicate that differentiation involves selective control of transcription in the nucleus, only a small part of total DNA being transcribed, while the remainder remains inaccessible for transcription.

One hypothesis postulates that selective gene expression involves "masking" and "unmasking" of different sections of the genome. Evidence for this view comes from studies with isolated chromatin. Chromatin is prepared by disruption of the tissues by homogenization, filtering and selective sedimentation by centrifugation. Isolated chromatin is capable of synthesizing RNA when supplied with the nucleoside triphosphates of the four RNA bases, guanine, adenine, cytosine and uracil, together with RNA polymerase. The sequence complementarity of the RNA thus produced is determined by hybridization with naked DNA, from which the proteins have been removed by digestion. In this way it is possible to determine what proportion of the DNA has been transcribed and what kinds of RNA have been made.

It is generally found that only a small fraction of the total repetitive DNA is transcribed into RNA, most of the DNA apparently being inaccessible, presumably because it is bound to histones and/or acidic proteins. If all the histones and acidic proteins are removed, the remaining naked DNA shows greatly increased transcriptive activity (Fig. 13.7). When chromatin is reconstituted from naked DNA by adding histones and acidic proteins, the transcription is now reduced to that of the original native chromatin; if only histones are added, transcription is nearly completely inhibited, but when acid proteins are also added the original template activity is restored. Thus, it appears that histones have a non-specific masking effect on DNA transcription, whereas the presence of acidic proteins reduces this masking effect. It has been suggested that histones combine with DNA and cause supercoiling or compaction, so making the DNA inaccessible to RNA polymerase. Acidic proteins could protect DNA from histones thereby making it more available for transcription.

As we have seen above (p. 305), the DNA in chromatin appears to be associated with histones

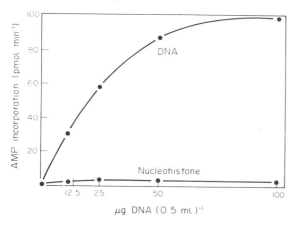

Fig. 13.7. The template activities of DNA and nucleohistone from *Pisum sativum* (garden pea). DNA and nucleohistone were used as templates in a reaction mixture containing radioactive ATP and non-radioactive CTP, GTP and UTP, together with essential co-factors, plus RNA polymerase. The incorporation of radioactivity into an acid-insoluble product was taken as a measure of the rate of RNA synthesis. (From J. Bonner, *The Molecular Biology of Development,* Clarendon Press, Oxford.)

in particles known as nucleosomes. There is some evidence that the nucleosomes of sections of the genome which are undergoing transcription are in a different state from those which are non-active, and that active nucleosomes may be "opened" or extended.

Most of the chromatin in interphase (resting nuclei) stains only lightly in response to cytological stains, but certain more heavily stained regions of the nucleus may often be recognized. This heavily staining material is known as *heterochromatin,* as distinct from the lightly staining *euchromatin.* It appears that heterochromatic regions remain contracted (coiled) in the interphase nuclei, when the remaining chromatin is uncoiled. There is good evidence that the genes in heterochromatic regions are inactive. Thus, the X or Y sex chromosomes of animals may be heterochromatic, and in such cases the genes are not expressed in somatic cells, but become active when the gametes differentiate. Thus, heterochromatin appears to result in the inactivation of large blocks of genes and the processes of chromosome contraction and uncoiling appear to provide a coarse means of regulating gene expression.

The foregoing discussion is based upon the hypothesis that selective gene expression in development may be regulated at the transcriptional level by the masking and unmasking of specific sections of the genome, i.e. by regulating template availability. However, other possible mechanisms for the control of transcription can be envisaged. Thus, eukaryotes contain a number of RNA polymerases with different properties and subcellular locations. In higher plants, the mitochondria and chloroplasts have distinct RNA polymerases and the nucleus has at least two distinct forms of these enzymes. RNA polymerases from microorganisms are complex proteins made up of several polypeptide chains, called alpha, beta and sigma chains. Sigma factors have a recognition function to initiate transcription on DNA templates. Several sigma factors have been identified in prokaryotes, each conferring a different specificity of action on alpha-beta subunits. Thus, some selective regulation of transcription of specific sections of the genome may be achieved in eukaryotes through the specificity of polymerase activity.

13.9 REGULATION BY PROCESSING OF NUCLEAR RNA

Most hypotheses for the regulation of gene expression in eukaryotes have been based upon the Jacob and Monod operon model for the regulation of gene expression in bacteria, and it is generally assumed that selective gene expression in eukaryotes involves selective transcription of certain sections of the genome, so that in any given tissue certain genes are transcribed while others are inactive, possibly because of selective masking and unmasking of various regions of the genome, as suggested in the preceding section. However, a very different model for the control of gene expression has recently been put forward by Britten and Davidson which does not assume that there is selective activation or repression of structural genes in different tissues.

Britten and Davidson point out that whereas large numbers of single copy sequences are present in nuclear (nRNA), only a small fraction of these are represented in the mRNA in the cytoplasm of a given cell. Moreover, whereas the sets of mRNA in a given tissue are mainly absent from other types of tissue, nevertheless, these sequences appear to be ubiquitously present in the heterogeneous nuclear RNA in all cell types. The implication of these observations is that each differentiated cell nucleus includes not only all of the genes ever utilized in the organism, but also transcripts of all or most of these genes.

Most heterogeneous nuclear RNAs turn over very rapidly, with a half-life of about 20 minutes in the sea-urchin egg. Britten and Davidson suggest that the regulation of gene expression is achieved by determining which of the potential mRNA precursors survive, are processed and are transported to the cytoplasm. It is postulated that certain mRNA precursors of structural genes are protected from degradation by forming RNA-RNA duplexes with transcripts of non-structural gene sequences. These non-structural gene sequences apparently consist of interspersed repetitive and single copy sequences and are complementary to the repeated sequences in the transcripts of the structural genes. Thus, according to this model, the survival of some mRNAs in any given cell type depends upon the production of specific regulatory non-structural gene sequences, but how this latter specificity is achieved in any given cell type remains unknown.

Whether or not the hypothesis of Britten and Davidson will prove correct remains to be seen, but studies on the tobacco plant are consistent with the hypothesis that control of gene expression is exercised at a post-transcriptional stage. Thus, while only 75 per cent of leaf mRNA sequences are represented in the polysomes of the stem, *all* of the leaf mRNAs are represented in the stem nuclear RNA. Thus, leaf structural genes, which are not utilized to produce proteins in the stem, are nevertheless transcribed into stem nuclear RNA, suggesting that "post-transcriptional processing or selection mechanisms play a central role in the regulation of plant gene expression" (Goldberg).

13.10 REGULATION OF GENE EXPRESSION AT THE TRANSLATIONAL LEVEL

We have already discussed the possibilities that gene expression may be regulated through masking and unmasking of the DNA (i.e. through regulation of template availability) or by regulation of the processing of nuclear DNA in the production of mRNA, but regulation may also be achieved at the translation, or even at a post-translation, stage.

An example of apparent control at the translation stage is seen in the aquatic fungus *Blastocladiella emersonii*. The zoospores of this species, which lack cell walls, have a single, posterior flagellum. There is a nuclear cap which surrounds one end of the nucleus and contains large numbers of pre-formed ribosomes. The zoospores are released into the water and settle on a solid substratum where they enter a short period of encystment, during which they undergo rapid and radical changes in structure. These changes include the breakdown of the nuclear cap and release of the ribosomes, which become dispersed throughout the cell. In about 10 minutes a small germ tube appears, indicating the commencement of germination (Fig. 13.8).

Studies on protein and nucleic acid synthesis have shown that RNA synthesis does not begin until much later (40-45 minutes after encystment) and this is followed in a further 30-40 minutes by protein synthesis and then DNA synthesis. Thus, germination takes place apparently without any RNA or protein synthesis. Moreover, inhibitors of RNA and protein synthesis do not prevent encystment and germination, so that it would appear that the necessary ribosomal, transfer and messenger RNAs are all present in the zoospores before they encyst, and polysome formation can be seen to occur during early germination. These structural changes occurring during encystment and germination apparently require only the protein and RNA present in the zoospore when it is released, but the later stages of germination do appear to require new protein synthesis.

Studies on the marine algae, *Acetabularia,* have been very instructive in showing, not only control

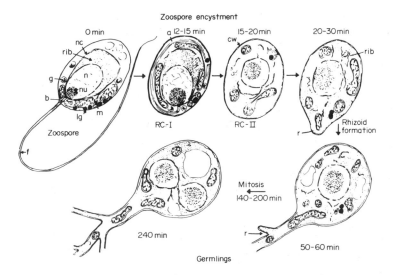

Fig. 13.8 Diagram of zoospore germination and early development in *Blastocladiella emersonii*. Symbols: b, basal body; g, gamma particle; m, mitochondrion; rib, ribosomes; nc, nuclear cap; nu, nucleolus; n, nucleus; lg, lipid granules; a, flagellar axoneme; cw, cell wall; r, rhizoid; f, flagellus; v, vacuole; RC-I, round cell I; RC-II, round cell II. (From C. J. Leaver and J. S. Lovett, *Perspectives in Experimental Biology*, Ed. N. Sunderland, Pergamon Press, Oxford, pp. 229-311, 1976.)

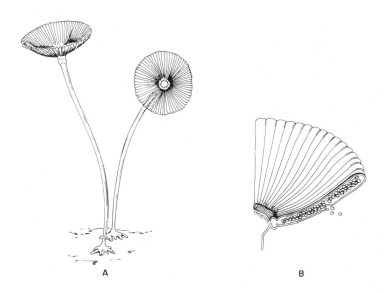

Fig. 13.9. Left: Mature plants of *Acetabularia mediterranea*. Right: Section through cap, showing production of cysts.

of gene expression at the translational level, but also the complex interaction between nucleus and cytoplasm in this process. *Acetabularia* consists of a single, giant cell, which has a rhizoid and a cylindrical stalk (Fig. 13.9). The stalk grows in length at its apical end and ultimately forms a "cap" in which large numbers of cysts are ultimately formed; after they are released the cysts give rise to isogametes. The shape and morphology of the cap varies considerably between different species. There is a large central vacuole, with a boundary layer of cytoplasm inside the wall, and the single nucleus lies in the cytoplasm at the rhizoidal end.

The large size of the cell makes it possible to remove the nucleus and transfer it to another individual of the same or a different species; when the nucleus of one species is transferred into an enucleate cell of another species, the cap that is formed later is characteristic of the species which supplied the nucleus. Grafting may also be carried out between different species of *Acetabularia,* by cutting off the apical part of the stalk of one species, and pushing it over the basal part of another part containing a nucleus. If an enucleate section of one species is grafted on to a nucleate base of another species, the first-formed cap of the regenerated plant is intermediate in its characteristics, but if this is removed the new cap resembles that of the species which supplied the nucleus.

If the stalk is cut off above the rhizoid the latter will regenerate a new stalk and cap. However, even parts lacking a nucleus can regenerate. For example, if a young stalk which has not yet developed a cap is separated from the rhizoid, so that it lacks a nucleus, it will survive for a long time and will ultimately form a cap.

These results seem to show that development is ultimately under the control of the nucleus, but that there are long-lasting effects of the nucleus on the cytoplasm, which can continue and complete morphogenesis even if the nucleus is removed. It would appear that (1) all the information necessary for the production of the cap passes from the nucleus to the cytoplasm long before the cap is normally formed; (2) information for the production of the cap may be present in the cytoplasm without being expressed; (3) the information for the production of the cap is very stable.

These conclusions suggest that the nucleus produces stable messenger RNA which persists in the cytoplasm for long periods. There is considerable evidence to support this hypothesis, including the following observations:

(1) Total protein synthesis continues for many months in the absence of a nucleus, whereas RNA synthesis stops immediately on removal of the nucleus.

(2) Any interference with RNA synthesis, such as treatment with the inhibitor Actinomycin D which inhibits the formation of mRNA, does not inhibit cap regeneration in enucleate fragments (which presumably already contain stable mRNA), but it does inhibit regeneration of the cap by nucleate fragments, which would presumably be dependent on the synthesis of new mRNA.

(3) Protein synthesis occurs in enucleated stalks. Inhibitors of protein synthesis, such as puromycin and cycloheximide, prevent cap regeneration, in both nucleated and enucleated stalks.

This and other indirect evidence suggests that morphogenesis in *Acetabularia* is dependent on the formation of stable, long-lived mRNA, but this hypothesis has not yet been demonstrated by direct experiment. The information passed from the nucleus to the cytoplasm can evidently remain unexpressed for several weeks, but the time of cap formation appears to be determined by events that take place in the cytoplasm. Control of gene expression in *Acetabularia* therefore appears to be at the *translational* level.

We have also seen that protein synthesis during the early stages of germination of seeds depends upon pre-formed mRNA (p. 280). Thus, although it seems very probable that development involves the production of specific mRNAs, translation of these mRNA does not necessarily occur immediately they are released into the cytoplasm, and the translation process offers another stage at which the *time* of gene expression may be controlled.

It is also possible for gene-expression to be regulated at a post-translational stage; for example, by controlling enzyme activity either by modifying the structure of the enzyme (as in allosteric effects, p. 311) or in other ways. Some enzymes are present in seeds in an inactive form (zymogen), but can be converted to an active state by appropriate treatments. Thus, protein granules have been isolated from pea seeds from which a glucosidase can be obtained by treatment with a proteolytic enzyme, and phosphatase activity in particulate fractions from lettuce seed can be greatly increased by treatment with the proteolytic enzyme trypsin.

13.11 NUCLEOCYTOPLASMIC INTERACTIONS IN DIFFERENTIATION

Studies on *Acetabularia,* in which experiments can be carried out with both nucleate and enucleate fragments of the cell, have provided a wealth of information on the nature of the interplay between nucleus and cytoplasm in the development of the plant, and have shown how events occurring in the cytoplasm are dependent upon factors (probably mRNAs) derived from the nucleus. However, there is also evidence of influences of the cytoplasm upon the behaviour of the nucleus. Thus, division of the nucleus to give large numbers of daughter nuclei normally occurs just before gamete formation, but if the cap is cut off division of the nucleus does not occur until a new cap is formed. How the control of nuclear division by the cytoplasm is achieved is not known.

The phenomenon of unequal (or polarized) division also indicates the powerful influence of the cytoplasmic environment upon the process of differentiation, which must in turn involve selective gene expression. A number of examples were described earlier (Chapter 1), in which the two daughter nuclei resulting from an unequal division subsequently followed very different patterns of differentiation, as in the formation of guard mother cells, and pollen grains. In such cases, mitosis is preceded by establishment of cytoplasmic differences within the parent cell, and the axis of the mitotic spindle is such that the two daughter nuclei come to lie in different cytoplasmic environments. If, by accident, the axis of the mitotic spindle is disturbed, so that as a result the new cell wall divides the cytoplasm equally, then differences in the pattern of differentiation of the two daughter cells do not occur.

We have no direct information as to the nature of the cytoplasmic components which are unequally distributed to the two daughter cells and which affect its subsequent differentiation. However, detailed studies have been carried out on the cytoplasmic changes preceding the polarized first division which occurs in the fertilized *Fucus* egg, and which gives rise to two daughter cells one of which forms the future thallus and the other the future rhizoid (p. 17). If the fertilized eggs are exposed to unilateral illumination, the shaded side of the cell begins to form a protuberance at about 14 h after fertilization and mitosis takes place so that the axis of the spindle is parallel to the direction of the incident light. Studies on the ultra-structural changes occurring following unilateral illumination have shown that 11-14 h after fertilization mitochondria, ribosomes and certain types of vesicle are more concentrated in the future rhizoidal half of the cell, which is thus more densely cytoplasmic than the other half. At about 12 hr after fertilization (before any visible signs of emergence of the rhizoidal protuberance) the nuclear surface becomes highly polarized with finger-like projections radiating towards the site at which the rhizoidal protuberance will arise. There are also changes in the cell wall at the site of the future rhizoidal protuberance, with the laying down of a sulphated polysaccharide, fucoidin.

It has been shown that the zygotes of *Fucus* become electrically polarized after unilateral illumination, the shaded side becoming electronegative to the illuminated side, as a result of which an electrical current flows through the egg, calcium and sodium ions entering the egg at the rhizoidal end and chloride ions at the thallus end.

This electrical potential difference may also lead to the electrophoretic movement of other molecules the distribution of which may become polarized within the egg and may influence the pattern of differentiation.

Studies of this sort have provided more detailed information about the types of change occurring in the cytoplasm before an unequal division, and may help ultimately to indicate how the state of the cytoplasm controls which genes in the nucleus are expressed.

13.12 SUCCESSIVE GENE EXPRESSION IN THE WHOLE PLANT

From the foregoing discussion it will be apparent that considerable progress has been made in the elucidation of the control of gene expression in eukaryotes, and the momentum which has been achieved is likely to be maintained, so that we may reasonably expect the nature of the processes involved will be established in the foreseeable future. This knowledge will undoubtedly be of the utmost value for our understanding of both general and specific aspects of development, but it is difficult to apply our present incomplete knowledge to the central problem of development, namely, how are the right genes expressed in the right cells at the right time? In the meantime, therefore, it is only possible to use a "phenomenological" approach to general aspects of development, but even though this is an unsatisfactory, though unavoidable situation, this type of approach is still useful in providing broad frames of reference in thinking about development.

We have seen that development of a flowering plant involves a series of successive stages, starting with the differentiation of the embryo into root and shoot, followed by the formation of organ initials, and ending with tissue differentiation within the individual organs. Superimposed on this pattern are the major changes in development represented by the transition to the flowering phase and the onset of dormancy. The orderly manner in which the successive stages follow each

other in a regular sequence is one of the most striking features of development. This phenomenon is well illustrated in the development of a flower, where not only is there a regular sequence in the initiation of the various flower parts (perianth, stamens and carpels), but within each of these there is an equally regular pattern of development.

The transition from one stage of development to the next would seem to involve a process of successive gene expression, in which certain previously unexpressed genes come into action and others become or remain unexpressed. Direct evidence of successive gene activation is provided by insect development. As is well-known, the cells in some organs of *Drosophila* and certain other flies have "giant" chromosomes, formed by the repeated replication of the DNA strands, without nuclear division. These giant chromosomes thus each consist of large numbers of single chromosome threads lying parallel to each other and aligned so that corresponding regions of the various threads are opposite to each other, to give a characteristic banded appearance. Each band appears to correspond to a single gene or operon. At certain stages of development of the insect one or more of these bands swells up and forms a "puff" of what is apparently RNA. It would seem that the occurrence of puffing indicates that a particular gene is active at that time. Different tissues have characteristic puffing patterns, and these occur at specific periods of development. Moreover, if the moulting hormone, ecdysone, is administered to a larva the puffing pattern changes rapidly, as the insect enters a new phase of development. We have no direct parallel evidence of successive gene activation during development in plants, but it seems very likely on *a priori* grounds that this does occur.

The regular sequence of changes seen in the development of an organ such as a leaf or a flower strongly suggests that once these particular pathways have been entered, then all the subsequent stages follow inexorably as a sort of "chain reaction" in which the attainment of one stage triggers off the next one. A mechanism of this type has been suggested for flower development. It is postulated that once the transition to the flowering stage has begun, a gene complex A is activated in the first formed primordia, and these genes produce an inducer X which moves to the next primordia and there activates a gene complex B and so on through the different flower parts. This hypothesis postulates the occurrence of short range inter-cellular "messengers" or "hormones", but the existence of such substances has yet to be demonstrated.

13.13 THE DIFFERENTIATED STATE

Ultimately the processes occurring during cell differentiation are terminated and the cell reaches the "steady state" of the mature condition, in which metabolism is maintained continuously (except, of course, in the case of non-living cells such as those of the xylem). The visible manifestations of differentiation include variations in the development of the cell wall and of certain cytoplasmic organelles, such as plastids. It is clear that differentiation must also extend to certain aspects of metabolism when it is remembered that some tissues are specially adapted for particular functions, such as photosynthesis, secretion and storage of reserve materials. Such differentiation almost certainly involves differences in enzyme production, which in turn implies the maintenance of differences in gene expression between various cells even in the mature state.

Many basic metabolic pathways are probably operative in all living cells of the plant. This must apply to the main pathways involved in the respiratory breakdown of carbohydrates, for example. On the other hand, there is much evidence that various tissues differ in their biosynthetic abilities. For example, it is found that the isolated roots of many species require the supply of certain vitamins, including thiamin, pyridoxine and nicotinic acid when grown in aseptic culture. It appears that in the intact plant these vitamins are synthesized in the shoot and are supplied to the roots. Similarly, callus cultures of certain plant tissues, including tobacco pith tissue, require to be supplied with auxin and cytokinin in order to

maintain cell division in culture (p. 155). Evidently tobacco pith cells do not have the ability to synthesize auxin and cytokinin and hence require an exogenous supply. However, this inability of pith cells to synthesize the two hormones is not due to any permanent loss of the potentiality to do so, as shown by the fact that cytokinin-requiring callus cultures can be "habituated" (p. 155), so that they become cytokinin-independent.

The inability of roots to synthesize certain vitamins, and of tobacco stem tissue to synthesize auxins and cytokinins, provides rather strong evidence that cell differentiation involves the activation of certain genes and the repression of others. It would be of interest to know whether the meristematic cells of the shoot apex of tobacco have the ability to synthesize cytokinins; if so, then it would seem that one of the processes occurring during cell differentiation in the stem is the repression of the enzymes responsible for auxin and cytokinin synthesis. Such a change in synthetic ability might, indeed, explain the transition from cell division to cell expansion seen in the apical region of both shoot and root.

We do not know how these permanent differences in biosynthetic ability are maintained but Jacob and Monod have shown that it is not difficult to construct "model" circuits, based upon the bacterial repression concepts, which would result in certain genes being permanently switched on. One very simple circuit is shown in Fig. 13.10, in which the inducer is not the substrate but the product of the controlled enzyme system. Such a system is known to occur in bacteria. In the absence of an exogenous inducer, the enzyme will not be produced, unless already present, but temporary contact with an inducer will cause the system to become "locked" indefinitely, at least so long as either substrate or product is present. Suppose, for example, that the operon in Fig. 13.10 is responsible for the biosynthesis of a hormone, such as cytokinin or gibberellin, and that in a dormant seed the operon is repressed. Then a single treatment with exogenous hormone will activate the operon and from then on the seedling would be able to synthesize its own hormone. The stimulation of endogenous auxin synthesis by a

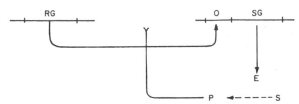

Fig. 13.10. Hypothetical model of a circuit for control of enzyme synthesis in which the inducing substance is the product of the controlled enzyme. Synthesis of enzyme E, genetically determined by the structural gene SG, is blocked by the repressor synthesized by the regulator gene R.G. The product P of the reaction catalysed by enzyme E acts as an inducer of the system by inactivating the repressor, O, operator; S, substrate. (From F. Jacob and J. Monod, *Cyto-differentiation and Molecular Synthesis* (Ed. M. Locke), Academic Press, New York and London, 1963).

single application of exogenous IAA in fruit set (p. 129) may provide a similar example. It is quite easy to construct other models which would account for the permanent suppression of certain enzymes. Thus, in the model illustrated in Fig. 13.11, the product of one enzyme acts as an inducer for the other, so that the two enzymes are mutually dependent. One could not be synthesized in the absence of the other, and inhibition of one enzyme or elimination of its substrate, even temporarily, would result in the permanent suppression of both.

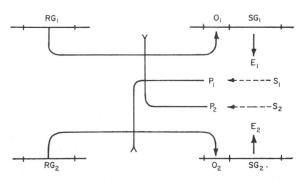

Fig. 13.11. Model circuit in which the product of one enzyme acts as an inducer of the other. Synthesis of enzyme E_1, genetically determined by the structural gene SG_1, is blocked by the repressor synthesized by the regulator gene RG_1. Synthesis of another enzyme E_2, controlled by structural gene SG_2, is blocked by another repressor, synthesized by regulator gene RG_2. The product P_1 of the reaction catalysed by enzyme E_1 acts as an inducer for the synthesis of enzyme E_1. O, operator; S_1,S_2, substrates. (From F. Jacob and J. Monod, *Cytodifferentiation and Molecular Synthesis* (Ed. M. Locke), Academic Press, New York and London, 1963.)

There is no evidence that determination and differentiation involve this type of mechanism, but these circuits do illustrate that it is possible to construct models which would account for the observed facts, within the concepts of bacterial repressor mechanisms.

13.14 DETERMINATION IN PLANT DEVELOPMENT

The successive stages of development can be regarded as a process in which there is divergence into alternative pathways of further development at various critical points of time and space. This divergence may occur at the cellular level, as when two daughter cells of a common mother cell show divergent patterns of differentiation following an unequal division, or it may occur in the differentiation of organs, or even in the shoot apex as a whole, as in the transition from the vegetative to the flowering phase. Furthermore, we saw that once an organ, such as a leaf primordium, passes a certain stage of development, it becomes irreversibly "determined" as a leaf (as opposed to a bud) and cannot normally then be converted into any other structure (p. 31).

This successive commitment of different parts of the developing organism to specific pathways of differentiation has been referred to as the "canalization of development". The development of the embryo appears to involve canalization into several major pathways, as a result of which there is the establishment of root and shoot regions, and with the establishment of an organized shoot apex the basis is laid for the initiation of various organs (leaf, bud, stem), which is maintained throughout the subsequent vegetative phase of the plant.

Once a group of cells has become committed to a particular pathway it usually follows the "normal" pattern of development through to completion and it is unusual for it to revert to an earlier stage or to make the transition into some other pathway. Thus, leaf primordia do not become buds or stems, and although abnormalities may sometimes occur during flower development, with reversion to a vegetative apex, such instances are relatively rare, and at certain critical stages a particular part of the organism is said to become "determined" in respect of its further differentiation. We have already seen an example of such determination in the development of leaf primordia (Fig. 2.12).

The development of an organ, in both animals and plants, is a process involving a series of changes which follow each other in regular succession, with no long time interval between them. However, in insects examples are known in which groups of cells become determined at an early stage of development, but the expression of this determination may not become apparent until many cell generations later. Thus, in *Drosophila*, the "imaginal discs", which are destined to give rise to the various body regions of the adult fly, are laid down at a very early stage in the larva, but they are already determined and they can retain this state of determination through many cell divisions without loss.

Exactly comparable examples are not known for plants, but evidence for the occurrence of determination in plant cells and tissues is not lacking and it is shown most clearly in shoot apices. It might seem paradoxical to suggest that meristems exhibit determination since meristematic cells are normally regarded as essentially cells which are undifferentiated and uncommitted. However, we have already considered two or three examples of determination in shoot apical meristems, viz. in vernalization and in phase change in woody plants. As we have seen, vernalization of rye embryos leads to intrinsic changes in the developmental potentialities of the cells of the shoot meristem so that they become capable of flower initiation. Once the apical meristem of rye has become fully vernalized all the cells derived from it are in the vernalized state, which is apparently transmitted through cell division without "dilution" (p. 243). Although the vernalized state is a very stable one, the new embryos produced by a vernalized plant are not themselves vernalized and hence "devernalization" must occur during normal gametogenesis.

"Phase-change" in woody plants (p. 256) provides another example of determination in apical meristems. There is no doubt that phase change involves intrinsic changes in meristematic cells, since if sterile callus cultures, derived from juvenile and adult stem tissue of ivy, are grown on the same nutrient medium, differences in the two cultures can be recognized, with "juvenile" cultures maintaining a higher rate of growth and more abundant root regeneration than the "adult" cultures. Evidently we are dealing with two relatively stable alternative states of determination which do not involve genetic change (since the adult shoots produce seeds from which juvenile seedlings develop), but which can be transmitted through cell division without loss of determination.

A further example of stable differences in shoot apical meristems is provided by the alternation of free living gametophytes and sporophytes in ferns and other lower plants. Normally the gametophyte is haploid and the sporophyte is diploid but the differences in morphology and life cycle between two generations cannot be ascribed to the difference in ploidy, since it is possible to produce experimentally diploid or tetraploid gametophytes and haploid sporophytes. These various phenomena must all depend upon the activity of different sections of the genome in the alternative states.

13.14.1 The nature of changes in states of determination

There is very little information as to what determination involves at the molecular level. It would seem likely that we are dealing with stable states of gene expression which can even be transmitted through cell division. It is difficult to envisage how differences in the states of selective gene activation could be transmitted through DNA replication and subsequent cytokinesis, but the transmission of the differentiated state through repeated cell division is well established for animal tissues.

Although we have little information on the molecular basis of determination there are close analogies between the phenomena of determination and that of "habituation" of plant cells in aseptic culture (p. 155). Callus cultures of tobacco pith tissue are normally found to require supplies of exogenous IAA and cytokinin in the culture medium. Thus, it would seem that tobacco pith cells either do not have the ability to synthesize auxin and cytokinin and hence require an exogenous supply, or possibly they have a high rate of catabolism of these growth substances. On the other hand, cultures which originally were heterotrophic for auxins and cytokinins may become "habituated" in culture so that they become self-sufficient for their requirements for one or other of these types of growth substance. Thus, it would appear that tobacco pith cultures can exist in stable, alternative states, in a manner closely analogous to tissues which show determination. It is significant that clones of habituated tissues which have been derived from

single cells are still habituated, showing that the change is a property of individual cells.

The question arises as to the nature of the changes involved in habituation. It would seem likely that these are changes in gene expression and are therefore "epigenetic", but the possibility remains that they may involve mutation. This problem has been investigated by Meins and his co-workers, using a strain of tobacco callus which habituates rapidly on a medium containing kinetin, or when maintained on a basal medium at 35°C. It is found that the rate of habituation per cell generation is at least 100-1000 times faster than the usual rate for somatic mutation in tobacco. In further experiments, whole tobacco plants were regenerated from habituated callus, and the pith from such plants was found to be non-habituated (Fig. 13.12), indicating that the habituated state had been reversed in the process of regeneration and subsequent differentiation; such reversal is again not characteristic of tissues which have undergone mutation, and their habituation does not affect the totipotency of the habituated cells.

As we saw above (p. 320), it is not difficult to construct model circuits, based upon bacterial systems for gene regulation, which could account for the stability of the differentiated state, in which certain genes are continuously expressed. Thus, it was suggested above that an exogenous hormone, such as a cytokinin or a gibberellin, might activate an operon for biosynthesis of endogenous hormone. Some evidence for this hypothesis has been obtained for habituated tobacco callus tissue. Habituated tissue was grown at 16°C, which apparently blocks the production of endogenous cytokinin. Some of the cultures were supplied with optimal concentrations of kinetin while others were supplied with low levels of kinetin which were only sufficient to support growth. When the two series of cultures were returned to 25°C, those previously maintained on low kinetin at 16°C were found to have become cytokinin requiring, whereas those supplied with higher cytokinin concentrations at 16°C were found still to be habituated when transferred to 25°C. These results are consistent with the

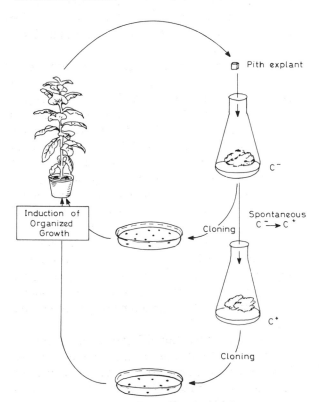

Fig. 13.12. Schematic representation of habituation reversal experiment. C⁻, cytokinin-habituated tissue able to produce cell division factors. (From F. Meins, 1977.)

hypothesis that the habituated state is maintained by a "positive feedback" circuit involving cell-division factors, such as cytokinins. The hypothesis therefore suggests that the determined state is maintained by products which induce their own synthesis (or block their own degradation).

13.14.2 Totipotency and determination

At the present time the capacity of the somatic cells of many plant species to regenerate whole plants by the application of appropriate aseptic techniques is exciting a great deal of attention, not least because of its great potential practical importance (Chapter 6). Great emphasis is thus laid upon the totipotent properties of many plant cells, and at first sight the concept of determination appears to conflict with that of totipotency. However, it is clear that determination is not necessarily irrevocable in plant cells in the way it is in many types of animal cell, where cells of developing embryos lose their capacity to regenerate whole animals at an early stage, and once this has occurred the change appears to be ir-reversible. Thus, "natural" regeneration in plant tissues, whether of adventitious roots, buds or of whole plantlets, normally occurs from cells (such as those of the pericycle) which have previously undergone vacuolation and maturation and hence must undergo "de-differentiation" before such regeneration can take place. Similarly, experimental regeneration of roots, buds or adventive embryos from callus or cell suspension cultures involves a stage of de-differentiation in the initial establishment of the cultures. The term "de-differentiation" is normally applied to the resumption of meristematic activity in cells which have ceased to divide and have undergone vacuolation, but it must also involve the "re-programming" of gene expression in these cells. It would appear, from the fact that some calluses retain certain characteristics of the differentiated tissues from which they were derived (such as the capacity for regeneration of roots in callus of juvenile *Hedera* (p.321)), that such "re-programming" may not be complete, so that the calluses still retain a degree of "determination". Thus, the resumption of meristematic activity to

form callus tissue does not necessarily result in complete "re-programming" of the constituent cells, although no doubt such re-programming is, in most cases, very extensive.

If we are correct in concluding that de-differentiation and "re-programming" largely obliterates any former state of selective masking of the genome, then the apparent conflict between the concepts of determination and totipotency is resolved and there would appear to be no conflict between the postulated state of determination in shoot and root apices and the production of adventitious roots and buds from the mature tissues of stems and roots respectively.

13.14.3 Differences in determination between shoot and root apices

The evidence for determination in shoot apices described above (p. 321) raises the question as to whether differences between shoot and root apices reflect differences in the state of determination in the two structures.

We have seen that the establishment of root and shoot poles in the embryo of higher plants is preceded by an unequal division in the zygote. Thus, the major division of the plant body into root and shoot is established at the earliest stages in development. Once established, the shoot and root meristems are very stable structures, and it is extremely rare for a shoot meristem to be converted into a root meristem and vice versa. Moreover, the stability of the root meristems is intrinsic in those organs themselves since isolated roots of tomato and other species can be maintained in aseptic culture for many years without the appearance of any buds, although they require to be supplied with thiamin and other vitamins, which are normally supplied by the shoot in intact plants.

From an "operational" standpoint, root and shoot apices behave as if they are determined. However, at first sight, this suggestion appears to conflict with the generally-accepted view that the cells in the shoot and root meristems are intrinsically uncommitted and that the different patterns of differentiation in the two organs lies in the structure and organization of the meristems themselves.

It is also widely accepted that the apex is a self-organizing region and that the pattern of differentiation occurring in the shoot apical region is controlled by the properties of the apex itself, which is independent of influences from existing, older differentiated tissues. However, these two concepts (non-commitment of the promeristem cells and independence of influences from older tissues) would seem incompatible. If we accept both these concepts, then the differing properties of the two types of apex must depend solely upon their different structure and organization, and it is then difficult to see how this difference could be maintained so tenaciously, so that a transition from root to shoot apices or vice versa rarely, if ever, occurs in higher plants. Moreover, if the properties of an apical meristem depend solely upon its structure and organization at the cellular level, it is difficult to conceive how adventitious buds and roots could arise *de novo* from unorganized callus tissue.

It seems to be logically inescapable that there are really only two possible alternative hypotheses to account for the stability of shoot and root apices and the maintenance of the differences between them, namely either (i) the meristematic cells of shoot and root apices are uncommitted, and development into either shoot or root pathways is controlled by influences from older cells which have undergone partial or complete differentiation; or (ii) the meristem cells are *inherently* programmed into either root or shoot channels and are *not* totally uncommitted, so that the different patterns of organization in shoot and root apices are a consequence of the inherent differences in their constituent cells. On this latter view, meristem cells are, in fact, specialized cells, just as gametes are specialized cells.

Let us consider, first, the possibility that meristems are subjected to influences from older tissues. We have seen that in some species the transition of the shoot apex from the vegetative to the floral condition may be regulated by a stimulus from the mature leaves. It is possible that endogenous growth hormones also play an important role in the activities of apical meristems. We

have seen that root initiation is promoted by high auxin levels, not only in the formation of adventitious roots on stems but also in callus cultures derived from tobacco pith, in which high auxin/cytokinin ratios produce roots and high cytokinin/auxin ratios produce buds. It seems possible that not only are root apices initiated under conditions of high auxin/cytokinin concentrations, but that a continued supply of auxin from neighbouring tissues is required also to *maintain* the structure and organization of a root apex, since it is now well-established that the transport of IAA in the root apex is acropetal, i.e. from older tissue towards the meristem. Thus, it would seem that the differentiating or mature tissues do, indeed, supply auxins to the root meristem.

Since high cytokinin/auxin ratios promote bud regeneration in certain callus cultures, there is the possibility that bud initiation and bud growth may depend upon the supply of cytokinins from other parts of the plant. There is indeed, evidence that young, expanding leaves may supply both auxins and cytokinin to the apical dome of shoot apices. Thus, excised meristem zones less than 0·1 mm in height can be induced to grow and develop into complete plants by using a simple nutrient medium containing only indole-3 acetic acid (IAA) as the hormonal addendum. Studies with *Coleus blumei* showed that no exogenous growth substances were required if the explant consisted of the apical meristem together with two or more pairs of primordial leaves. But the addition of IAA was essential when leaf primordia were not present on the explant. In experiments with excised apical meristems of *Dianthus caryophyllus,* it was shown that although a medium containing kinetin alone enabled a few apical dome explants to develop into rooted plants, the highest frequency of plants was obtained when both IAA and kinetin were supplied. On the other hand, explants that had two pairs of primordial and a pair of expanding leaves were able to continue their development without supplementary growth substances. Thus, it would appear that the shoot apical meristem in a number of species is dependent upon the supply of auxin from the expanding leaves, and that in some species these leaves also supply cytokinin.

It is possible that the stability and homeostatic properties of root and shoot meristems may be dependent upon and controlled by the continued supply of growth hormones from the neighbouring tissues, and that the stability of apical meristems resides in the neighbouring mature tissues which produce and supply the hormones. However, an alternative hypothesis is that the meristematic cells of the shoot and root apices have become determined along "shoot" and "root" pathways of development respectively, i.e. that *intrinsic* differences in gene activity have been established, which are not caused primarily by the cellular environment in which they occur. This suggestion is in conflict with the current opinion that meristematic cells are completely uncommitted with respect to their future pattern of differentiation, but we have seen that in the phenomenon of phase-change, shoot apices can exist in two alternative but stable states (juvenile and adult). Since callus tissues derived from juvenile and adult shoots clearly exhibit intrinsic differences even when grown in isolation in aseptic culture (p. 321), it is possible that the differences in organization and structure between juvenile and adult apices arises from the intrinsic, "determined" differences in the potentialities of their constituent cells.

On this basis it is logical to suggest that, by analogy, the characteristic structures and properties of root and shoot apices may arise from intrinsic differences in the developmental potentialities (though not in their *genetic* potentialities) of their constituent cells, rather than that the differences in the patterns of differentiation in root and shoot apices is the *result* of the organization and biochemical activities of the apices.

The most direct method of testing whether the cells of the meristem are intrinsically determined would be to establish aseptic cultures from the meristem regions of root and shoot apices and to compare their behaviour and growth factor requirements, but although shoot apices of a number of species have been successfully cultured they have always been large enough to include

cells which have begun to undergo vacuolation and maturation. However, in experiments with maize root tips it was possible to isolate the quiescent centre and to demonstrate that the quiescent centre itself is able to give rise to new roots. The initial divisions following excision gave rise to unorganized cell proliferations from which organized root tissues later arose. The quiescent centre was itself polarized, regenerating a root cap on its (normally) distal side and root tissue on the proximal side. Both IAA and kinetin were required by the quiescent centre for growth. This result would seem to indicate that the cells of the quiescent centre are, in fact, determined as root cells.

Evidence which at first sight appears to conflict with the view that shoot and root apical cells are intrinsically determined along "shoot" and "root" pathways of development, appears to be provided by the fact that adventitious roots can develop from stem tissue and adventitious buds from roots. This evidence indicates that some mature stem and root cells are not irrevocably determined and can give rise to root and shoot apices, respectively. However, the development of adventitious roots and buds from pericycle or cortical tissue entails a degree of "de-differentiation" of mature cells and the renewal of meristematic activity (p. 165). Thus, it is possible that during the process of maturation and subsequent de-differentiation a state of determination of the apical meristem cells from which they were derived may be obliterated. In some species adventitious roots arise in close proximity to the shoot apical meristems. However, close scrutiny of such cases seems to indicate that in all such cases the root initials arise from tissues which have already undergone a degree of maturation and differentiation.

13.15 FURTHER DIFFERENTIATION IN APICAL REGIONS

We have seen that development involves the successive commitment of cells to specific pathways of differentiation which we have referred to as the "canalization of development". The canalization of development requires (1) that cells of a common lineage diverge into alternative pathways of differentiation and (2) that once they have so diverged they become committed to that pathway, i.e. they are "determined". We have discussed one process by which the two daughter cells of a common parent cell are diverted into different patterns of differentiation, i.e. by unequal or asymmetric division, and we have discussed the evidence for determination in plants, especially in apical meristems. We have, therefore, to consider further how far these processes can account for the regular patterns of differentiation of tissues and organs observed in the apical regions of roots and shoots, in both lower and higher plants.

In alga, such as *Chara,* which have apical cells, and in mosses and many ferns, the single apical cell divides unequally, the outer daughter cell continuing as the apical cell and the other giving rise to differentiated thallus cells after further unequal divisions. In certain mosses and a few pteridophytes, such as certain species of *Selaginella,* the various tissues of the mature stem can apparently be traced back to precise divisions of the single apical cell and its immediate derivatives. This problem has recently been studied in greater detail in the root of the aquatic fern, *Azolla,* using the electron microscope to follow the process at the ultra-structural levels. There is a single tetrahedral apical cell, from the three proximal faces of which derivative cells are cut off in succession. Each of these derivative cells undergoes further divisions and the pattern of subsequent divisions is extremely regular, so that the precise cell lineage leading to each cell of the mature root can be traced (Fig. 13.13). Some of these divisions are asymmetric (unequal) and others are symmetric. However, differentiation apparently does not necessarily depend upon unequal division since in some instances the daughter cells derived from an unequal division are similar, whereas in other instances symmetric divisions produce sister cells which undergo very different patterns of differentiation (e.g. outer sieve element and pericycle cells). Thus, it would appear that cell differentiation in the *Azolla* root is not necessarily preceded by an unequal division.

It is difficult to identify the initial (promeristem) cells in roots of higher plants, but it is nevertheless possible, by studying the patterns of cell division, to trace the cell lineages back to the margins of the quiescent centre. Thus, the cell lineages which ultimately form specific tissues in the mature root (central cylinder, cortex, epidermis and root cap) can be seen to be established very close to the promeristem itself. However, we do not know whether the cells of any given lineage are already *determined* to give rise to certain tissues; that is to say, we do not know whether (1) the developmental potentialities of the early derivatives of the initial cells of the promeristem are already different and that these differences are transmitted through further divisions and so give rise to the various tissue zones of the mature root, or (2) the cell lineages are, at first, still capable of giving rise to any of the main zones of the root, and only later is the pattern of differentiation imposed on them by some unknown process.

The fact that in some roots cell lineages can apparently be traced back to specific layers of cells in the meristem region is consistent with the Histogen Theory put forward by Hanstein in 1868 according to which three meristematic zones or histogens can be recognized in both shoot and root apices, known as plerome, periblem and dermatogen, which were held to give rise to the vascular cylinder, cortex and epidermis, respectively. Thus, according to the Histogen Theory, initiating cells of the various tissue regions are separate, and destined to give rise to different regions of the mature root. It is possible, therefore, that the quiescent centre does, indeed, contain cells of different developmental potential.

Although the histogen concept is still a tenable hypothesis for root meristems, it does not appear to be applicable to shoot apices of higher plants for the following reasons. Firstly, the vascular tissue of leaf traces is formed from cortical cells and the pattern of development of these traces appears to cut across any potential histogen boundaries that might be held to occur in the apical region of the shoot. Secondly, we have seen (p. 34) that the shoot apex has the properties of a self-organizing entity, with the capacity for self-

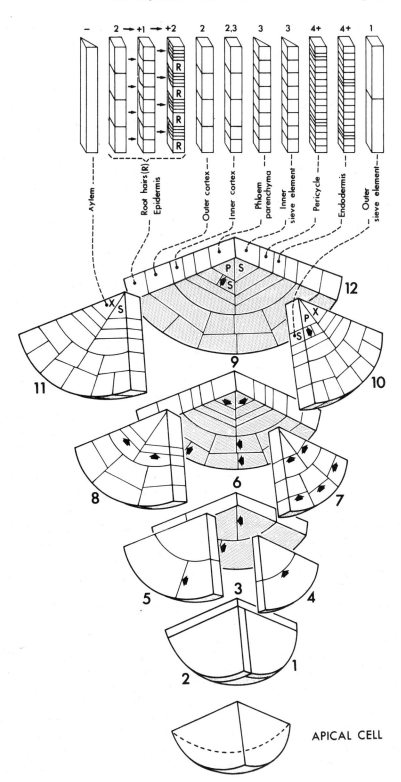

Fig. 13.13. Patterns of cell division within the root apical region of the water fern *Azolla pinnata*. Division of the single tetrahedral apical cell cuts off daughter cells from each of its 3 inner faces, and each of these undergoes a sequence of regular further divisions. (Each new cell wall is indicated by →.) Each of the cells in the uppermost sextant is destined to give rise to a specific tissue of the mature root, there being longitudinal files of cells for each such tissue, as indicated at the top of the diagram. (Figure kindly supplied by Prof. B. E. Gunning.)

regulation and regeneration. If this is so, then it would appear that the pattern of differentiation within the shoot apex is imposed on its constituent cells by the properties of the structure as a whole, whereas the hypothesis that differentiation involves an unequal division and the establishment of lineages of determined cells implies that the organization and properties of the apical region are determined by the intrinsic properties of its constituent cells.

Although it has been argued above that the difference between shoot and root apices may be dependent upon a degree of determination in the meristematic cells, it should be noted that this is not incompatible with the view that, so far as the differentiation of organs and tissues within the shoot apical region is concerned, it is the properties of the apex as a whole which regulate the pattern of differentiation rather than unequal division and the establishment of histogens. That is to say, even if the meristem cells of shoot and root apices are determined as suggested, differentiation into a variety of cells and tissue types takes place within each type of meristem indicating that the cells are still undetermined at these levels of organization. The commonly-accepted view that the pattern of differentiation within shoot and root apices is controlled by the overall structure and organization of the apices may thus be valid at this level of organization.

The rather complex interactions between what might be referred to as the "intrinsic" and "organizational" influences affecting differentiation in shoot and root apices may be summarized as follows:

(1) It is possible that the differences in structure between root and shoot apices depends upon intrinsic (determined) differences between their meristematic cells.

(2) Organogenesis and tissue differentiation in the shoot apex are regulated by the organization and properties of the apex as a whole.

(3) It is not clear whether differentiation in the root apex is partly regulated by its organization properties, but the patterns of cell lineage are not inconsistent with the possible

existence of histogens established close to the quiescent centre.

13.16 HORMONES AND DIFFERENTIATION

So far we have discussed mainly the role of intrinsic cell factors in differentiation, but we now have to consider an alternative situation, namely where the pattern of differentiation is controlled by factors *external* to the cell, and especially the role of hormones. By definition, the term hormone can be applied to a growth regulating substance only where it leaves the cell in which it is formed and affects another cell.

We have already met a number of instances where a hormone, or a combination of hormones, brings about a qualitative change which can be regarded as an aspect of differentiation, including:

(1) the induction of buds and roots in callus tissue by the combined action of IAA and a cytokinin (p. 157);

(2) the induction of vascular strands in callus tissue and in cortical and pith tissue in stems by IAA (p. 121);

(3) the interaction of IAA and GA_3 in the differentiation of vascular tissue (p. 122);

(4) the initiation of adventitious roots in stem tissue in response to IAA (p. 128);

(5) the induction of flowering and control of sex expression by gibberellins, auxins and ethylene (pp. 247-249).

Although these appear to be valid instances in which hormones appear to stimulate differentiation, most of the effects of hormones appear not to control selective gene expression directly, but rather to stimulate growth and differentiation in cells and tissues which are already "programmed" to differentiate in a specific manner. We have seen that each type of hormone has a wide range of effects and that the same hormone may have quite different effects in different tissues. For example, IAA may stimulate cell vacuolation in young fruits and in developing internodes, but it stimulates cell division in the cambium and it is necessary for the formation of the secondary wall in xylem differentiation. Again, GA_3 will stimulate internode elongation in many plants, and the synthesis of α-amylase in barley aleurone. Thus, *in plants the specificity of hormone action resides in the "target" tissue itself* and the hormone does not seem to determine the pattern of selective gene expression. There are a number of well-authenticated examples in which application of a hormone results in the synthesis of a specific enzyme protein, as in the stimulation of α-amylase synthesis in the barley aleurone layer by GA_3 (p. 96), or of cellulase synthesis by auxin. No doubt the hormone stimulates the production of specific mRNAs for such enzyme proteins, but there is no evidence to suggest that the hormone determines which parts of the genome are expressed in a given cell or tissue.

It also has to be borne in mind that only five main types of endogenous hormone have so far been identified, whereas during the life cycle of the plant differentiation must involve the regulation of large numbers of genes in the right cells in the right sequence, and it seems unlikely that the same small number of hormones could regulate the expression of so many genes. However, it is possible that the establishment of the major developmental pathways involves the regulation of certain "master genes" controlling the activities of a large number of subordinate genes, which become activated during the subsequent stages of differentiation. Indeed, it is a striking feature of certain aspects of differentiation that it appears to involve the co-ordinated expression of blocks of genes, as in the development of a leaf or a flower. The number of major steps in the development of a higher plant involving master genes is probably quite small and it is possible that interaction between the known hormones may play a controlling role at some of these steps.

For the reasons already given, it seems unlikely that the known hormones act as specific "effector" substances in gene expression and possibly the finer control of gene activity involves substances with a higher specificity of action such as protein or RNA molecules. Indeed, there is evidence that hormones bind with specific proteins and, if this is the case, then the specificity of hormone action in any given type of target cell may depend upon the receptor protein rather than on the hormone molecule. It is possible, therefore, that determination involves the production of specific receptor proteins and that differentiation will subsequently follow a predetermined pattern when the hormones become available to bind with the proteins.

13.17 SHORT-RANGE CELL INTERACTIONS IN PLANTS

Cell interactions mediated via hormones may occur over short or very long distances, but we now have to consider short-range, cell-to-cell interactions in plants. In animals, short-range cell interactions are well established. Firstly, since animal cells are motile and different cell types may become intermingled it is necessary that they should have the capacity for *cell recognition,* to enable them to distinguish between cells of the same and of a different type from themselves. This function appears to be fulfilled by the occurrence of recognition substances, especially proteins, in the cell surfaces. Secondly, in animal embryos certain tissue types have the capacity for inducing differentiation in neighbouring cells with which they are in contact and there is strong evidence that this effect involves diffusible substances capable of moving from the inducing to the induced cells. By contrast, somatic plant cells are not motile and the need for the capacity for cell recognition is not apparent, but sexual reproduction in plants, on the other hand, does involve motile cells and there is now increasing evidence for cell recognition mechanisms not only in pollen/stigma interactions in higher plants, but also in gamete fusion in unicellular algae, such as *Chlamydomonas.*

The discovery of cell-recognition processes in plants arose from studies on incompatibility in flowering plants. Many plant species have genetically determined incompatibility systems which prevent self-fertilization and hence promote outbreeding. In *gametophytic* self-incompatibility there are one or more gene loci each with multiple alleles (S_1, S_2, S_3, etc.) and the S allele borne by each pollen grain is expressed phenotypically when it alights on the diploid stigma, rejection occurring when the same allele occurs in both pollen and stigma (Fig. 13.14). In this type the pollen germinates and the tube penetrates the stigma, but growth is arrested either in the stigma or in the transmitting tissue of the style before fertilization can occur. In *sporophytic* self-incompatibility the behaviour of the pollen grains is determined by the genotype of the parent plant, and dominance may be demonstrated (Fig. 13.14). In this system, the "self" pollen grain starts to germinate and the tube may even penetrate the stigma surface, but growth is arrested thereafter, so fertilization is again prevented.

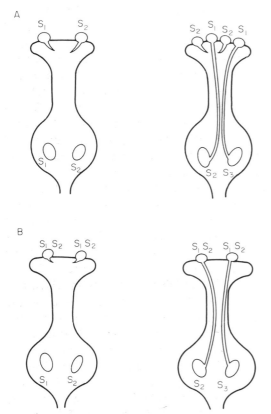

Fig. 13.14. Gametophytic (A) and sporophytic (B) self-incompatibility. In the gametophytic system, the S allele expresses its own phenotype during pollination on the diploid stigma, whereas in the sporophytic system, the phenotype of all pollen grains is determined by the genotype of the parent $(S_1 S_2)$. In both systems illustrated, the left-hand pollinations are self-incompatible and the right-hand ones compatible for (a) S_1 pollen grains of the gametophytic system, and (b) for all pollen grains of the sporophytic system, assuming that S_1 is dominant.

The walls of pollen grains consist of two layers, an outer *exine* and an inner *intine*. In many species the exine is complex and includes cavities containing proteins, and the intine also contains proteins. The exine proteins are derived from the tapetum of the anther and so are sporophytic in origin, whereas the intine proteins are formed later in the pollen grain itself. When a pollen grain alights on a moist stigma surface, the exine proteins are released and diffuse into the stigma surface within a few minutes. It can be shown that the pollen exine proteins bind with proteins in the stigma surface. In the families Cruciferae and Compositae, with sporophytic system, the pollen grain swells and germinates, but in incompatible matings growth of the tube quickly stops, and the tip becomes occluded with callose (a ß-1,3-linked glucan). A counterpart reaction is induced in the contiguous stigma cells, which also produce callose deposits on the inner surface of the wall at the point of contact with the incompatible grain. The callose depositions are not formed in compatible matings, and the tube continues growth to effect fertilization. Even non-viable pollen can produce the callose responses, but washed pollen does not do so. Moreover extracts of pollen will produce the callose response and such extracts have been shown to contain glycoproteins. Thus, the reaction at the stigma surface can discriminate between wall proteins of different origin.

In plants showing the gametophytic system of self-incompatibility the response depends on interactions in the stylar canal or transmitting tissue, or in some cases, such as the Gramineae, in the stigma. Where the response is early, the intine-borne proteins of the pollen grain may be involved, but where it is delayed the synthesis of the incompatibility factors probably takes place during the later growth of the tube. The interaction is with the interstitial material of the transmitting tract, which is rich in glycoproteins. The arrest of the tube in the Gramineae is accompanied by the deposition of callose, but there is no counterpart reaction in the stigma. In species with delayed tube inhibition, such as in the Solanaceae, the tube tip may become occluded, or it may burst with the release of particles containing callose precursors.

Little is known as to the nature of events occurring between the recognition reactions involving binding between pollen and stigma/style proteins and the subsequent rejection responses including the inhibition of tube growth and the deposition of callose. In the sporophytic system as exemplified by the Cruciferae, however, it seems likely that a diffusible substance is produced at the binding site on the stigma surface which stimulates the synthesis of callose, since this compound would not otherwise be formed in the stigma papillae. The response is in some respects similar to that induced in plant tissues by invading pathogens.

Although much remains to be elucidated regarding the nature of the pollen/stigma interactions it seems clear that recognition of "self" from "not self" pollen is achieved through the initial binding reactions between pollen and stigma proteins.

It is not yet possible to say whether analogous recognition reactions occur between somatic cells of the plant body, but it may be significant that grafts between certain genotypes are successful and others unsuccessful. Little is known as to the molecular basis of acceptance or rejection between grafted tissues in plants, although a great deal is known regarding the corresponding phenomena in animals. It is possible, however, that cell recognition responses, analogous to those in pollen/stigma interactions, occur between plant somatic cells.

There is increasing evidence that glycoproteins play an essential role in cell recognition phenomena in both plants and animals. Glycoproteins have simple or branched carbohydrate chains covalently attached at their surface. Such glycoproteins are of widespread occurrence in plants, especially in seeds, but their function has hitherto remained unknown. The *lectins* are another group of proteins which probably play an important role in cell recognition responses. The molecules of lectins have pockets that can accommodate sugar residues, and consequently they are capable of binding to the sugar chains of glycoproteins. Some plant lectins show specificity in their binding properties, and attach to specific combinations of sugars. Thus, we have a possible molecular basis for cell recognition mechanisms, and indeed experiments with the gametes of *Chlamydomonas* suggest that this type of molecular interaction may be involved in the recognition responses and adhesion between gametes of opposite mating types.

13.18 NON-GENIC FACTORS IN DEVELOPMENT

So far, we have regarded development as a process of "selective gene expression", involving the controlled activity of specific groups of genes, which in turn control the synthesis of enzyme and structural proteins characteristic of specialized cells. The information encoded in the DNA determines the amino acid sequence in the polypeptide chains of proteins, constituting their *primary* structure. Once the primary structure is established, the polypeptide chain can usually take on higher configurations without further directions from the genome. That is, genetic control of primary structure determines the secondary, tertiary and quarternary structure.

However, the cell is clearly not simply a random mixture of enzyme and other proteins, but has a highly ordered structure, in the form of organelles, membranes, etc., the structure of which appears not to be controlled by enzymic processes and hence not directly controlled by the information in the DNA. One way in which certain subcellular structures are formed is by the process of *self-assembly,* in which larger units, such as ribosomes, microtubules and membranes, are formed by the spontaneous assembly of smaller sub-units, frequently protein or nucleic acid macromolecules. This self-assembly is not directly enzyme controlled, but results directly from the physicochemical properties of the constituent sub-units.

Apart from the well-established process of self-assembly at the molecular level, there is other evidence of the importance of non-genic factors in development at the organelle and cell levels of organization. Thus, a considerable number of instances of "cytoplasmic inheritance" (in which certain characters of the offspring are transmitted through the maternal cytoplasm) are known for plants, and we now know that certain organelles, notably the plastids and mitochondria, have some degree of autonomy and indeed contain their own DNA.

Other non-genic factors probably include certain "physical" properties of cells and tissues. We have already seen the importance of external environmental factors, such as light and temperature, in plant development, but we must also consider the effects of other physical factors, such as surface tension and diffusion gradients of oxygen, arising and operating within the cells and tissues themselves. For example, attempts have been made to account for the position of cell walls in terms of physical laws. Earlier workers drew a parallel between the shapes and arrangements of cells in a tissue and those of groups of soap bubbles. Now, the shapes of soap bubbles can be interpreted in terms of the effects of surface tension,

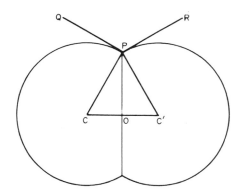

Fig. 13.15. Stable partition and walls of minimum surface assumed by two equal bubbles which are in contact. Angles OPQ and OPR are 120°. The distance between the centres equals the radii. (From D. A. W. Thompson, *Growth and Form,* Cambridge University Press, 1942.)

which tends to make them adopt forms with the least possible surface area (Fig. 13.15). Errera suggested that the shapes of cells can be similarly interpreted (Fig. 13.16), although it is questionable how far cell walls can be regarded as equivalent to the almost weightless films of soap bubbles. Moreover many cell types, such as cambium cells, do not appear to adopt a form with a minimal surface area. It is possible, however, that Errera's

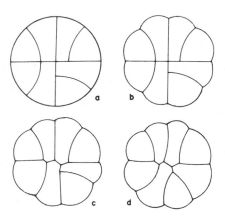

Fig. 13.16. Plate of eight cells (or bubbles) assuming a position of equilibrium where cell surfaces are of minimum area. (From D. A. W. Thompson, *Growth and Form,* Cambridge University Press, 1942.)

laws are applicable to isolated cells, or group of cells, which will be free of the pressures and other influences from surrounding cells which occur in a tissue mass. Thus, the position of the walls formed during the early development of plant embryos may follow Errera's law and be determined by surface tension effects.

Although we cannot elaborate further on this theme here, it is clear that we should avoid the danger of assuming that all aspects of growth and development can be accounted for in terms of the information encoded in the DNA of the nucleus.

FURTHER READING

General

Bryant, J. A. (Ed.). *Molecular Aspects of Gene Expression in Plants,* Academic Press, 1976.
Cohen, S. H. and J. A. Shapiro. Transposable genetic elements, *Sci. Amer.* **242**(2), 36, 1980.
Graham, C. and P. F. Wareing. *The Developmental Biology of Plants and Animals,* Blackwell Scientific Publications, Oxford, 1976.
Grant, P. *Biology of Developing Systems,* Holt, Rinehard & Winston, New York, 1978.
Heslop-Harrison, J. *Cellular Recognition Systems in Plants,* Edward Arnold, London, 1978.
Smith, H. (Ed.). *The Molecular Biology of Plant Cells,* Blackwell Scientific Publications, London, 1977.
Thomas, H. The structure of nucleic acids and proteins. In: *Plant Structure Function and Adaptation* (Ed. M. A. Hall), MacMillan, London, 1976.

More Advanced Reading

Brachet, J. Synthesis of macromolecules and morphogenesis in *Acetabularia, Current Topics in Devel. Biol.* **3**, 1, 1968.

Brink, R. A. Phase change in higher plants and somatic cell heredity, *Quart. Rev. Biol.* **37**, 1, 1962.

Britten, E. H and R. J. Davidson. Regulation of gene expression: Possible role of repetitive sequences, *Science* **204**, 1052, 1979.

Causton, D. R. *A Biologist's Mathematics,* Edward Arnold, 1977.

Flavell, R. The molecular characterisation and organisation of plant chromosomal DNA sequences, *Ann. Rev. Plant Physiol.* **31**, 569, 1980.

Gunning, B. E. S., J. E. Hughes and A. R. Hardham. Formative and proliferative cell divisions, cell differentiation and developmental changes in the meristem of *Azolla* roots, *Planta* **143**, 121, 1978.

Halperin, W. Organogenesis at the shoot apex, *Ann. Rev. Plant Physiol.* **29**, 239, 1978.

Hämmerling, J. Nucleo-cytoplasmic interactions in *Acetabularia* and other cells, *Ann. Rev. Plant Physiol.* **14**, 65, 1963.

Jacob, F. and J. Monod. Chapter in: *Cytodifferentiation and Macromolecular Synthesis,* Academic Press, New York and London, 1963.

Klein, R. M. The physiology of bacterial tumors in plants and of habituation, *Encycl. Plant Physiol.* **15**(2), 209, 1965.

Kung, S. D. Expression of chloroplast genomes in higher plants, *Ann. Rev. Plant Physiol.* **28**, 401, 1977.

Leaver, C. L. and J. S. Lovett. An analysis of protein and RNA synthesis during encystment and outgrowth (germination) of *Blastocladiella* zoospores, *Cell Differentiation* **3**, 165, 1974.

Leaver, C. J. (Ed.). *Genome Organisation in Plants,* NATO Advanced Study Institute Series, Series A, Life Sciences A29, Plenum Press, 1980.

Meins, F. Cell determination in plant development. *Bio Science,* **29**, 221-225, 1979.

Puiseux-Dao, S. *Acetabularia and Cell Biology,* Logos Press Ltd., London, 1970.

Quatrano, R. S. Development of cell polarity, *Ann. Rev. Plant Physiol.* **29**, 487, 1978.

Sinnott, E. W. Cell polarity and the development of form in cucurbit fruits, *Amer. J. Bot.* **31**, 338, 1944.

Wareing, P. F. Some aspects of differentiation in plants. In: *Control Mechanisms of Growth and Differentiation* (ed. D. D. Davies and M. Balls), *Symp. Soc. Exp. Biol.,* 25, 1971.

Wareing, P. F. Determination in plant development, *Bot. Mag. Tokyo,* special edition 1, 3-18, 1978.

INDEX

335